Introduction to Fluid Power

James L. Johnson

DELMAR

THOMSON LEARNING ™

Australia Canada Mexico Singapore Spain United Kingdom United States

Introduction to Fluid Power
James L. Johnson

Business Unit Director:
Alar Elken

Executive Editor:
Sandy Clark

Acquisitions Editor:
Mark Huth

Developmental Editor:
Dawn Daugherty

Executive Marketing Manager:
Maura Theriault

Channel Manager:
Mary Johnson

Marketing Coordinator:
Brian McGrath

Executive Production Manager:
Mary Ellen Black

Production Manager:
Larry Main

Production Editor:
Tom Stover

Art and Design Coordinator:
Rachel Baker

Cover Design:
Cummings Advertising Art, Inc.

COPYRIGHT © 2002 by Delmar,
a division of Thomson Learning, Inc.
Thomson Learning™ is a trademark
used herein under license

Printed in the United States
1 2 3 4 5 XXX 05 04 03 02 01

For more information contact Delmar,
3 Columbia Circle, PO Box 15015,
Albany, NY 12212-5015.
Or find us on the World Wide Web at
http://www.delmar.com

ALL RIGHTS RESERVED. No part of this
work covered by the copyright hereon
may be reproduced or used in any
form or by any means—graphic, electronic, or mechanical, including pho-
tocopying, recording, taping, Web distribution or information storage and
retrieval systems—without written
permission of the publisher.

For permission to use material from
this text or product, contact us by
Tel   (800) 730-2214
Fax   (800) 730-2215
www.thomsonrights.com

Library of Congress
Cataloging-in-Publication Data

Johnson, James, 1969-
    Introduction to fluid power /
James Johnson.
        p. cm.
    ISBN 0-7668-2365-2
    1. Fluid power technology.   I. Title.
TJ844 .J64 2001
620.1'06--dc21

                                    2001047320

NOTICE TO THE READER

Publisher does not warrant or guarantee any of the products described herein or perform any independent analysis in connection with any of the product information contained herein. Publisher does not assume, and expressly disclaims, any obli-
gation to obtain and include information other than that provided to it by the manufacturer.

The reader is expressly warned to consider and adopt all safety precautions that might be indicated by the activities herein
and to avoid all potential hazards. By following the instructions contained herein, the reader willingly assumes all risks in connection with such instructions.

The Publisher makes no representation or warranties of any kind, including but not limited to, the warranties of fitness for
particular purpose or merchantability, nor are any such representations implied with respect to the material set forth herein,
and the publisher takes no responsibility with respect to such material. The publisher shall not be liable for any special, con-
sequential, or exemplary damages resulting, in whole or part, from the readers' use of, or reliance upon, this material.

Contents

Preface.. **viii**

1   *An Introduction to Fluid Power*

1.1   Fluid Power Defined.. **2**
1.2   Hydraulics versus Pneumatics **2**
1.3   Standards.. **3**
1.4   Applications **3**
1.5   The Fluid Power Industry **4**
1.6   Units **5**
1.7   Review Questions **5**

2   *Basic Principles of Hydraulics*

2.1   Introduction **8**
2.2   Pascal's Law **8**
2.3   Transmission and Multiplication of Force.............................. **14**
2.4   Basic Properties of Hydraulic Fluids **19**
2.5   Liquid Flow **26**
2.6   Static Head Pressure.. **45**
2.7   Pressure Losses **47**
2.8   Power... **48**
2.9   Hydraulic Systems.. **51**
2.10 Equations.. **52**
2.11 Review Questions and Problems... **53**

3   *Hydraulic Pumps*

3.1   Introduction **58**
3.2   Pump Flow and Pressure... **58**
3.3   Pump Drive Torque and Power **63**
3.4   Pump Efficiency.. **67**
3.5   Pump Types **73**
3.6   Pressure-Compensated Pumps **87**
3.7   Cavitation and Aeration.. **88**
3.8   Graphic Symbols **88**
3.9   Pump Specifications **90**
3.10 Equations.. **98**
3.11 Review Questions and Problems... **99**

4   Hydraulic Cylinders

4.1	Introduction	104
4.2	Cylinder Force	107
4.3	Cylinder Speed	112
4.4	Cylinder Power	118
4.5	Differential Flow	121
4.6	Cylinder Types	124
4.7	Graphic Symbols	130
4.8	Cylinder Applications	132
4.9	Cylinder Specifications	142
4.10	Equations	146
4.11	Review Questions and Problems	147

5   Hydraulic Motors

5.1	Introduction	154
5.2	Motor Types	154
5.3	Motor Torque	161
5.4	Motor Speed	163
5.5	Motor Power	166
5.6	Motor Efficiency	167
5.7	Graphic Symbols	172
5.8	Motor Applications	174
5.9	Motor Specifications	176
5.10	Equations	182
5.11	Review Questions and Problems	183

6   Hydraulic Directional Control

6.1	Introduction	188
6.2	Check Valves	188
6.3	Shuttle Valves	190
6.4	Two-way Directional Control Valves	193
6.5	Three-way Directional Control Valves	198
6.6	Four-way Directional Control Valves	204
6.7	Directional Control Valve Actuation	218
6.8	Circuits	224
6.9	Directional Control Valve Mounting	234
6.10	Directional Control Valve Specifications	238
6.11	Equations	240
6.12	Review Questions and Problems	241

7 Hydraulic Pressure Control

7.1	Introduction	244
7.2	Pressure Relief Valves	244
7.3	Unloading Valves	253
7.4	Pressure Reducing Valves	255
7.5	Sequence Valves	258
7.6	Counterbalance Valves	265
7.7	Brake Valves	268
7.8	Pressure-compensated Pumps	269
7.9	Pressure Control Valve Mounting	275
7.10	Pressure Control Valve Specifications	276
7.11	Review Questions and Problems	276

8 Hydraulic Flow Control

8.1	Introduction	280
8.2	Flow Control Valve Types	280
8.3	Flow Coefficient	284
8.4	Circuits	287
8.5	Cushioned Cylinders	298
8.6	Flow Dividers	300
8.7	Flow Control Valve Specifications	304
8.8	Equations	306
8.9	Review Questions and Problems	307

9 Ancillary Hydraulic Components

9.1	Introduction	310
9.2	Accumulators	310
9.3	Intensifiers	321
9.4	Reservoirs	325
9.5	Heat Exchangers	329
9.6	Filters	333
9.7	Instrumentation and Measurement	342
9.8	Conduits and Fittings	347
9.9	Seals and Bearings	361
9.10	Hydraulic Fluids	369
9.11	Equations	371
9.12	Review Questions and Problems	372

10   Basic Principles of Pneumatics

10.1	Introduction	376
10.2	Absolute Pressure and Temperature	378
10.3	Gas Laws	379
10.4	Gas Flow	389
10.5	Vacuum	390
10.6	Pneumatic Systems	394
10.7	Equations	396
10.8	Review Questions and Problems	396

11   Pneumatic Power Supply

11.1	Introduction	400
11.2	Compressor Types	401
11.3	Compressor Sizing	407
11.4	Vacuum Pumps	412
11.5	Equations	413
11.6	Review Questions and Problems	414

12   Pneumatic Components

12.1	Introduction	416
12.2	Pneumatic Cylinders	417
12.3	Pneumatic Motors	424
12.4	Other Pneumatic Actuators	426
12.5	Pneumatic Directional Control Valves	427
12.6	Pneumatic Flow Control Valves	445
12.7	Air Preparation	448
12.8	Air Distribution	456
12.9	Equations	457
12.10	Review Questions and Problems	457

13   Electronic Control of Fluid Power

13.1	Introduction	460
13.2	Solenoid Valves	462
13.3	Proportional and Servo Valves	467
13.4	Pump Controls	472
13.5	Review Questions	473

Appendix A
Nomenclature and Common Units **475**

Appendix B
Metric (SI) Prefixes and Conversion Factors **477**
    B.1 Metric (SI) Prefixes **477**
    B.2 Conversion Factors **477**
    B.3 How To Use Conversion Factors **478**

Appendix C
Equations **480**

Appendix D
Graphic Symbols **487**

Glossary **497**

Index **501**

Preface

Goals and Methods

This textbook provides a comprehensive introduction to fluid power, including both hydraulics and pneumatics, by focusing on the following learning objectives. Upon completion of a course using this textbook, the student will:

1. Understand the underlying theoretical concepts,
2. Be familiar with the construction and function of the components,
3. Know how the components are selected and integrated into a system,
4. Understand the operation of basic circuits, and
5. Know how to read basic schematics.

An understanding of the theoretical concepts is achieved through easy-to-understand explanations and numerous examples. Knowledge of the construction and function of fluid power components is accomplished with the aid of color cut-away drawings and color photographs of the components from the manufacturers. Many schematics are provided, with the flow paths traced in color, so that the student can learn to read schematics and gain an understanding of the operation of basic circuits.

To help the student develop system design skills and understand component selection criteria, specifications sections are located in each of the components chapters. Throughout the text, realistic numbers are used and the problems are stated in such a way as to develop system design skills.

Who Is This Textbook For?

This textbook is designed primarily for use in an introductory fluid power course in Engineering Technology or Industrial Technology programs. This text may also be used by technologists, sales people, engineers, and others who have completed their formal education, but wish to learn more about fluid power. It can be used for this purpose in a formal course, or for individual study.

A working knowledge of algebra is required to understand the theoretical material in this text. Trigonometry is used in only one section (Chapter 4: Hydraulic Cylinders) and may be omitted with no loss in continuity. The theoretical material is explained so that the instructor may choose to place more or less emphasis on the mathematics.

Organization

Chapter 1 is a brief introduction to the subject of fluid power and the fluid power industry. Chapter 2 covers the basic principles of hydraulics. Chapters 3 through 9 cover virtually all of the components one will find in a hydraulic system. The emphasis is on understanding their construction, function, and how they are selected. Schematic symbols of all components covered are shown throughout the text, allowing the student to learn how to read schematics gradually. The instructor may cover the chapters on components in any order to correspond to laboratory exercises.

Chapter 10 covers the basic principles of pneumatics and highlights the differences between hydraulics and pneumatics. Chapter 11 covers pneumatic power supply, including compressors and receiver tanks. Chapter 12 covers pneumatic valves, actuators, and other miscellaneous pneumatic components. Chapter 13 is an introduction to the electronic control of fluid power, including both hydraulics and pneumatics.

Several useful appendices are provided at the end of the text, including sections on nomenclature, conversion factors, equations, and graphic symbols. A glossary is included to aid the student in learning fluid power terminology.

Units

Both U.S. customary and metric (SI) units are used throughout the text. Equations and examples for each unit system are given separately so that the instructor can cover both systems, or place more emphasis on one or the other.

Note to the Instructor

Although this text works best when used in conjunction with hands-on lab exercises, those without equipment can still effectively use the text. Whatever the level of equipment, industrial tours can greatly improve a student's understanding of the subject. Search for companies in your area that are willing to give tours and demonstrate the operation of machinery that utilizes fluid power. It has been the experience of the author that students return from these tours with increased motivation and a greater appreciation for the relevance of the course.

Because fluid power is a very practical and applied subject area, having actual components on hand for the students to manipulate and disassemble is an extremely helpful supplement to the text. Many companies are willing to donate old components that may no longer function but are of great educational value. Finally, it is useful to provide students with actual manufacturers' catalogs that can be used to do component selection exercises. Many manufacturers have catalogs that are readily available for downloading at their web sites.

Acknowledgments

I would like to acknowledge some of the people who made a significant contribution to this text. First, I thank my wife, Holly Johnson, for carefully and thoroughly reviewing the manuscript and making many helpful suggestions. I thank the helpful and knowledgeable staff at Delmar, particularly Mark Huth and Dawn Daugherty. I thank the reviewers, who provided thoughtful and constructive comments and suggestions. Thanks also to the many manufacturers who provided photographs and illustrations.

Special thanks to Bill Doebele, for suggesting engineering as a career and for providing good advice throughout the years, and to Arnold von Engelbrechten, a talented engineer and valued mentor.

Dedication

This book is dedicated to my Mom and Dad for their love, support, and dedication to family.

The author and Delmar wish to acknowledge and thank the individuals who participated on the review panel for their suggestions and contributions to the text. The review panel members were:

Ray Greb
Mesa State College
Grand Junction, CO

Robert Williams
Spokane Community College
Spokane, WA

Skip Davis
Kennebec Valley Technical College
Fairfield, ME

John Seim
Alexandria Technical College
Alexandria, MN

Richard Minch
Waukesha Technical College
Waukesha, WI

Jan Stenberg
Greenville Technical College
Henderson, SC

Michael Renzoni
Gateway Technical College
Kenosha, WI

Stuart Hilton
Chattanooga State Technical Community College
Chattanooga, TN

Richard Dettloff
San Joaquin Delta Community College
Stockton, CA

Karen Hardesty
Stark State Technical College
Canton, OH

CHAPTER 1

An Introduction to Fluid Power

OUTLINE

1.1 Fluid Power Defined

1.2 Hydraulics versus Pneumatics

1.3 Standards

1.4 Applications

1.5 The Fluid Power Industry

1.6 Units

1.7 Review Questions

1.1 Fluid Power Defined

Fluid power is the use of a confined fluid flowing under pressure to transmit power from one location to another. It is one of three commonly used methods of transmitting power in an industrial setting; the others are electrical and mechanical power transmission: Electrical power transmission uses an electric current flowing through a wire to transmit power. Its main advantage is its ability to transmit power over large distances very quickly. The most obvious example is the use of electricity to transmit power from the power plant to our homes. Mechanical power transmission uses gears, pulleys, chains, and other such devices to transmit power over short distances with a large degree of rigidity. A simple example is the use of a chain on a bicycle to transmit power from the pedals to the rear wheel. It is not uncommon to find all three forms of power transmission on a single machine.

Two of the most important advantages of fluid power transmission are its ability to multiply force and its flexibility to change direction quickly without damage to the system. Multiplication of force is discussed in detail in Chapter 2. The flexibility of fluid power results because the medium of transmission is a flowing fluid, which allows flexible hoses to be used. This makes it very easy to change direction and transmit the power through angles.

1.2 Hydraulics versus Pneumatics

Fluid power is divided into two areas: hydraulics and pneumatics. Hydraulics is the transmission of power through a liquid, most commonly petroleum-based oil. Pneumatics is the use of a gas, usually air. The fundamental difference between air and oil is that air is *compressible,* while oil is *relatively incompressible*. This leads to advantages and disadvantages that determine which should be used in a given application. The compressibility of air causes a pneumatic system to behave in a "springy" fashion. When a valve is shifted and the air is allowed to flow, it expands very quickly, resulting in quick response time. In addition to this quickness, however, the compressibility of air makes it more difficult to control a pneumatic system with precision. Due to the incompressibility of oil, hydraulic systems move more slowly but are capable of higher precision. In general, hydraulic systems are usually preferred for applications that require:

1. high power/large load capacity,
2. precise positioning, and
3. smooth movement.

Pneumatics is well suited for applications that require:

1. low power/light to moderate load capacity,
2. low to moderate precision, and
3. quick response.

Many of the basic concepts of fluid power are similar whether we are using air or oil, while others are affected by the difference in compressibility. The first part of this text covers hydraulics because the compressibility of air makes a pneumatic system slightly more complicated to analyze. Pneumatics is covered in the latter part of the text, where the similarities and differences between hydraulics and pneumatics are discussed in detail.

1.3 Standards

As with most other industries, the fluid power industry has governing bodies that promote standardization. The *National Fluid Power Association* (NFPA, www.nfpa.com) is the most prominent organization in this industry. The NFPA sets standards such as cylinder sizes, mounting styles for components, testing methods for fluid power devices, and numerous others. It also promotes the fluid power industry and gathers statistical information on industry sales and forecasts. Other standards organizations in this field are the *American Society for Testing and Materials* (ASTM, www.astm.org) and the *Society of Automotive Engineers* (SAE, www.sae.org). ASTM and SAE are involved with setting standards in many segments of industry, including the fluid power industry.

The *American National Standards Institute* (ANSI, www.ansi.org) coordinates the efforts of organizations, such as those just mentioned, whose responsibilities overlap. It does not develop standards, but ensures that there is a truly national standard by establishing a consensus among the concerned parties. The *International Standards Organization* (ISO, www.iso.ch) performs a similar function on the international level. It is important to note that these standards are not mandatory, but it is generally in everyone's best interest to have standardization. Standardization allows compatibility, interchangeability, and components to be compared with one another on a level playing field.

1.4 Applications

There are actually two types of fluid systems used in industry: fluid transfer and fluid power. The purpose of a fluid transfer system is to simply move a fluid from one location to another. We encounter this type of system regularly when we stop at the gas station. The purpose of the gasoline pump is simply to transfer the gas from the holding tank to our gas tank. Other examples include the water and oil pumps used in our vehicles. Their objective is simply to keep the fluid circulating, which requires only minimal pressure because the resistance is small. Fluid power systems have as their objective the transmission of power. They move large loads, which creates large resistances, and therefore must be capable of withstanding high pressures. Although this text is concerned primarily with fluid power, many of the concepts discussed will apply to the fluid transfer field as well.

The majority of fluid power applications fall into one of three industry segments: *mobile hydraulic, industrial hydraulic,* and *industrial pneumatic.* Mobile hydraulic applications include excavating equipment such as backhoes and bulldozers, farm equipment, and other vehicles such as garbage trucks that use hydraulics to compact trash. Industrial hydraulic applications include machining equipment, robots, presses, stamping equipment, and other machines used in factories and machine shops. Components used in mobile hydraulics tend to be of lighter construction than those used in industrial hydraulic systems because weight is a significant factor and they usually undergo fewer cycles over a given time period. Industrial hydraulic components are often more robust because they must operate almost continuously over long periods, with as little downtime as possible, and weight is not normally a concern. Industrial pneumatic applications include air tools, automated manufacturing and assembly, and others. According to the NFPA, mobile hydraulics is the largest segment, comprising 50% of fluid power equipment sales. Industrial hydraulics and pneumatics each account for 25% of total sales.

1.5 The Fluid Power Industry

Most fluid power companies fall into one of three broad categories: *component manufacturers, original equipment manufacturers (OEMs),* and *distributors.* Component manufacturers produce components such as valves, cylinders, pumps, hydraulic fluids, hoses, fittings, and seals. OEMs manufacture mobile and industrial machinery, such as excavating equipment and hydraulic presses that utilize fluid power. OEMs purchase items from the component manufacturers for use in their equipment. Distributors serve as a one-stop shop for purchasing components from a variety of manufacturers. They sell to smaller OEMs, custom machine builders, and end users of equipment. They also provide application engineering support and can usually answer most questions regarding the implementation of the components they sell. Many distributors also do repairs and build custom hydraulic systems.

1.6 Units

Due to the international nature of today's marketplace, it is important to be familiar with both the U.S. customary and metric (SI) units. These unit systems are used side by side throughout the text. Common units for both systems and conversion factors are given in the appendix for quick reference.

1.7   Review Questions

1. Define *fluid power.*
2. Differentiate between hydraulics and pneumatics.
3. Describe the difference between fluid transfer and fluid power applications.
4. What are the three principal application areas for fluid power?
5. What are the three categories that most fluid power companies fit into?
6. List the governing bodies that set the standards for the fluid power industry. How do ANSI and ISO differ from the others?

CHAPTER 2

Basic Principles of Hydraulics

OUTLINE

2.1 Introduction

2.2 Pascal's Law

2.3 Transmission and Multiplication of Force

2.4 Basic Properties of Hydraulic Fluids

2.5 Liquid Flow

2.6 Static Head Pressure

2.7 Pressure Losses

2.8 Power

2.9 Hydraulic Systems

2.10 Equations

2.11 Review Questions and Problems

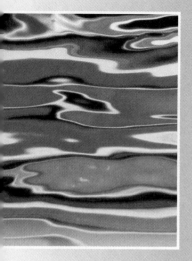

2.1   Introduction

As stated in Chapter 1, hydraulics is well suited for applications that require large forces, precise positioning, and smooth movement. Hydraulic systems typically operate at pressures from 500 to 5000 pounds per square inch (psi), and can easily generate thousands of pounds of force. They can also be very precise in terms of speed and positioning when coupled with electronic controls. Chapter 2 focuses on the basic principles needed to understand the operation of a hydraulic system. Subsequent chapters will focus on components.

2.2   Pascal's Law

Pascal's Law is the most fundamental principle in fluid power. It deals with *hydrostatics,* the transmission of force through a confined fluid under pressure. Pascal's Law states that:

> *The pressure exerted on a confined fluid is*
> *transmitted undiminished in all directions and*
> *acts at right angles to the containing surfaces.*

This is shown graphically in Figure 2-1. Here, a force is being applied to a piston, which in turn exerts a pressure on the confined fluid. The pressure is equal everywhere, and at right angles to the containing surfaces. Red arrows will be used throughout this text to represent fluid pressure.

The fluid is often in motion in a hydrostatic system, but it is the *pressure,* and not the motion of the fluid, that transmits the force and energy. A system in which the *motion* is the mode of transmittal is known as a *hydrodynamic* system. In these systems, the velocity or kinetic energy of the fluid is converted into mechanical energy, usually in rotational form. The fluid is not pressurized to any significant degree, as it is with a hydrostatic system. A simple and ancient example is the water wheel. In this device, the velocity of the falling fluid is used to turn a wheel, which powers a grain mill or other device. A more modern example is a steam-driven turbine, which uses high-velocity steam to generate electricity at power plants. Most industrial hydraulic and pneumatic systems use pressurized fluids and are therefore considered hydrostatic.

To relate force and pressure for specific values, Pascal's Law must be quantified. Pressure is defined mathematically as a force distributed over an area. In equation form:

$$p = \frac{F}{A}$$

(2-1)

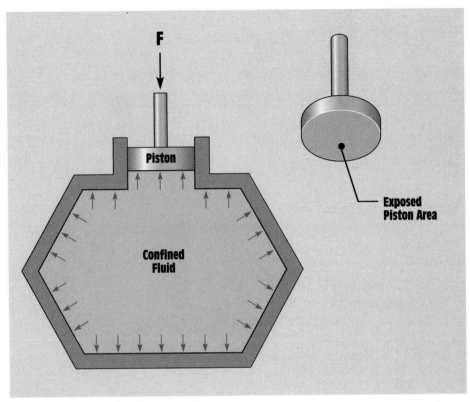

FIGURE 2-1   Pascal's law.

where: F = force

   A = area

   p = pressure

   The most common units for these quantities in both the U.S. customary and metric (SI) unit systems are shown in Table 2-1.

   The unit lbs/in² is commonly abbreviated psi. The unit N/m² is also known as a *pascal* (abbreviated Pa). Another unit for pressure that is often used by European manufacturers is a *bar*, which is equal to 100,000 Pa. Conversion factors between these and other units are shown in Appendix B for easy reference.

   The diameter of the piston, rather than the area, is usually provided. The piston area exposed to the fluid is circular, as shown in Figure 2-1, so we can use the equation for the area of a circle to calculate the piston area:

$$A = \frac{\pi \cdot D^2}{4}$$ **(2-2)**

QUANTITY	U.S. CUSTOMARY UNIT	METRIC (SI) UNIT
Force (*F*)	Pounds (lbs)	Newtons (N)
Area (*A*)	Square inches (in²)	Square meters (m²)
Pressure (*p*)	Pounds per square inch $\left(\frac{lbs}{in^2}, psi\right)$	Newtons per square meter $\left(\frac{N}{m^2}, Pa\right)$

TABLE 2-1   Force and Pressure Values

where: D = diameter
π (pi) = a mathematical constant equal to approximately 3.142

EXAMPLE 2-1.

The piston and cylinder shown in Figure 2-2 have a diameter (*D*) of 2 in and are loaded with a force (*F*) of 1000 lbs. What is the pressure (*p*) inside the cylinder?

1. Calculate the piston area:

$$A = \frac{\pi \cdot D^2}{4} = \frac{3.142 \cdot (2 \text{ in})^2}{4} = 3.142 \text{ in}^2$$

2. Calculate the pressure:

$$p = \frac{F}{A} = \frac{1000 \text{ lbs}}{3.142 \text{ in}^2} = 318 \frac{\text{lbs}}{\text{in}^2} \quad (318 \text{ psi})$$

EXAMPLE 2-1M.

The piston and cylinder shown in Figure 2-2 have a diameter of 100 mm (0.100 m) and are loaded with a force of 2500 N. What is the pressure inside the cylinder?

SOLUTION:

1. Calculate the piston area:

$$A = \frac{\pi \cdot D^2}{4} = \frac{3.142 \cdot (0.1 \text{ m})^2}{4} = 0.007854 \text{ m}^2$$

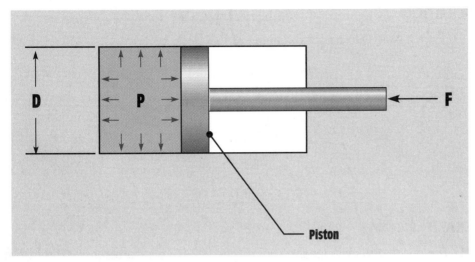

FIGURE 2-2   Example 2-1.

2. Calculate the pressure:

$$p = \frac{F}{A} = \frac{2500 \text{ N}}{0.007854 \text{ m}^2} = 318,300 \frac{\text{N}}{\text{m}^2} \quad (318.3 \text{ kPa}, \ 3.183 \text{ bar})$$

Cylinder sizes are often given in millimeters (mm), as in the previous example. One millimeter is one thousandth of a meter (1 mm = 0.001 m). Because the unit N/m² is so small, it is often expressed in kPa (1 kPa = 1000 Pa = 1000 N/m²) or bars (1 bar = 100,000 Pa). When doing calculations, it is best to use the units listed in Table 2-1. Converting to N, m, and N/m² before beginning a calculation reduces the chances of making a mistake.

Because Equation 2-1 is an equation with three unknowns, we could calculate any one of the three quantities p, F, or A, provided the other two are known. In Example 2-1, we knew the force and cylinder size and we calculated the pressure. We look next at examples in which F and A are the unknowns.

EXAMPLE 2-2.

The piston and cylinder shown in Figure 2-2 have a diameter of 1.5 in. We do not want to exceed a pressure of 500 psi. What is the maximum force this system can withstand?

SOLUTION:

1. Calculate the piston area:

$$A = \frac{\pi \cdot D^2}{4} = \frac{3.142 \cdot (1.5 \text{ in})^2}{4} = 1.767 \text{ in}^2$$

2. Calculate the force:

$$F = p \cdot A = 500 \frac{\text{lbs}}{\text{in}^2} \cdot 1.767 \text{ in}^2 = 884 \text{ lbs}$$

EXAMPLE 2-2M.

The piston and cylinder shown in Figure 2-2 have a diameter of 40 mm. We do not want to exceed a pressure of 3500 kPa. What is the maximum force this system can withstand?

SOLUTION:

1. Calculate the piston area:

$$A = \frac{\pi \cdot D^2}{4} = \frac{3.142 \cdot (0.04 \text{ m})^2}{4} = 0.001257 \text{ m}^2$$

2. Convert to N/m²:

$$3500 \text{ kPa} = 3,500,000 \text{ Pa} = 3,500,000 \frac{\text{N}}{\text{m}^2}$$

3. Calculate the force:

$$F = p \cdot A = 3,500,000 \frac{\text{N}}{\text{m}^2} \cdot 0.001257 \text{ m}^2 = 4398 \text{ N}$$

EXAMPLE 2-3.

The piston and cylinder shown in Figure 2-2 are required to support a force of 2500 lbs. We do not want to exceed a pressure of 1000 psi. What size cylinder is required?

SOLUTION:

1. Calculate the piston area:

$$A = \frac{F}{P} = \frac{2500 \text{ lbs}}{1000 \frac{\text{lbs}}{\text{in}^2}} = 2.5 \text{ in}^2$$

2. As stated previously, cylinders are usually sized by their diameter. To calculate the diameter, we simply rearrange Equation 2-2:

$$A = \frac{\pi \cdot D^2}{4} \Rightarrow D = \sqrt{\frac{4 \cdot A}{\pi}} = \sqrt{\frac{4 \cdot (2.5 \text{ in}^2)}{3.142}} = 1.784 \text{ in}$$

If we do not want to exceed 1000 psi, we would need a cylinder this size or larger. A cylinder of this exact size is not available as a standard cylinder from a catalog, so we would choose the next larger available size.

EXAMPLE 2-3M.

The piston and cylinder shown in Figure 2-2 are required to support a force of 10 kN (10,000 N). We do not want to exceed a pressure of 70 bar. What size cylinder is required?

SOLUTION:

1. Convert to N/m²:

$$70 \text{ bar} = 7,000,000 \text{ Pa} = 7,000,000 \frac{\text{N}}{\text{m}^2}$$

2. Calculate the piston area:

$$A = \frac{F}{p} = \frac{10,000 \text{ N}}{7,000,000 \frac{\text{N}}{\text{m}^2}} = 0.001429 \text{ m}^2$$

3. Calculate the diameter:

$$D = \sqrt{\frac{4 \cdot A}{\pi}} = \sqrt{\frac{4 \cdot (0.001429 \text{ m}^2)}{3.142}} = 0.04265 \text{ m} \quad (42.65 \text{ mm})$$

2.3   Transmission and Multiplication of Force

The previous examples demonstrate how pressure, force, and area are related in a hydraulic system. In order for this to be of any practical use, we must be able to transmit the pressure to a second location where it can be converted into an output force. Consider the system shown in Figure 2-3. In this system, we have an input cylinder on the left and an output cylinder on the right with a confined fluid in between. We apply an input force of 100 lbs, which is distributed over an area of 10 in². This results in a pressure of:

$$p = \frac{F}{A} = \frac{100 \text{ lbs}}{10 \text{ in}^2} = 10 \frac{\text{lbs}}{\text{in}^2}$$

We are dealing with a confined fluid under pressure, so Pascal's Law applies. The pressure at the output cylinder is therefore also 10 psi. The area of the output piston is also 10 in², so the output force is:

$$F = p \cdot A = 10 \frac{\text{lbs}}{\text{in}^2} \cdot 10 \text{ in}^2 = 100 \text{ lbs}$$

Therefore, we have simply transmitted the force of 100 lbs from the input to the output cylinder through the fluid.

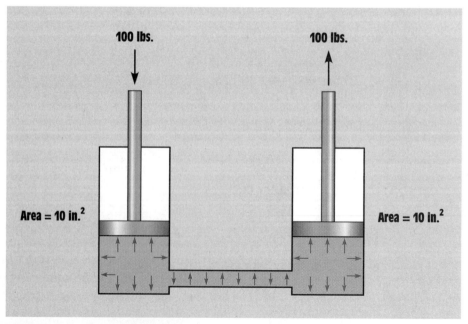

FIGURE 2-3   Transmission of force.

One of the most useful features of fluid power is the ease with which it is able to multiply force. This is accomplished by using an output piston that is larger than the input piston. Such a system is shown in Figure 2-4. Here we have an input cylinder that has a piston area of 10 in² and an output cylinder with a piston area of 100 in². As in the previous problem, we have an input force of 100 lbs that is distributed over 10 in², so the pressure is 10 psi. We have a confined fluid under pressure once again, therefore a pressure of 10 psi is also applied to the output piston. The output piston is distributing the pressure over an area of 100 in², so the force is:

$$F = p \cdot A = 10 \frac{\text{lbs}}{\text{in}^2} \cdot 100 \text{ in}^2 = 1000 \text{ lbs}$$

The force was multiplied by a factor that was equal to the ratio of the output piston area to the input piston area. The output piston area was ten times larger, therefore the output force was ten times larger. Stated mathematically:

$$F_{OUT} = \frac{A_{OUT}}{A_{IN}} \cdot F_{IN} \qquad\qquad \textbf{(2-3)}$$

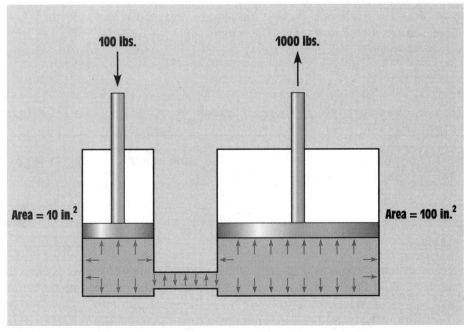

100 lbs.   1000 lbs.

Area = 10 in.²   Area = 100 in.²

FIGURE 2-4   Multiplication of force.

This increase in force may seem remarkable. We have increased our load capacity by a factor of ten simply by using a larger piston on the output. However, we know that in physics, as in life, we never get something for nothing. If we gain force we must lose something else in return. In explanation of this it will be useful to introduce the concept of *work*. Work is a measure of energy expended. In a physical system, it is defined as an applied force multiplied by a distance traveled due to that force. In equation form:

$$W = F \cdot d \qquad\qquad \textbf{(2-4)}$$

where: F = force (lbs, N)
$\quad\quad d$ = distance traveled (ft, m)
$\quad\quad W$ = work (ft · lbs, N · m)

For example, suppose we are pushing a wheelbarrow full of gravel that requires us to exert a force of 50 lbs. If we move the wheelbarrow 100 ft, then the work we have done is equal to:

$$W = 50 \text{ lbs} \cdot 100 \text{ ft} = 5000 \text{ ft} \cdot \text{lbs}$$

One of the fundamental laws of nature is that we cannot get more energy out of a system than we put into it, so the work done must remain constant. In the system shown in Figure 2-4, the force was increased by a factor of ten, so in order for the work to have remained constant the distance must have decreased by the same factor. This means that we have to move the input piston 10 inches in order to move the output piston 1 inch. In equation form:

$$d_{IN} = \frac{A_{OUT}}{A_{IN}} \cdot d_{OUT} \qquad\qquad \textbf{(2-5)}$$

The pistons move these distances in the same amount of time, so we may substitute velocity for distance to obtain:

$$v_{IN} = \frac{A_{OUT}}{A_{IN}} \cdot v_{OUT} \qquad\qquad \textbf{(2-6)}$$

This gain in force, loss of distance trade-off is also apparent if we look at the volume of fluid flow. We have a closed system, so the amount of fluid must remain constant. Whatever fluid flows out of the input piston must flow into the output piston. The output piston is ten times larger, so it will move one-tenth the amount of the input piston for an equal amount of fluid.

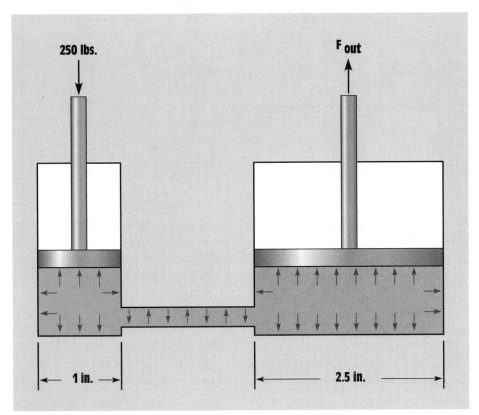

FIGURE 2-5   Example 2-4.

EXAMPLE 2-4.

Figure 2-5 shows an input cylinder with a diameter of 1 in and an output cylinder with a diameter of 2.5 in. A force of 250 lbs is applied to the input cylinder. What is the output force? How far would we need to move the input cylinder to move the output cylinder 1 in?

SOLUTION:

1. Calculate the input piston area:

$$A = \frac{\pi \cdot D^2}{4} = \frac{3.142 \cdot (1 \text{ in})^2}{4} = 0.7854 \text{ in}^2$$

2. Calculate the output piston area:

$$A = \frac{\pi \cdot D^2}{4} = \frac{3.142 \cdot (2.5 \text{ in})^2}{4} = 4.909 \text{ in}^2$$

3. Calculate the output force:

$$F_{OUT} = \frac{A_{OUT}}{A_{IN}} \cdot F_{IN} = \frac{4.909 \text{ in}^2}{0.7854 \text{ in}^2} \cdot (250 \text{ lbs}) = 1563 \text{ lbs}$$

4. Calculate the input distance:

$$d_{IN} = \frac{A_{OUT}}{A_{IN}} \cdot d_{OUT} = \frac{4.909 \text{ in}^2}{0.7854 \text{ in}^2} = (1 \text{ in}) = 6.25 \text{ in}$$

EXAMPLE 2-4M.

An input cylinder with a diameter of 30 mm is connected to an output cylinder with a diameter of 80 mm. A force of 1000 N is applied to the input cylinder. What is the output force? How far would we need to move the input cylinder to move the output cylinder 100 mm?

SOLUTION:

1. Calculate the input piston area:

$$A = \frac{\pi \cdot D^2}{4} = \frac{3.142 \cdot (0.030 \text{ m})^2}{4} = 0.0007069 \text{ m}^2$$

2. Calculate the output piston area:

$$A = \frac{\pi \cdot D^2}{4} = \frac{3.142 \cdot (0.080 \text{ m})^2}{4} = 0.005027 \text{ m}^2$$

3. Calculate the output force:

$$F_{OUT} = \frac{A_{OUT}}{A_{IN}} \cdot F_{IN} = \frac{0.005027 \text{ m}^2}{0.0007069 \text{ m}^2} \cdot (1000 \text{ N}) = 7111 \text{ N}$$

4. Calculate the input distance:

$$d_{IN} = \frac{A_{OUT}}{A_{IN}} \cdot d_{OUT} = \frac{0.005027 \text{ m}^2}{0.0007069 \text{ m}^2} \cdot (100 \text{ mm}) = 711.1 \text{ mm}$$

EXAMPLE 2-5.

The output cylinder in Example 2-4 is required to move at 4 in/s. At what speed must the input cylinder move?

SOLUTION:

$$v_{IN} = \frac{A_{OUT}}{A_{IN}} \cdot v_{OUT} = \frac{4.909 \text{ in}^2}{0.7854 \text{ in}^2} \cdot \left(4\,\frac{\text{in}}{\text{s}}\right) = 25.0\,\frac{\text{in}}{\text{s}}$$

EXAMPLE 2-5M.

The output cylinder in Example 2-4M is required to move at 0.5 m/s. At what speed must the input cylinder move?

SOLUTION:

$$v_{IN} = \frac{A_{OUT}}{A_{IN}} \cdot v_{OUT} = \frac{0.005027 \text{ m}^2}{0.0007069 \text{ m}^2} \cdot \left(0.5\,\frac{\text{m}}{\text{s}}\right) = 3.556\,\frac{\text{m}}{\text{s}}$$

It is important to note that the amount of force multiplication is not equal to the ratio of the diameters, but the ratio of the areas. We could substitute $A = \frac{\pi}{4} \cdot D^2$ into Equations 2-3 and 2-5 to obtain:

$$F_{OUT} = \frac{D^2_{OUT}}{D^2_{IN}} \cdot F_{IN}$$

$$d_{IN} = \frac{D^2_{OUT}}{D^2_{IN}} \cdot d_{OUT}$$

These equations show that the amount of force multiplication is proportional to the ratio of the *square* of the diameters. Thus, the effect of doubling the diameter of the output piston is to increase the force output by a factor of four ($2^2 = 4$). The distance traveled by the output piston would decrease by a factor of four.

2.4   Basic Properties of Hydraulic Fluids

An understanding of the most basic properties of hydraulic fluids is required to understand the operation of a hydraulic system. More information on fluid properties and specifications is provided in Chapter 9, along with information on some commonly used hydraulic fluids.

2.4.1 MASS VERSUS WEIGHT

The distinction between mass and weight can be confusing. The *mass* of an object can be thought of simply as how much matter it contains. This is independent of where we measure it, whether it is on the Earth, the moon, or anywhere else in the universe. *Weight,* on the other hand, is a measure of how much force an object's mass creates due to gravity. Gravity on Earth is equal to 32.2 ft/s² (9.81 m/s²), which means that any object near the Earth will accelerate toward it at this rate. Newton's second law states that force equals mass times acceleration ($F = m \cdot a$). The force of gravity, or weight, is therefore given by:

$$w = m \cdot g \qquad \text{(2-7)}$$

where: w = weight (lbs, N)
   m = mass (slugs, kg)
   g = gravity = 32.2 ft/s² = 9.81 m/s²

Gravity changes depending on what planet we are on (or if we are in space), so the weight of an object depends on its location.

EXAMPLE 2-6.

◼ An object weighs 1000 lbs. What is its mass?

SOLUTION:

$$m = \frac{w}{g} = \frac{1000 \text{ lbs}}{32.2 \dfrac{\text{ft}}{\text{s}^2}} = 31.06 \frac{\text{lb} \cdot \text{s}^2}{\text{ft}} \ (31.06 \text{ slugs})$$

Note: The units $\frac{\text{lb} \cdot \text{s}^2}{\text{ft}}$ and *slugs* are equivalent because $1 \text{ lb} = 1 \frac{\text{slug} \cdot \text{ft}}{\text{s}^2}$.

EXAMPLE 2-6M.

◼ An object weighs 2000 *N.* What is its mass?

SOLUTION:

$$m = \frac{w}{g} = \frac{2000 \text{ N}}{9.81 \dfrac{\text{m}}{\text{s}^2}} = 203.9 \frac{\text{N} \cdot \text{s}^2}{\text{m}} \ (203.9 \text{ kg})$$

Note: The units $\frac{\text{N} \cdot \text{s}^2}{\text{m}}$ and kg are equivalent because $1 \text{ M} = 1 \frac{\text{kg} \cdot \text{m}}{\text{s}^2}$.

The term pounds is commonly used to express both mass and weight. In this context, pounds mass should be written "lbm," while pounds force or weight should be written "lbf." Both have the same value, however, because on Earth an object with 1 pound of mass exerts 1 pound of force. Wherever "lb" is used, it is assumed to mean "lbf" or pounds force.

2.4.2 DENSITY, SPECIFIC WEIGHT, AND SPECIFIC GRAVITY

Density, specific weight, and specific gravity are all different ways of measuring the same quantity. *Density* is a measure of a substance's mass per unit volume. In equation form:

$$\rho = \frac{m}{V} \qquad\qquad \textbf{(2-8)}$$

where: ρ = the Greek letter rho = density (slugs/ft^3, kg/m^3)
   m = mass (slugs, kg)
   V = volume (ft^3, m^3)

A typical density for hydraulic oil is 1.74 slugs/ft^3 (897 kg/m^3). The *specific weight* of a substance is the weight per unit volume:

$$\gamma = \frac{w}{V} \qquad\qquad \textbf{(2-9)}$$

where: γ = the Greek letter gamma = specific weight (lbs/ft^3, N/m^3)
   w = weight (lbs, N)
   V = volume (ft^3, m^3)

Notice that specific weight is simply the density multiplied by gravity ($\gamma = \rho \cdot g$), which is a constant, so both convey the same information. Petroleum-based hydraulic oils typically have a specific weight of around 56 lbs/ft^3 (8800 N/m^3).

Specific gravity is yet another way to represent the same quantity. Specific gravity is the ratio of the specific weight of a liquid to the specific weight of water:

$$sg_X = \frac{\gamma_X}{\gamma_{WATER}} \qquad\qquad \textbf{(2-10)}$$

where: sg_X = specific gravity of fluid x (no units)
   γ_X = specific weight of fluid x,
   γ_{WATER} = specific weight of water

Water has a specific weight of 62.4 lbs/ft^3. Thus, the specific gravity for petroleum oil is $sg_{OIL} = \frac{56 \text{ lbs/ft}^3}{62.4 \text{ lbs/ft}^3} \approx 0.9$. Because specific gravity has no units associated with it, the specific gravity of petroleum oil would be 0.9 regardless of the unit system being used.

2.4.3 VISCOSITY

Viscosity is a measure of a fluid's thickness or resistance to flow. Viscosity is of critical importance in a hydraulic system because the fluid is not only the medium for transmitting power, it is also the lubricant. Lubrication essentially provides a cushion of fluid between mating components. The thickness of the fluid is critical to this function.

The physical mechanism behind viscosity can best be explained by looking at the behavior of a fluid film between two parallel plates, as shown in Figure 2-6. The top plate is moving at a velocity v, while the bottom plate is stationary. They are separated by a distance, y, which is the thickness of the fluid film. In this situation, a very thin layer of fluid actually adheres to the surfaces of each plate. This means that the velocity of the fluid at the top plate is v, and the velocity of the fluid at the bottom plate is zero. A series of very thin layers of fluid that are moving at different velocities, as indicated by the arrows, occurs between the plates. These layers are resisting the shearing between them due to their mutual attraction (on the molecular level). This resistance causes what we think of as "thickness" in a fluid. Viscosity can be defined mathematically by the following equation:

$$\mu = \frac{F \cdot y}{v \cdot A} \qquad\qquad \textbf{(2-11)}$$

where: μ = the Greek letter mu = dynamic viscosity $\left(\dfrac{\text{lb} \cdot \text{s}}{\text{ft}^2} , \dfrac{\text{N} \cdot \text{s}}{\text{m}^2} \right)$

F = the force required to pull the plate at velocity v (lbs, N)
y = the thickness of the fluid film (ft, m)

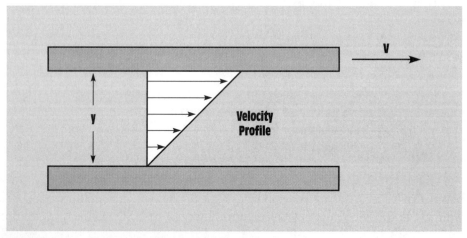

FIGURE 2-6   Parallel plates separated by an oil film.

v = the velocity of the moving plate (ft/s, m/s)

A = the area of the moving plate (ft², m²)

The area (A) is the *wetted* area, or the area in contact with the fluid. The following example will clarify this concept.

EXAMPLE 2-7.

The top plate shown in Figure 2-6 has a wetted area of 0.5 ft² and is moving at a velocity of 10 ft/s. The force required to maintain this speed is found to be 5 lbs. The fluid film thickness is 0.05 in. What is the viscosity of the fluid?

SOLUTION:

1. Convert the thickness to feet:

$$0.05 \text{ in} \cdot \left(\frac{1 \text{ ft}}{12 \text{ in}}\right) = 0.00417 \text{ ft}$$

2. Calculate the viscosity:

$$\mu = \frac{F \cdot y}{v \cdot A} = \frac{5 \text{ lbs} \cdot (0.00417 \text{ ft})}{10 \frac{\text{ft}}{\text{s}} \cdot (0.5 \text{ ft}^2)} = 0.00417 \frac{\text{lb} \cdot \text{s}}{\text{ft}^2}$$

EXAMPLE 2-7M.

The top plate shown in Figure 2-6 has a wetted area of 0.1 m² and is moving at a velocity of 3 m/s. The force required to maintain this speed is found to be 25 N. The fluid film thickness is 1 mm. What is the viscosity of the fluid?

SOLUTION:

$$\mu = \frac{F \cdot y}{v \cdot A} = \frac{25 \text{ N} \cdot (0.001 \text{ m})}{3 \frac{\text{m}}{\text{s}} \cdot (0.1 \text{ m}^2)} = 0.0833 \frac{\text{N} \cdot \text{s}}{\text{m}^2}$$

The viscosity discussed previously is known as the *dynamic viscosity*. The *kinematic viscosity* is also used in calculations. This is simply the dynamic viscosity divided by the density:

$$v = \frac{\mu}{\rho} \tag{2-12}$$

where: ν = the Greek letter nu = kinematic viscosity $\left(\dfrac{ft^2}{s}, \dfrac{m^2}{s} \right)$

μ = the dynamic viscosity $\left(\dfrac{lb \cdot s}{ft^2}, \dfrac{N \cdot s}{m^2} \right)$

ρ = density $\left(\dfrac{slugs}{ft^3}, \dfrac{kg}{m^3} \right)$

As stated earlier, the density of a typical hydraulic oil is about 1.74 slugs/ft³, which can also be written as 174 $\frac{lb \cdot s^2}{ft^4}$ because a slug is equivalent to $\frac{lb \cdot s^2}{ft}$. The kinematic viscosity of the oil in Example 2-7 would then be:

$$\nu = \frac{0.00417 \dfrac{lb \cdot s}{ft^2}}{1.74 \dfrac{lb \cdot s^2}{ft^4}} = 0.00240 \frac{ft^2}{s}$$

In metric units, hydraulic oil typically has a density of 897 kg/m³, which can also be written as 897 $\frac{N \cdot s^2}{m^4}$ because a kg is equivalent to $\frac{N \cdot s^2}{m}$. The kinematic viscosity of the oil in Example 2-7M would then be:

$$\nu = \frac{0.0833 \dfrac{N \cdot s}{m^2}}{897 \dfrac{N \cdot s^2}{m^4}} = 0.0000929 \frac{m^2}{s}$$

Another unit that can be used in the metric system to measure kinematic viscosity is a *stoke* (St), which is equal to 1 cm²/s. Because the stoke is a relatively large unit, it is common to measure viscosity in *centistokes* (cSt), which is 1/100 of a stoke. A centistoke is equal to 0.000001 m²/s (1 m²/s = 1,000,000 cSt). The kinematic viscosity just calculated would therefore be equal to 92.9 cSt.

In industry, the viscosity of a fluid is usually determined using a Saybolt viscometer or similar device. This apparatus measures the time for a 60-milliliter sample of oil to drain through a standard orifice. The resulting time is a measurement of viscosity known as *saybolt seconds universal* (abbreviated SSU or SUS). This test is typically performed at two temperatures: 100°F and 210°F. SSU is the preferred unit for use in component specifications, while ft²/s is preferred when discussing the theory of viscosity or when doing calculations. No matter which units we use for viscosity, a higher viscosity represents a thicker fluid.

Viscosity *decreases* with *increasing* temperature, so a fluid becomes thinner as its temperature increases. Because viscosity changes with temperature, it is also

useful to know how much a particular fluid's viscosity changes due to temperature changes. *Viscosity index* (V.I.) does this. V.I. was originally devised on a scale
from 0 to 100, with 100 being less change with temperature. Thus, the higher
this number, the more stable the fluid's viscosity with varying temperature. Modern chemistry has provided us with some fluids with a V.I. much greater than
100. Viscosity index is an important property to consider when selecting an oil
for a system that will operate under large temperature variations, such as with
mobile equipment.

Pumps are the components that are most sensitive to viscosity variations. Great
care should be taken to choose an oil that is within the manufacturer's specification
for a particular pump at the operating temperature of system. If the viscosity of a
fluid is too high, it becomes difficult to move through the system, causing increased
power loss and heat buildup due to excessive fluid friction. If the viscosity is too
low, excessive wear and heat buildup results because the oil is not thick enough to
provide good lubrication. Low viscosity also causes increased leakage because a
thinner fluid will flow more easily through an opening. This leakage results in power
loss. Note that low and high viscosities have similar results: increased power loss
and overheating. This will be discussed further in Chapter 3, which covers pumps.

2.4.4 BULK MODULUS

As stated in Chapter 1, hydraulic fluids are *relatively* incompressible. They are
not, however, completely incompressible. The *bulk modulus* of a liquid is a measure of its incompressibility, or "stiffness." It is defined by the following equation:

$$B = \frac{-\Delta p}{\Delta V / V}$$

(2-13)

where: B = bulk modulus (psi, Pa),
Δp = change in pressure (psi, Pa),
ΔV = change in volume (in³, m³),
V = volume (in³, m³).

The $\Delta V / V$ term is a proportional change in volume. The Δp term has a negative sign because an increase in pressure will result in compression or a
decrease in volume (ΔV is negative). This ensures that B is always positive. The
following example illustrates the meaning of this equation.

EXAMPLE 2-8.

In a system using hydraulic oil (B = 250,000 psi), the pressure is 3000 psi.
By what percentage is the oil being compressed relative to the unpressurized
state (p = 0 psi)?

SOLUTION:

1. Calculate the change in pressure:

$$\Delta p = 3000 \text{ psi} - 0 \text{ psi} = 3000 \text{ psi}$$

2. Calculate the proportional change in volume by solving Equation 2-13 for $\Delta V/V$:

$$\Delta V/V = \frac{-\Delta p}{B} = \frac{-3000 \text{ psi}}{250,000 \text{ psi}} = -0.012$$

3. Multiply this by 100 to obtain the % change:

$$-0.012 \times 100 = -1.2\%$$

The previous example tells us that at 3000 psi, the oil is being compressed by only 1.2% of its original volume. This justifies our earlier statement that oil is relatively incompressible. This small amount of compressibility only becomes an issue in systems in which fast response time is a critical factor.

2.5 Liquid Flow

2.5.1 FLOW RATE VERSUS FLOW VELOCITY

The *flow rate* is the volume of fluid that moves through a system in a given period of time. Flow rates determine the speed at which the output device (e.g., a cylinder) will operate. The *flow velocity* of a fluid is the distance the fluid travels in a given period of time. These two quantities are often confused, so care should be taken to note the distinction. The following equation relates the flow rate and flow velocity of a liquid to the size (area) of the conduit (i.e., a pipe or tube) through which it flows:

$$Q = v \cdot A \qquad \qquad \textbf{(2-14)}$$

where: Q = flow rate $\left(\dfrac{\text{in}^3}{\text{min}}, \dfrac{\text{m}^3}{\text{min}} \right)$,

v = flow velocity $\left(\dfrac{\text{in}}{\text{min}}, \dfrac{\text{m}}{\text{min}} \right)$,

A = area (in^2, m^2).

This is shown graphically in Figure 2-7. Light-blue arrows will be used throughout this text to represent fluid flow. It is important to note that the area

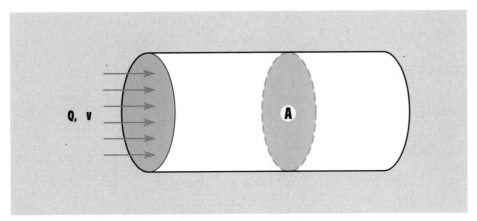

FIGURE 2-7   Flow velocity versus flow rate.

is the *inside* area of the conduit. We therefore must know the inside diameter (*ID*) of the conduit to determine this area.

EXAMPLE 2-9.

A fluid flows at a velocity of 1000 in/min through a conduit with an *ID* of 1 in. Determine the flow rate.

SOLUTION:

1. Calculate the conduit area:

$$A = \frac{\pi \cdot D^2}{4} = \frac{3.142 \cdot (1 \text{ in})^2}{4} = 0.7854 \text{ in}^2$$

2. Calculate the flow rate:

$$Q = v \cdot A = 1000 \frac{\text{in}}{\text{min}} \cdot 0.7854 \text{ in}^2 = 785.4 \frac{\text{in}^3}{\text{min}}$$

Flow rates are commonly expressed in gallons per minute (written gal/min or gpm). The conversion factor from in³ to gallons is 1 gal = 231 in³.

3. Convert to gpm:

$$785.4 \frac{\text{in}^3}{\text{min}} \cdot \left(\frac{1 \text{ gal}}{231 \text{ in}^3} \right) = 3.4 \frac{\text{gal}}{\text{min}}$$

Many of the common conversion factors used in fluid power are compiled in Appendix B.

EXAMPLE 2-9M.

A fluid flows at a velocity of 25 m/min through a conduit with an *ID* of 30 mm. Determine the flow rate.

SOLUTION:

1. Calculate the conduit area:

$$A = \frac{\pi \cdot D^2}{4} = \frac{3.142 \cdot (0.030 \text{ m})^2}{4} = 0.0007070 \text{ m}^2$$

2. Calculate the flow rate:

$$Q = v \cdot A = 25\,\frac{\text{m}}{\text{min}} \cdot 0.0007069 \text{ m}^2 = 0.01767\,\frac{\text{m}^3}{\text{min}}$$

Flow rates are commonly expressed in liters per minute (written l/min or lpm). The conversion factor from m³ to liters is 1 m³ = 1000 l.

3. Convert to gpm:

$$0.01767\,\frac{\text{m}^3}{\text{min}} \cdot \left(\frac{1000 \text{ l}}{1 \text{ m}^3}\right) = 17.67\,\frac{\text{l}}{\text{min}}$$

EXAMPLE 2-10.

A fluid flows at a rate of 10 gpm through a conduit with an *ID* of 1.5 in. Determine the flow velocity in ft/s.

SOLUTION:

1. Calculate the conduit area:

$$A = \frac{\pi \cdot D^2}{4} = \frac{3.142 \cdot (1.5 \text{ in})^2}{4} = 1.767 \text{ in}^2$$

2. Convert to in³/min:

$$10\,\frac{\text{gal}}{\text{min}} \cdot \left(\frac{231 \text{ in}^3}{1 \text{ gal}}\right) = 2310\,\frac{\text{in}^3}{\text{min}}$$

3. Calculate the flow velocity:

$$v = \frac{Q}{A} = \frac{2310 \dfrac{\text{in}^3}{\text{min}^3}}{1.767 \ \text{in}^2} = 1307 \ \frac{\text{in}}{\text{min}}$$

4. Convert to ft/s:

$$1307 \ \frac{\text{in}}{\text{min}} \cdot \left(\frac{1 \ \text{ft}}{12 \ \text{in}} \right) \cdot \left(\frac{1 \ \text{in}}{60 \ \text{s}} \right) = 1.815 \ \frac{\text{ft}}{\text{s}}$$

EXAMPLE 2-10M.

A fluid flows at a rate of 30 lpm through a conduit with an *ID* of 40 mm. Determine the flow velocity in m/s.

SOLUTION:

1. Calculate the conduit area:

$$A = \frac{\pi \cdot D^2}{4} = \frac{3.142 \cdot (0.040 \ \text{m})^2}{4} = 0.001257 \ \text{m}^2$$

2. Convert to m³/min:

$$30 \ \frac{1}{\text{min}} \cdot \left(\frac{1 \ \text{m}^3}{1000 \ 1} \right) = 0.030 \ \frac{\text{m}^3}{\text{min}}$$

3. Calculate the flow velocity:

$$v = \frac{Q}{A} = \frac{0.030 \dfrac{\text{m}^3}{\text{min}}}{0.001257 \ \text{m}^2} = 23.87 \ \frac{\text{m}}{\text{min}}$$

4. Convert to m/s:

$$23.87 \ \frac{\text{m}}{\text{min}} \cdot \left(\frac{1 \ \text{min}}{60 \ \text{s}} \right) = 0.3978 \ \frac{\text{m}}{\text{s}}$$

EXAMPLE 2-11.

A conduit size needs to be determined for a system in which the flow rate will be 15 gpm. Determine the conduit *ID* if the flow velocity is not to exceed 20 ft/s.

SOLUTION:

1. Convert gpm to in³/min:

$$Q = 15 \frac{\text{gal}}{\text{min}} \cdot \left(\frac{231 \text{ in}^3}{1 \text{ gal}} \right) = 3465 \frac{\text{in}^3}{\text{min}}$$

2. Convert ft/s to *in/min:*

$$v = 20 \frac{\text{ft}}{\text{s}} \cdot \left(\frac{12 \text{ in}}{1 \text{ ft}} \right) \cdot \left(\frac{60 \text{ s}}{1 \text{ min}} \right) = 14,400 \frac{\text{in}}{\text{min}}$$

3. Calculate the conduit area:

$$A = \frac{Q}{v} = \frac{3465 \dfrac{\text{in}^3}{\text{min}}}{14,400 \dfrac{\text{in}}{\text{min}}} = 0.2406 \text{ in}^2$$

4. Calculate the conduit *ID:*

$$D = \sqrt{\frac{4 \cdot A}{\pi}} = \sqrt{\frac{4 \cdot (0.2406 \text{ in}^2)}{3.142}} = 0.5535 \text{ in}$$

    If we do not want the flow velocity to exceed 20 ft/s, we must choose a conduit that has an *ID* this size or larger. We would therefore select the next larger available size.

EXAMPLE 2-11M.

A conduit size needs to be determined for a system in which the flow rate will be 100 lpm. Determine the conduit *ID* if the flow velocity is not to exceed 6 m/s.

SOLUTION:

1. Convert lpm to m³/min:

$$Q = 100 \frac{\text{l}}{\text{min}} \left(\frac{1 \text{ m}^3}{1000 \text{ l}} \right) = 0.100 \frac{\text{m}^3}{\text{min}}$$

2. Convert m/s to m/min:

$$v = 6\,\frac{\text{m}}{\text{s}} \cdot \left(\frac{60\ \text{s}}{1\ \text{min}}\right) = 360\,\frac{\text{m}}{\text{min}}$$

3. Calculate the conduit area:

$$A = \frac{Q}{v} = \frac{0.100\,\dfrac{\text{m}^3}{\text{min}}}{360\,\dfrac{\text{m}}{\text{min}}} = 0.0002778\ \text{m}^2$$

4. Calculate the conduit *ID*:

$$D = \sqrt{\frac{4 \cdot A}{\pi}} = \sqrt{\frac{4 \cdot (0.0002778\ \text{m}^2)}{3.142}} = 0.01881\ \text{m}\ \ (18.81\ \text{mm})$$

If we do not want the flow velocity to exceed 6 m/s, we must choose a conduit that has an *ID* this size or larger. We would therefore select the next larger available size.

2.5.2 THE CONTINUITY EQUATION

Hydraulic systems commonly have a pump that produces a constant flow rate. If we assume the fluid is incompressible, this situation is referred to as *steady flow*. This simply means that whatever volume of fluid flows through one section of the system must also flow through any other section. Figure 2-8 shows a system

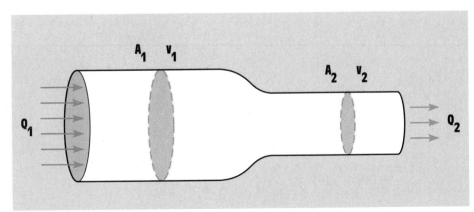

FIGURE 2-8   Continuity of flow.

in which the flow rate is constant and the diameter varies. The following equation applies in this system:

$$Q_1 = Q_2$$

for any two points in the system. If we plug $Q = v \cdot A$ into this equation, the continuity equation results:

$$v_1 \cdot A_1 = v_2 \cdot A_2 \qquad \textbf{(2-15)}$$

The following examples illustrate the significance of this equation.

EXAMPLE 2-12.

A fluid flows at a velocity of 120 in/min at point 1 in the system shown in Figure 2-8. The diameter at point 1 is 2 in and the diameter at point 2 is 1.5 in. Determine the flow velocity at point 1. Also determine the flow rate in gpm.

SOLUTION:

1. Calculate the conduit areas:

$$A_1 = \frac{\pi \cdot D_1^2}{4} = \frac{3.142 \cdot (2 \text{ in})^2}{4} = 3.142 \text{ in}^2$$

$$A_2 = \frac{\pi \cdot D_2^2}{4} = \frac{3.142 \cdot (1.5 \text{ in})^2}{4} = 1.767 \text{ in}^2$$

2. Solve Equation 2-15 for v_2:

$$v_2 = v_1 \cdot \frac{A_1}{A_2} = 120 \frac{\text{in}}{\text{min}} \cdot \frac{3.142 \text{ in}^2}{1.767 \text{ in}^2} = 213.4 \frac{\text{in}}{\text{min}}$$

3. Calculate the flow rate:

$$Q = v_1 A_1 = 120 \frac{\text{in}}{\text{min}} \cdot (3.142 \text{ in}^2) = 377.0 \frac{\text{in}^3}{\text{min}}$$

4. Convert to gpm:

$$Q = 377.0 \frac{\text{in}^3}{\text{min}} \cdot \left(\frac{1 \text{ gal}}{231 \text{ in}^3} \right) = 1.632 \frac{\text{gal}}{\text{min}}$$

EXAMPLE 2-12M.

A fluid flows at a velocity of 10 m/min at point 1 in the system shown in Figure 2-8. The diameter at point 1 is 50 mm and the diameter at point 2 is 30 mm. Determine the flow velocity at point 1. Also determine the flow rate in lpm.

SOLUTION:

1. Calculate the conduit areas:

$$A_1 = \frac{\pi \cdot D_1^{\,2}}{4} = \frac{3.142 \cdot (0.050 \text{ m})^2}{4} = 0.001964 \text{ m}^2$$

$$A_2 = \frac{\pi \cdot D_2^{\,2}}{4} = \frac{3.142 \cdot (0.030 \text{ m})^2}{4} = 0.0007070 \text{ m}^2$$

2. Solve Equation 2-15 for v_2:

$$v_2 = v_1 \cdot \frac{A_1}{A_2} = 10 \, \frac{\text{m}}{\text{min}} \cdot \frac{0.001964 \text{ m}^2}{0.0007070 \text{ m}^2} = 27.78 \, \frac{\text{m}}{\text{min}}$$

3. Calculate the flow rate:

$$Q = v_1 A_1 = 10 \, \frac{\text{m}}{\text{min}} \cdot (0.001964 \text{ m}^2) = 0.01964 \, \frac{\text{m}^3}{\text{min}}$$

4. Convert to lpm:

$$Q = 0.01964 \, \frac{\text{m}^3}{\text{min}} \cdot \left(\frac{1000 \text{ l}}{1 \text{ m}^3} \right) = 19.64 \, \frac{\text{l}}{\text{min}}$$

These examples show that in a system with a steady flow rate, a reduction in area (conduit size) corresponds to an increase in flow velocity by the same factor. If the conduit size increases, the flow velocity is reduced by the same factor. This is an important concept to understand because in an actual hydraulic system, the conduit size changes repeatedly as the fluid flows through hoses, fittings, valves, and other devices.

The increase in velocity that occurs when the conduit size is reduced represents an increase in energy. No energy has been added to the system, so we must have a corresponding reduction in energy in some other area. This leads us to Bernoulli's equation, which describes the energy balance in a fluid system.

2.5.3 BERNOULLI'S EQUATION

Bernoulli's equation is a relationship that describes the total energy of an incompressible fluid. Hydraulic fluids are nearly incompressible, so this is applicable to a hydraulic system. Energy in fluids appears in three forms:

    1. Potential energy (due to elevation and gravity) $= w \cdot h$

    2. Pressure energy (due to pressure) $= w \cdot \dfrac{p}{\gamma}$

    3. Kinetic energy (due to velocity) $= w \cdot \dfrac{v^2}{2 \cdot g}$

where: w = weight (lbs, N)
      h = height or elevation (in, m)
      p = pressure (lbs/in², N/m²)
      γ = specific weight (lbs/in³, N/m³)
      v = velocity (in/s, m/s)
      g = gravity = 386.4 in/s² = 9.81 m/s²

    All three energy terms have units of in · lbs. If no energy is added or removed from the system by some external source, then the energy at any two points must be equal. In equation form:

$$w \cdot h_1 + w \cdot \frac{p_1}{\gamma} + w \cdot \frac{v_1^{\,2}}{2 \cdot g} = w \cdot h_2 + w \cdot \frac{p_2}{\gamma} + w \cdot \frac{v_2^{\,2}}{2 \cdot g}$$

Because w appears in each term, it can be divided out to obtain:

$$h_1 + \frac{p_1}{\gamma} + \frac{v_1^{\,2}}{2 \cdot g} = h_2 + \frac{p_2}{\gamma} + \frac{v_2^{\,2}}{2 \cdot g} \tag{2-16}$$

    This is Bernoulli's equation. It expresses the energy a fluid contains per unit of weight. It tells us that any increase in energy in any one of the three areas must be balanced by a reduction in one or more of the other two areas of an equal amount. The units for each term are inches (or feet) and are referred to as *head*. The first term is called the *elevation head,* the second is the *pressure head,* and the third is the *velocity head.* The sum of the three is referred to as the *total head.* Head is simply a measure of the energy a fluid possesses per pound. Following are a few examples.

EXAMPLE 2-13.

A fluid ($\gamma = 0.0324$ lbs/in³) flows at a constant flow rate of 150 in³/s through the system shown in Figure 2-9. The areas at points 1 and 2 are equal. The pressure at point 1 is 100 psi and $h = 200$ in. Determine the pressure at point 2.

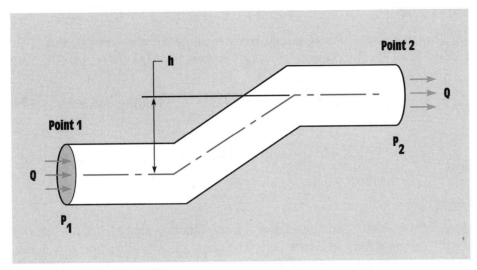

FIGURE 2-9   Example 2-13.

SOLUTION:

The areas at points 1 and 2 are equal and the flow rate is constant, so the flow velocities at points 1 and 2 are equal, as shown by the continuity equation. The velocity terms can be ignored because they are equal and on both sides of the equation. Bernoulli's equation reduces to: $h_1 + \dfrac{p_1}{\gamma} = h_2 + \dfrac{p_2}{\gamma}$.

1. Solve for p_2:

$$p_2 = p_1 - \gamma\,(h_2 - h_1)$$

2. Calculate p_2:

$$p_2 = 100\,\frac{\text{lbs}}{\text{in}^2} - 0.0324\,\frac{\text{lbs}}{\text{in}^3}\,(200\ \text{in}) = 93.5\,\frac{\text{lbs}}{\text{in}^2}$$

We can interpret this result as follows: The cost of raising the fluid 200 inches in elevation is to sacrifice 6.5 psi of pressure. The total energy of the fluid remains the same.

EXAMPLE 2-13M.

A fluid ($\gamma = 8800$ N/m³) flows at a constant flow rate of 10 lpm through the system shown in Figure 2-9. The areas at points 1 and 2 are equal. The pressure at point 1 is 700 kPa and $h = 5$ m. Determine the pressure at point 2.

SOLUTION:

Recall that a pascal (Pa) is simply another name for the unit N/m² and that 1 kPa = 1000 Pa. The pressure at point 1 is therefore 700 kPa = 700,000 N/m².

$$p_2 = p_1 - \gamma(h_2 - h_1) = 700,000 \frac{N}{m^2} - 8800 \frac{N}{m^3}(5m)$$

$$= 656,000 \frac{N}{m^2} \quad (656 \text{ kPa})$$

EXAMPLE 2-14.

A fluid ($\gamma = 0.0324$ lbs/in³) flows at a constant flow rate of 300 in³/s through the system shown in Figure 2-10. The areas are $A_1 = 2$ in² and $A_2 = 1$ in². If the pressure at point 1 is 150 psi, determine the pressure at point 2.

SOLUTION:

Because $h_1 = h_2$, the elevation terms drop out. Bernoulli's equation reduces to:

$$\frac{p_1}{\gamma} + \frac{v_1^2}{2 \cdot g} = \frac{p_2}{\gamma} + \frac{v_2^2}{2 \cdot g}$$

Solving for p_2, we obtain: $p_2 = p_1 + \dfrac{\gamma}{2 \cdot g}(v_1^2 - v_2^2)$

1. Calculate the flow velocities:

$$v_1 = \frac{Q}{A_1} = \frac{300 \frac{in^3}{s}}{2 \text{ in}^2} = 150 \frac{in}{s}$$

$$v_2 = \frac{Q}{A_2} = \frac{300 \frac{in^3}{s}}{1 \text{ in}^2} = 300 \frac{in}{s}$$

2. Calculate p_2:

$$p_2 = 150 \frac{lbs}{in^2} + \frac{0.0324 \frac{lbs}{in^3}}{2 \cdot \left(386.4 \frac{in}{s^2}\right)} \left(\left(150 \frac{in}{s}\right)^2 - \left(300 \frac{in}{s}\right)^2 \right) = 147.2 \frac{lbs}{in^2}$$

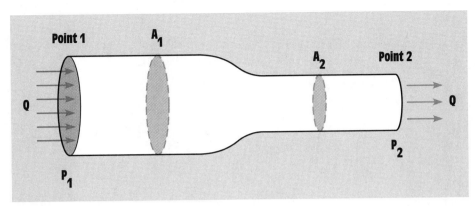

Point 1 A_1 A_2 Point 2

Q

Q

P_2

P_1

FIGURE 2-10   Example 2-14.

EXAMPLE 2-14M.

A fluid ($\gamma = 8800$ N/m³) flows at a constant flow rate of 0.005 m³/s through the system shown in Figure 2-10. The areas are $A_1 = 0.002$ m² and $A_2 = 0.001$ m². If the pressure at point 1 is 1000 kPa, determine the pressure at point 2.

SOLUTION:

1. Calculate the flow velocities:

$$v_1 = \frac{Q}{A_1} = \frac{0.005\,\dfrac{m^3}{s}}{0.002\,m^2} = 2.5\,\frac{m}{s}$$

$$v_2 = \frac{Q}{A_2} = \frac{0.005\,\dfrac{m^3}{s}}{0.001\,m^2} = 5\,\frac{m}{s}$$

2. Calculate p_2:

$$p_2 = p_1 + \frac{\gamma}{2 \cdot g}(v_1^{\,2} - v_2^{\,2})$$

$$p_2 = 1,000,000\,\frac{N}{m^2} + \frac{8800\,\dfrac{N}{m^3}}{2 \cdot \left(9.81\,\dfrac{m}{s^2}\right)}\left(\left(2.5\,\frac{m}{s}\right)^2 - \left(5\,\frac{m}{s}\right)^2\right)$$

$$= 991,600\,\frac{N}{m^2} \quad (991.6\ kpa)$$

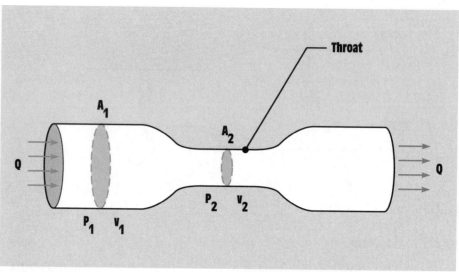

FIGURE 2-11   A venturi.

The previous examples illustrate that reducing the size of the conduit causes an *increase* in velocity and a *decrease* in pressure. As stated previously, this is an important principle to understand because the size of the conduit through which the hydraulic fluid flows changes repeatedly as it moves through the system.

Certain applications take advantage of Bernoulli's principle. One example is a *venturi,* a conduit in which the diameter of a section has been reduced to intentionally produce a low-pressure area (Figure 2-11). The reduced diameter section is known as the *throat.* One application for a venturi is in the measurement of flow velocity and flow rate. In this situation, the heights are equal and the elevation terms drop out. Bernoulli's equation reduces to:

$$\frac{p_1}{\gamma} + \frac{v_1^{\,2}}{2 \cdot g} = \frac{p_2}{\gamma} + \frac{v_2^{\,2}}{2 \cdot g}$$

We then plug in $v_1 = v_2 \cdot \dfrac{A_2}{A_1}$ (from the *continuity equation*), and solve for v_2. This results in the following equation:

$$v_2 = \sqrt{\frac{2 \cdot g \cdot (p_1 - p_2)}{\gamma \cdot \left[1 - \left(\dfrac{A_2}{A_1}\right)\right]}} \qquad \textbf{(2-17)}$$

Equation 2-17 is applied as follows: We know the quantities A_1 and A_2, as they can easily be calculated from the diameters. The quantities p_1 and p_2 are measured using pressure sensors. We can then calculate the velocity, v_2, using Equation 2-17. Finally, the flow rate is calculated using $q = v_2 \cdot A_2$. This is one of the methods used to determine fluid flow rates in industry. Others will be discussed in Chapter 9 when flowmeters are discussed. The following example illustrates the preceding calculation.

EXAMPLE 2-15.

A fluid ($\gamma = 0.0324$ lbs/in³) flows at a constant flow rate through the system shown in Figure 2-11. The areas are $A_1 = 3$ in² and $A_2 = 1.5$ in². The pressures were measured to be: $p_1 = 100$ psi and $p_2 = 96$ psi. Determine the flow rate of this fluid in gpm.

SOLUTION:

1. Calculate v_2:

$$v_2 = \sqrt{\dfrac{2\left(386.4\,\dfrac{\text{in}}{\text{s}^2}\right)\left(100\,\dfrac{\text{lbs}}{\text{in}^2} - 96\,\dfrac{\text{lbs}}{\text{in}^2}\right)}{0.0324\,\dfrac{\text{lbs}}{\text{in}^3}\cdot\left[1-\left(\dfrac{1.5\ \text{in}^2}{3\ \text{in}^2}\right)\right]}} = 436.8\,\dfrac{\text{in}}{\text{s}}$$

2. Calculate Q:

$$Q = v_2 \cdot A_2 = 436.8\,\frac{\text{in}}{\text{s}}\,(1.5\ \text{in}^2) = 655.2\,\frac{\text{in}^3}{\text{s}}$$

3. Convert to gpm:

$$655.2\,\frac{\text{in}^3}{\text{s}}\cdot\left(\frac{60\ \text{s}}{1\ \text{min}}\right)\cdot\left(\frac{1\ \text{gal}}{231\ \text{in}^3}\right) = 170\,\frac{\text{gal}}{\text{min}}$$

EXAMPLE 2-15M.

A fluid ($\gamma = 8800$ N/m³) flows at a constant rate through the system shown in Figure 2-11. The areas are $A_1 = 0.002$ m² and $A_2 = 0.001$ m². The pressures were measured to be: $p_1 = 900$ kPa and $p_2 = 800$ kPa. Determine the flow rate of this fluid in lpm.

SOLUTION:

1. Calculate v_2:

$$v_2 = \sqrt{\frac{2\left(9.81\,\frac{m}{s^2}\right)\left(900,000\,\frac{N}{m^2} - 800,000\,\frac{N}{m^2}\right)}{8800\,\frac{N}{m^3}\cdot\left[1-\left(\frac{0.001\,m^2}{0.002\,m^2}\right)\right]}} = 21.12\,\frac{m}{s}$$

2. Calculate Q:

$$Q = v_2 \cdot A_2 = 21.12\,\frac{m}{s}\,(0.001\,in^2) = 0.02112\,\frac{m^3}{s}$$

3. Convert to lpm:

$$0.02112\,\frac{m^3}{s}\cdot\left(\frac{60\,s}{1\,min}\right)\cdot\left(\frac{1000\,l}{1\,m^3}\right) = 1267\,\frac{l}{min}$$

Because of the low-pressure area it creates, a venturi can also be used to draw another fluid into the fluid stream. A carburetor uses this method to draw gasoline into the air stream before it is fed into the inlet valve of the engine.

2.5.4 TORRICELLI'S THEOREM

Torricelli's theorem is a special case of Bernoulli's equation that applies to a liquid draining from a tank, as shown in Figure 2-12. In this situation, we may want to know the velocity of the flow out of the tank at the outlet. This can be determined by applying Bernoulli's equation to the surface of the tank (point 1) and at the outlet (point 2). The pressure at both points is equal (0 psi), so these terms drop out. It is also valid to assume that the velocity at point 1 (v_1) is zero because it will be very small as compared to the velocity at point 2 (v_2). Bernoulli's equation then simplifies to:

$$h_1 = h_2 + \frac{v_2^2}{2\cdot g}$$

Solving for v_2, we obtain:

$$v_2 = \sqrt{2\cdot g(h_1 - h_2)}$$

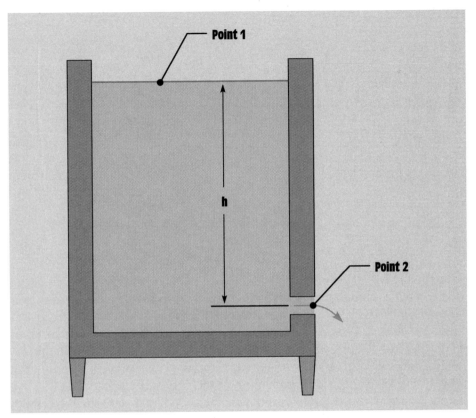

FIGURE 2-12   Liquid draining from a tank.

We may refer to $h_1 - h_2$ simply as h, the height of the fluid level above the outlet. This results in the following equation, known as Torricelli's theorem:

$$v_2 = \sqrt{2 \cdot g \cdot h} \qquad \textbf{(2-18)}$$

The preceding equation shows that the discharge velocity depends *only* on the height of the fluid level above the outlet and not on the size of the tank. As the fluid drains out of the tank, the height is reduced, so the velocity decreases constantly. We are therefore calculating the velocity only at a particular instant in time.

EXAMPLE 2-16.

The tank shown in Figure 2-12 is filled with liquid to a height of 5 feet. Determine the velocity at the outlet.

SOLUTION:

$$v_2 = \sqrt{2 \cdot \left(32.2\,\frac{ft}{s^2}\right) \cdot 5\,ft} = 17.94\,\frac{ft}{s}$$

EXAMPLE 2-16M.

The tank shown in Figure 2-12 is filled with liquid to a height of 3 m. Determine the velocity at the outlet.

SOLUTION:

$$v_2 = \sqrt{2 \cdot \left(9.81\,\frac{m}{s^2}\right) \cdot 3\,m} = 7.67\,\frac{m}{s}$$

2.5.5 LAMINAR VERSUS TURBULENT FLOW

As stated earlier, fluid often flows in thin layers, each having a different velocity. We now apply this concept to flow through a cylindrical conduit. Figure 2-13 shows the velocity profile in this situation. Just as with the parallel plates example (Figure 2-6), a thin layer of fluid adheres to the stationary surface, in this case, the conduit wall. As we move away from the wall, the velocity increases with each layer, until it reaches its maximum at the center of the conduit. This type of flow is known as *laminar* flow, and is characterized by its smooth, layered flow.

Earlier we computed the flow velocity using the equation $v = Q/A$. This equation actually computes the *average* velocity through a conduit, which in most cases is exactly what we want. If the average velocity is increased, it will

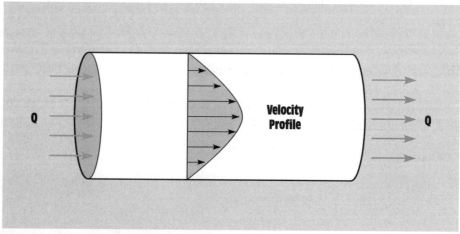

FIGURE 2-13   Laminar flow velocity profile.

eventually cause the flow to transition from laminar to *turbulent* flow. In this type of flow, the fluid does not flow in a smooth, layered pattern as it does with laminar flow. Instead, it flows in a rough, chaotic fashion with parcels of fluid flowing in all directions, as shown in Figure 2-14. This gives turbulent flow its characteristic churning action. It is always desirable to avoid turbulent flow in fluid power systems because the chaotic flow is inefficient and causes more energy losses than does laminar flow. The type of flow present in a given situation can be predicted with the Reynolds number, which is given by the following equation:

$$Re = \frac{v \cdot D}{\nu}$$ **(2-19)**

where: Re = Reynolds number (no units)
v = velocity (ft/s, m/s)
D = diameter (ft, m)
ν = kinematic viscosity (ft^2/s, m^2/s)

The units in Equation 2-19 cancel out so that the Reynolds number has no units. Quantities without units, such as this, are known as *dimensionless parameters*. Note the distinction between velocity and kinematic viscosity; both use symbols that are similar in appearance.

Experimentation has determined that the flow through a conduit in practical situations will be laminar for $Re < 2000$. The flow will be turbulent for $Re > 4000$. The range between 2000 and 4000 is a transitional range in which the type of flow is difficult to predict.

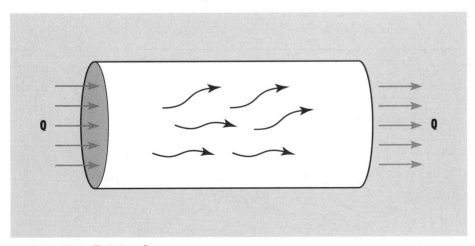

FIGURE 2-14   Turbulent flow.

EXAMPLE 2-17.

A fluid with a kinematic viscosity of 0.00240 ft²/s is flowing through a 2-in diameter conduit at a flow rate of 200 gpm. Determine if the flow will be laminar or turbulent.

SOLUTION:

1. Convert gpm to in³/s:

$$Q = 200\,\frac{\text{gal}}{\text{min}} \cdot \left(\frac{231\ \text{in}^3}{1\ \text{gal}}\right) \cdot \left(\frac{1\ \text{min}}{60\ \text{s}}\right) = 770\,\frac{\text{in}^3}{\text{s}}$$

2. Calculate the area:

$$A = \frac{\pi \cdot D^2}{4} = \frac{3.142 \cdot (2\ \text{in})^2}{4} = 3.1416\ \text{in}^2$$

3. Calculate the flow velocity:

$$v = \frac{Q}{A} = \frac{770\,\frac{\text{in}^3}{\text{s}}}{3.1416\ \text{in}^2} = 245.1\,\frac{\text{in}}{\text{s}}$$

4. Convert in/s to ft/s:

$$v = 245.1\,\frac{\text{in}}{\text{s}} \cdot \left(\frac{1\ \text{ft}}{12\ \text{in}}\right) = 20.43\,\frac{\text{ft}}{\text{s}}$$

5. Convert in to ft:

$$D = 2\ \text{in} \cdot \left(\frac{1\ \text{ft}}{12\ \text{in}}\right) = 0.1667\ \text{ft}$$

6. Calculate *Re*:

$$Re = \frac{v \cdot D}{\nu} = \frac{20.43\,\frac{\text{ft}}{\text{s}} \cdot (0.1667\ \text{ft})}{0.00240\,\frac{\text{ft}^2}{\text{s}}} = 1419$$

EXAMPLE 2-17M.

A fluid with a kinematic viscosity of 0.000223 m²/s is flowing through a 50-mm diameter conduit at a flow rate of 800 lpm. Determine if the flow will be laminar or turbulent.

SOLUTION:

1. Convert lpm to m³/s:

$$Q = 800 \, \frac{l}{min} \cdot \left(\frac{1 \, m^3}{1000 \, l} \right) \cdot \left(\frac{1 \, min}{60 \, s} \right) = 0.01333 \, \frac{m^3}{s}$$

2. Calculate the area:

$$A = \frac{\pi \cdot D^2}{4} = \frac{3.142 \cdot (0.050 \, m)^2}{4} = 0.001964 \, m^2$$

3. Calculate the flow velocity:

$$v = \frac{Q}{A} = \frac{0.01333 \, \frac{m^3}{s}}{0.001964 \, m^2} = 6.787 \, \frac{m}{s}$$

4. Calculate *Re*:

$$Re = \frac{v \cdot D}{v} = \frac{6.787 \, \frac{m}{s} \cdot (0.050 \, m)}{0.000223 \, \frac{m^2}{s}} = 1522$$

   In both of the previous examples, the Reynolds number is below the critical number of 2000, so the flow will be laminar in each case. As stated previously, turbulent flow should be avoided because of its inefficient nature. This can be done by choosing conduits, fittings, and valves that are appropriately sized for the flow rate of the system.

2.6   Static Head Pressure

As we submerge into a body of water, we are subjected to a pressure that increases as we dive deeper. This pressure is due to the weight of the fluid above our location pressing down on us. The technical term for this is *static head pressure,* and its magnitude can be determined using the following equation:

$$p_H = \gamma \cdot h \tag{2-20}$$

where: p_H = head pressure (lbs/in³, N/m³)
  γ = specific weight (lbs/in³, N/m³),
  h = height (in, m)

EXAMPLE 2-18.

The tank shown in Figure 2-15 is filled with oil with a specific weight of 0.0324 lbs/in³ to a depth of 20 ft. What is the pressure at the bottom of the tank?

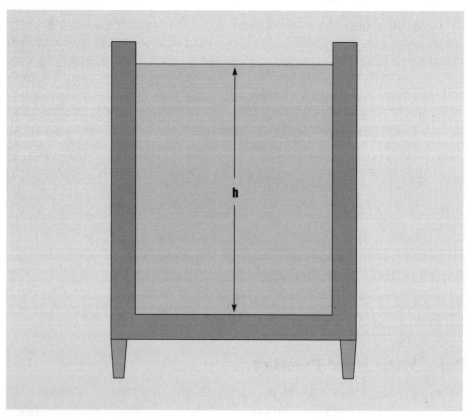

FIGURE 2-15   Example 2-18.

SOLUTION:

1. Convert feet to inches:

$$h = 20 \text{ ft} \cdot \left(\frac{12 \text{ in}}{1 \text{ ft}}\right) = 240 \text{ in}$$

2. Calculate the head pressure:

$$p_H = \gamma \cdot h = 0.0324 \frac{\text{lbs}}{\text{in}^3} \cdot (240 \text{ in}) = 7.78 \frac{\text{lbs}}{\text{in}^2}$$

Notice that the pressure at the bottom of the tank does not depend on the size of the tank. It depends only on the height, or depth, of the fluid. Whether the tank in the previous example has a diameter of 2 ft or 10 ft, the pressure at the bottom is 7.78 psi.

EXAMPLE 2-18M.

The tank shown in Figure 2-15 is filled with oil with a specific weight of 8800 N/m³ to a depth of 5 m. What is the pressure at the bottom of the tank?

SOLUTION:

$$p_H = \gamma \cdot h = 8800 \frac{\text{N}}{\text{m}^3} \cdot (5\text{m}) = 44{,}000 \frac{\text{N}}{\text{m}^2} \quad (44 \text{ kPa})$$

2.7   Pressure Losses

Pascal's Law states that in a pressurized, confined fluid, the pressure is equal everywhere and acts at right angles to the containing surfaces. When we have a *flowing* fluid, however, the pressure drops as the fluid flows through the system due to energy losses. These energy losses are primarily due to the viscosity of the fluid resisting the flow. Whenever the flow stops, the pressure very quickly equalizes again and Pascal's Law holds. The pressure continues to act perpendicular to the containing surfaces.

Pressure drops are a major concern not only because they represent energy losses and inefficiency, but also because of where the energy goes. The lost energy is transformed into heat, which raises the temperature of the fluid. As stated earlier, when a fluid's temperature increases, its viscosity goes down (it becomes thinner). When a fluid becomes thinner, its ability to lubricate is reduced, resulting in increased component wear and reduced life.

2.8   Power

Work is defined as a measure of energy expended and is calculated by multi-plying an applied force times a distance moved due to that force ($W = F \cdot d$). Previously in this chapter, an example of a wheelbarrow that was moved 100 ft and required 50 lbs of force to push was given. In this case the work done is: $W = 50 \text{ lbs} \cdot 100 \text{ ft} = 5000 \text{ ft} \cdot \text{lbs}$. What work does not tell us, however, is how *fast* the wheelbarrow was moved. Was it done in 10 seconds or 5 seconds? *Power* takes into account how fast energy is expended. By dividing the work by the time (t) in which that work was done, we have an equation for power:

$$P = \frac{W}{t} = \frac{F \cdot d}{t} \qquad \textbf{(2-21)}$$

If the work is in ft $\cdot$ lbs and the time is in seconds (s), the power will be in $\frac{\text{ft} \cdot \text{lbs}}{\text{s}}$. Note that uppercase P is used for power, while lowercase p is used for pressure. We can replace $\frac{d}{t}$ in Equation 2-21 with the velocity, v, to obtain:

$$P = F \cdot v \qquad \textbf{(2-22)}$$

If the force is in lbs and the velocity is in ft/s, we will again have units of $\frac{\text{ft} \cdot \text{lbs}}{\text{s}}$ for the power. It is most common, however, to express power in *horsepower* (hp). The conversion factor between these units is:

$$1 \text{ hp} = 550 \, \frac{\text{ft} \cdot \text{lbs}}{\text{s}}$$

This is a common conversion that is useful to incorporate into Equation 2-22. The following equation results:

$$HP = \frac{F \cdot v}{550} \qquad \textbf{(2-23)}$$

HP signifies that the power is in horsepower. By inserting the conversion factor into the equation, we have locked ourselves into using lbs for force and ft/s for velocity. *No other units may be used.* These types of equations with built-in conversions are commonly used in industry because of their obvious convenience. There are important differences between these equations, in which only specific units may be used, and equations such as Equation 2-22, in which any *consistent units* (those that cancel properly) may be used.

In the metric system of units, the force will be in newtons (N) and the velocity will be in m/s. Equation 2-22 will therefore result in units of $\frac{N \cdot m}{s}$ for power. This unit is also known as a *watt* (W). Because this is a small unit, power is most commonly expressed in kilowatts (kW) in the metric system. Inserting the appropriate conversion factor (1 kW = 1000 W) into Equation 2-22 results in the following equation, which uses the most common metric units:

$$kW = \frac{F \cdot v}{1000} \qquad \textbf{(2-23M)}$$

We use kW to signify that the resulting power is in kilowatts. For Equation 2-23M, the required units are F in N and v in m/s.

EXAMPLE 2-19.

A 2000-lb load must be moved at a velocity of 2 ft/s. How much horsepower is required?

SOLUTION:

$$HP = \frac{F \cdot v}{550} = \frac{2000 \text{ lbs} \cdot \left(2\frac{\text{ft}}{\text{s}}\right)}{550} = 7.27 \text{ hp}$$

EXAMPLE 2-19M.

A 10,000-N load must be moved at a velocity of 1 m/s. How much power is required?

SOLUTION:

$$kW = \frac{F \cdot v}{1000} = \frac{10,000 \text{ N} \cdot \left(1\frac{\text{m}}{\text{s}}\right)}{1000} = 10 \, kW$$

For a hydraulic system, it is useful to express the power in terms of pressure and flow. If we substitute $F = p \cdot A$ (Equation 2-1) into Equation 2-22, we obtain:

$$P = p \cdot A \cdot v$$

We can then insert $Q = A \cdot v$ (Equation 2-14) to obtain:

$$P_H = p \cdot Q \qquad \textbf{(2-24)}$$

P_H is the hydraulic power of a flowing fluid. If we have the pressure in lbs/in² and the flow in in³/min, we will get units of $\frac{\text{in} \cdot \text{lbs}}{\text{min}}$ for the power. Again, the preferred unit in the U.S. customary system for power is horsepower, so we will use the following conversion factor:

$$1 \text{ hp} = 396,000 \, \frac{\text{in} \cdot \text{lbs}}{\text{min}}$$

As stated previously, the preferred unit in industry for flow is gpm and 1 gal = 231 in³. To incorporate these units into Equation 2-24, we must multiply by 231 and divide by 396,000. The following commonly used equation for hydraulic horsepower results:

$$HP_H = \frac{p \cdot Q}{1714} \qquad \textbf{(2-25)}$$

Equation 2-25 is similar to Equation 2-23 in that specific units are required. The pressure must be in psi, the flow in gpm, and the power in hp.

If metric units are used in Equation 2-24 and the pressure is in N/m² (Pa), and the flow is in m³/s, the power will be in $\frac{\text{N} \cdot \text{m}}{\text{s}}$, which is known as a watt (W). The preferred unit for flow in the metric system is lpm. To convert from lpm to m³/s, we must use the following conversion factor:

$$1 \frac{\text{m}^3}{\text{s}} \cdot \left(\frac{1000 \text{ l}}{1 \text{ m}^3} \right) \cdot \left(\frac{60 \text{ s}}{1 \text{ min}} \right) = 60,000 \, \frac{1}{\text{min}}$$

The preferred unit for pressure is kPa (1 kPa = 1000 Pa) and the preferred unit for power is kW (1 kW = 1000 W). If we plug these three conversions into Equation 2-24, the following useful equation results:

$$kW_H = \frac{p \cdot Q}{60,000} \qquad \textbf{(2-25M)}$$

We must use the following units for Equation 2-25M: pressure in kPa, flow in lpm, and the hydraulic power in kW.

EXAMPLE 2-20.

▎ Oil is flowing through a hydraulic system at 10 gpm and 2500 psi. What is the hydraulic power of this system?

SOLUTION:

$$HP_H = \frac{p \cdot Q}{1714} = \frac{2500 \text{ psi} \cdot (10 \text{ gpm})}{1714} = 14.6 \text{ hp}$$

EXAMPLE 2-20M.

▎ Oil is flowing through a hydraulic system at 40 lpm and 15,000 kPa. What is the hydraulic horsepower of this system?

SOLUTION:

$$kW_H = \frac{p \cdot Q}{60,000} = \frac{15,000 \text{ kPa} \cdot (40 \text{ lpm})}{60,000} = 10 \text{ kW}$$

2.9 Hydraulic Systems

Hydraulic systems can be divided into three main segments: (1) *power supply,* (2) *control,* and (3) *output.* The power supply segment supplies flow to the system. Its primary components include the *prime mover* and the *pump.* The prime mover is the device that provides the mechanical input power by driving the pump. In an industrial system, this device is usually an electric motor. In mobile systems, such as excavators and farm equipment, the prime mover is the engine of the vehicle. The pump takes the mechanical power from the prime mover and converts it to fluid power (a fluid flowing under pressure).

The output segment of the system includes the *actuator* and the *load.* The actuator converts the fluid power back into mechanical power, which is used to move the load. The actuator may be a *linear actuator* (cylinder), which produces straight-line motion, or a *rotary actuator* (motor), which produces rotational motion.

The control segment is in the middle of the system. This segment includes directional control valves, pressure control valves, and flow control valves. By controlling the direction of the fluid flow, we can control the direction of the actuator, allowing a cylinder to extend or retract, for example. Controlling the pressure of the fluid allows us to control the output force, as stated earlier in this chapter in the discussion of Pascal's Law. Controlling the flow rate of the fluid regulates the speed of the actuator.

Chapters 3 through 8 cover the components discussed previously. Chapter 9 covers the ancillary hydraulic components such as seals, filters, reservoirs, etc.

2.10 Equations

EQUATION NUMBER	EQUATION	REQUIRED UNITS
2-1	$p = \dfrac{F}{A}$	Any consistent units
2-2	$A = \dfrac{\pi \cdot D^2}{4}$	Any consistent units
2-3	$F_{OUT} = \dfrac{A_{OUT}}{A_{IN}} \cdot F_{IN}$	Any consistent units
2-4	$W = F \cdot d$	Any consistent units
2-5	$d_{IN} = \dfrac{A_{OUT}}{A_{IN}} \cdot d_{OUT}$	Any consistent units
2-6	$v_{IN} = \dfrac{A_{OUT}}{A_{IN}} \cdot v_{OUT}$	Any consistent units
2-7	$w = m \cdot g$	Any consistent units
2-8	$\rho = \dfrac{m}{V}$	Any consistent units
2-9	$\gamma = \dfrac{w}{V}$	Any consistent units
2-10	$sg_x = \dfrac{\gamma_x}{\gamma_{WATER}}$	Any consistent units
2-11	$\mu = \dfrac{F \cdot y}{v \cdot A}$	Any consistent units
2-12	$v = \dfrac{\mu}{\rho}$	Any consistent units
2-13	$B = \dfrac{-\Delta p}{\Delta V / V}$	Any consistent units
2-14	$Q = v \cdot A$	Any consistent units
2-15	$v_1 \cdot A_1 = v_2 \cdot A_2$	Any consistent units
2-16	$h_1 + \dfrac{p_1}{\gamma} + \dfrac{v_1^2}{2 \cdot g} = h_2 + \dfrac{p_2}{\gamma} + \dfrac{v_2^2}{2 \cdot g}$	Any consistent units
2-17	$v_2 = \sqrt{\dfrac{2 \cdot g \cdot (p_1 - p_2)}{\gamma \cdot \left[1 - \left(\dfrac{A_2}{A_1} \right) \right]}}$	Any consistent units

EQUATION NUMBER	EQUATION	REQUIRED UNITS
2-18	$v_2 = \sqrt{2 \cdot g \cdot h}$	Any consistent units
2-19	$Re = \dfrac{v \cdot D}{v}$	Any consistent units
2-20	$P_H = \gamma \cdot h$	Any consistent units
2-21	$P = \dfrac{W}{t} = \dfrac{F \cdot d}{t}$	Any consistent units
2-22	$P = F \cdot v$	Any consistent units
2-23	$HP = \dfrac{F \cdot v}{550}$	F in lb, v in ft/s, HP in hp
2-23M	$kW = \dfrac{F \cdot v}{1000}$	F in N, v in m/s, kW in kW
2-24	$P_H = p \cdot Q$	Any consistent units
2-25	$HP_H = \dfrac{p \cdot Q}{1714}$	p in psi, Q in gpm, HP_H in hp
2-25M	$kW_H = \dfrac{p \cdot Q}{60,000}$	p in kPa, Q in lpm, kW_H in kW

2.11 Review Questions and Problems

1. State Pascal's Law in your own words.
2. Differentiate between a hydrostatic system and a hydrodynamic system.
3. Define the terms *work* and *power*.
4. Define *viscosity*. Why is it such an important fluid property?
5. Define *density, specific gravity,* and *specific weight*.
6. What does the *bulk modulus* of a fluid measure?
7. A cylinder with a diameter of 3 in is loaded with a force of 10,000 lbs. What is the pressure inside the cylinder?
8. A cylinder with a diameter of 50 mm is loaded with a force of 15,000 N. What is the pressure (in kPa) inside the cylinder?
9. A cylinder with a diameter of 2 in is loaded with a force of 15,000 lbs. What is the pressure inside the cylinder?
10. A system with a 2-in diameter cylinder has a maximum operating pressure of 1500 psi. What is the maximum load the cylinder can move?

11. A system with a 100-mm diameter cylinder has a maximum operating pressure of 25,000 kPa. What is the maximum load the cylinder can move?

12. An application requires a load of 10,000 lbs to be moved with a maximum pressure of 2500 psi. What is the minimum diameter cylinder?

13. A cylinder is required to move a load of 20 kN with a pressure of 250 bar. What is the minimum diameter cylinder (in mm)?

14. Considering an arrangement similar to that shown in Figure 2-5, calculate the output force for the following parameters: input diameter = 2 in, output diameter = 4 in, input force = 5000 lbs. Calculate the distance the input piston must be moved to move the output piston 3 in.

15. Considering an arrangement similar to that shown in Figure 2-5, calculate the output force for the following parameters: input diameter = 25 mm, output diameter = 100 mm, input force = 15 kN. Calculate the distance the input piston must be moved to move the output piston 100 mm.

16. Considering an arrangement similar to that shown in Figure 2-5, calculate the output force for the following parameters: input diameter = 1 in, output diameter = 8 in, input force = 2000 lbs. Calculate the velocity of the input piston if the output piston must move at 2 in/s.

17. A cylinder must move a 2000-lb load 2 ft. How much work must it perform?

18. A cylinder must move a 30-kN load 1.5 m. How much work must it perform?

19. An oil has a specific gravity of 0.86. What is its specific weight? Water has a specific weight of 62.4 lbs/ft³.

20. A liquid with a bulk modulus of 275,000 psi is subjected to a pressure of 5000 psi. By how much is it being compressed relative to the unpressurized state?

21. Oil flows at a velocity of 8 ft/s through a conduit with an ID of 0.5 in. Determine the flow rate in gpm.

22. Oil flows at a velocity of 5 m/s through a conduit with an ID of 30 mm. Determine the flow rate in lpm.

23. A pump is producing a constant flow rate of 12 gpm in a system that has piping with an ID of 1 in. What is the flow velocity in ft/s?

24. A pump is producing a constant flow rate of 100 lpm in a system that has piping with an ID of 40 mm. What is the flow velocity in m/s?

25. In a system that will have a flow rate of 20 gpm, it has been determined that the flow velocity should not exceed 15 ft/s. What is the minimum conduit ID (in inches)?

26. In a system that will have a flow rate of 75 lpm, it has been determined that the flow velocity should not exceed 6 m/s. What is the minimum conduit ID (in mm)?

27. Standard hydraulic oil ($\gamma = 0.0324$ lbs/in^3) is flowing at a constant rate of 25 gpm through a conduit that has a 2-in ID. What is the flow velocity? If the conduit ID is reduced to 0.5 inch in one section of the system, what will the velocity be at this point?

28. In problem 27, the pressure in the 1.5-in section is 500 psi. What will the pressure be in the 0.5-in section? Assume equal elevations at both points.

29. Standard hydraulic oil ($\gamma = 8800$ N/m^3) is flowing at a constant rate of 80 lpm through a conduit that has a 60-mm ID. What is the flow velocity? If the conduit ID is reduced to 20 mm in one section of the system, what will the velocity be at this point?

30. In problem 29, the pressure in the 60-mm section is 5000 kPa. What will the pressure be in the 20-mm section? Assume equal elevations at both points.

31. Standard hydraulic oil is flowing at 10 gpm and 100 psi at the pump outlet. The system then flows uphill to an elevation of 100 ft. What is the pressure at this point?

32. Standard hydraulic oil is flowing at 50 lpm and 1000 kPa at the pump outlet. The system then flows uphill to an elevation of 50 m. What is the pressure at this point?

33. What is the flow velocity at the outlet of the tank shown in Figure 2-16 if the height is 10 ft?

34. What is the flow velocity at the outlet of the tank shown in Figure 2-16 if the height is 5 m?

35. A fluid with a kinematic viscosity of 0.00250 ft^2/s is flowing through a 0.5-in diameter conduit at a flow rate of 250 gpm. Determine the Reynolds number and state if the flow will be laminar or turbulent.

36. A fluid with a kinematic viscosity of 0.000223 m^2/s is flowing through a 30-mm diameter conduit at a flow rate of 1000 lpm. Determine the Reynolds number and state if the flow will be laminar or turbulent.

37. A 8-ft diameter cylindrical tank is filled with oil ($\gamma = 0.0324$ lbs/in^3) to a height of 35 ft. What is the pressure at the bottom of the tank?

38. A 2-m diameter cylindrical tank is filled with oil ($\gamma = 8800$ N/m^3) to a height of 10 m. What is the pressure (in kPa) at the bottom of the tank?

39. A 20,000-lb load must be moved at a velocity of 2 ft/s. How much horsepower is required?

40. A 100-kN load must be moved at a velocity of 1 m/s. How much power (in kW) is required?

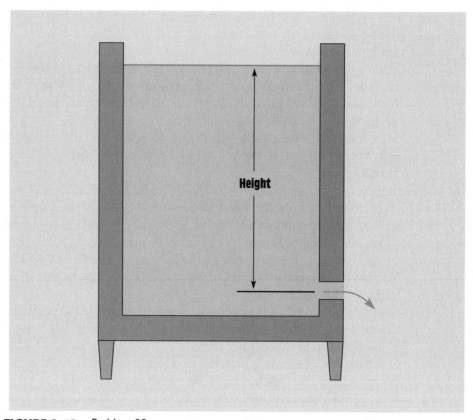

Height

FIGURE 2-16 Problem 33.

41. A machine that moves a load at 3 ft/s has a maximum power rating of 50 hp. What is the maximum load?

42. A machine that moves a load at 2.5 m/s has a maximum power rating of 50 kW. What is the maximum load?

43. A hydraulic system operates at 10 gpm and 3500 psi. How much hydraulic horsepower is generated?

44. A hydraulic system operates at 50 lpm and 20,000 kPa. How much hydraulic power (in kW) is generated?

45. A power unit (pump and electric motor) has a power rating of 25 hp and supplies flow at a rate of 15 gpm. What is the maximum pressure this unit can withstand?

46. A power unit has a power rating of 30 kW and supplies flow at a rate of 60 lpm. What is the maximum pressure this unit can withstand?

CHAPTER 3

Hydraulic Pumps

OUTLINE

3.1 Introduction

3.2 Pump Flow and Pressure

3.3 Pump Drive Torque and Power

3.4 Pump Efficiency

3.5 Pump Types

3.6 Pressure-Compensated Pumps

3.7 Cavitation and Aeration

3.8 Graphic Symbols

3.9 Pump Specifications

3.10 Equations

3.11 Review Questions and Problems

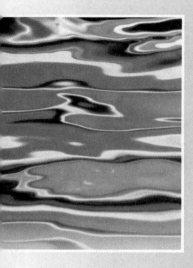

3.1   Introduction

As stated in Chapter 2, the power supply segment of a fluid power system consists of the *prime mover* and the *pump.* The prime mover supplies the input power to the pump. It may be an electric motor (industrial applications) or a gasoline engine (mobile applications). The pump takes this mechanical input and converts it into fluid power (a fluid flowing under pressure).

Chapter 2 also discussed the difference between a hydrostatic system, which uses fluid pressure to transmit power, and a hydrodynamic system, which uses fluid motion to transmit power. The type of pump used in a hydrostatic system is a *positive displacement* pump. Positive displacement pumps operate by opening up a cavity very quickly to create a vacuum, which allows atmospheric pressure to push fluid in from a reservoir. They then close the cavity very quickly to force the fluid back out in the direction of the system. These two steps are known as one *cycle* of the pump and usually correspond to one revolution of the pump shaft. Positive displacement pumps have very close-fitting mating components. They therefore have only a small amount of leakage that increases slightly as the pressure increases. There are three main types of construction for positive displacement pumps: gear, vane, and piston. Each is discussed in detail in Section 3.5.

Hydrodynamic systems often use *nonpositive displacement* pumps, which typically operate by spinning the fluid with an impeller, as shown in Figure 3-1. The fluid is "thrown" toward the outlet by centrifugal force. In this type of pump, the mating components (i.e., the impeller and housing) *do not* make a tight fit with one another. This causes a large amount of leakage to occur if the pressure at the outlet becomes too great. In addition, they are usually not *self-priming,* which means they cannot generate enough vacuum to draw in fluid on their own and must be provided with flow at their inlet. This can be accomplished by positioning the pump inlet below the reservoir to allow gravity to feed in the fluid. In spite of these limitations, nonpositive displacement pumps are well suited for fluid transfer applications that do not require high pressure because their objective is simply to move the fluid from one location to another. This type of design also has two important advantages: It is inexpensive to manufacture and can pump nonlubricating fluids because of the lack of close-fitting components in the pumping mechanism.

3.2   Pump Flow and Pressure

Many beginning fluid power students, and even some people who work with hydraulic systems on a daily basis, assume incorrectly that the purpose of a hydraulic pump is to create pressure. It is not. The purpose of a pump is to create *flow.* Pressure is created only when the flow is resisted, by putting a load on

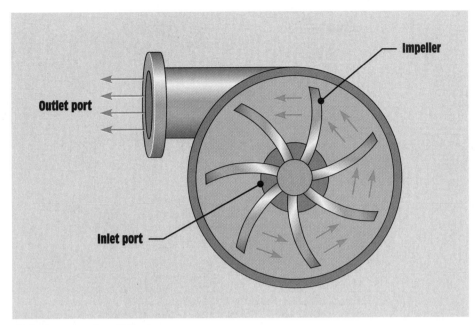

FIGURE 3-1   Impeller pump.

a cylinder, for example. Pressure is directly proportional to the load or force that is applied to the system. If there is no resistance to flow, there is no pressure.

Pump flow will always go to the path of least resistance (lowest pressure), as illustrated in Figure 3-2. Here we have a pump connected to two cylinders, A and B. Cylinder A is loaded with a 1000-lb load and has an area of 1 in². The pressure required to lift cylinder A is therefore:

$$P_A = \frac{F_A}{A_A} = \frac{1000 \text{ lb}}{1 \text{ in}^2} = 1000 \text{ psi}$$

Cylinder B is loaded with a 1500-lb load and has an area of 2 in². The pressure required to lift cylinder B is:

$$P_B = \frac{F_B}{A_B} = \frac{1500 \text{ lb}}{2 \text{ in}^2} = 750 \text{ psi}$$

The pump flow will therefore go to cylinder B, the path of least resistance, at a pressure of 750 psi. Only after cylinder B is fully raised ("bottoms out") and this flow path is no longer available will the pressure build to 1000 psi and lift cylinder A.

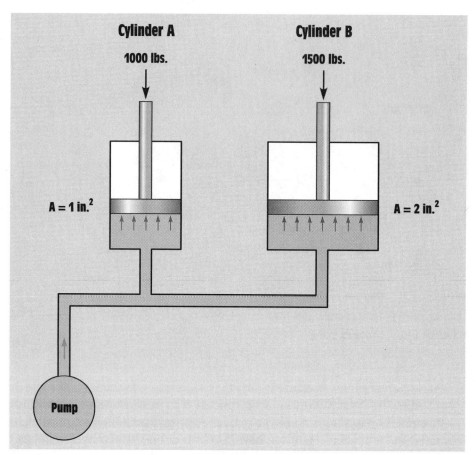

FIGURE 3-2   Pump connected to two cylinders.

Fluid power pumps all apply the same basic principle: opening a cavity to draw in fluid during half of the cycle, then closing the cavity to expel the fluid during the other half of the cycle. The volume of fluid that is discharged per cycle is called the *displacement*. A cycle is most commonly one revolution of the pump shaft. Displacement is usually expressed in cubic inches per revolution (in³/rev) in the U.S. customary unit system and in cubic centimeters per revolution (cm³/rev) in the metric (SI) system.

The flow rate of a pump is the volume of fluid that it outputs per unit time, also known as the *delivery* or *output*. This is determined by the displacement of the pump and how fast it is driven. The drive speed of a pump is expressed in revolutions per minute, which may be written as rev/min or rpm. The most commonly used drive speeds for hydraulic pumps are 1200 and 1800 rpm. We can determine the flow rate by multiplying the displacement by the drive speed, as given by the following equation:

$$Q_T = V_p \cdot N \qquad\qquad \textbf{(3-1)}$$

where: Q_T = theoretical flow rate of the pump $\left(\dfrac{in^3}{min}, \dfrac{cm^3}{min}\right)$

V_p = pump displacement $\left(\dfrac{in^3}{rev}, \dfrac{cm^3}{rev}\right)$

N = drive speed $\left(\dfrac{rev}{min}\right)$

The flow rate that is calculated with Equation 3-1 is termed *theoretical* because the actual flow will be less due to a small amount of leakage. This issue is addressed in Section 3.4.

The units that result for the flow rate when using Equation 3-1 are in³/min. In most pump literature, however, the flow rate is expressed in gallons per minute (gpm). We can insert the conversion factor of 1 gal = 231 in³ in Equation 3-1 so that it contains the most common units:

$$Q_T = \frac{V_p \cdot N}{231} \qquad\qquad \textbf{(3-2)}$$

The following units must be used for Equation 3-2: V_p in in³/rev, N in rpm, and Q_T in gpm.

In metric units, the displacement is typically given in cm³/rev, which will result in units of cm³/min for Q when using Equation 3-1. The preferred unit for flow in the metric system, however, is liters per minute (lpm). We can insert the conversion factor of 1 l = 1000 cm³ into Equation 3-1 so that it contains the most common metric units:

$$Q_T = \frac{V_p \cdot N}{1000} \qquad\qquad \textbf{(3-2M)}$$

The following units must be used for Equation 3-2M: V_p in cm³/rev, N in rpm, and Q_T in lpm.

EXAMPLE 3-1.

A pump has a displacement of 2 in³/rev and is driven at 1200 rpm. What is its theoretical flow rate?

SOLUTION:

$$Q_T = \frac{V_p \cdot N}{231} = \frac{2\,\frac{in^3}{rev} \cdot \left(1200\,\frac{rev}{min}\right)}{231\,\frac{in^3}{gal}} = 10.4\,\frac{gal}{min}$$

EXAMPLE 3-1M.

A pump has a displacement of 30 cm³/rev and is driven at 1200 rpm. What is its theoretical flow rate?

SOLUTION:

$$Q_T = \frac{V_p \cdot N}{1000} = \frac{30\,\frac{cm^3}{rev} \cdot \left(1200\,\frac{rev}{min}\right)}{100\,\frac{cm^3}{l}} = 36.0\,\frac{l}{min}$$

EXAMPLE 3-2.

A flow rate of 8 gpm is required for a new system. If we use an 1800-rpm electric motor to drive the pump, what should the displacement be to achieve this flow rate? Assume 100% efficiency.

SOLUTION:

$$V_p = \frac{231 \cdot Q_T}{N} = \frac{231\,\frac{in^3}{gal} \cdot \left(8\,\frac{gal}{min}\right)}{1800\,\frac{rev}{min}} = 1.03\,\frac{in^3}{rev}$$

EXAMPLE 3-3.

For the system in Example 3-2, we have located a standard pump with a displacement of 1.15 in³/rev. What will the flow rate be with this pump?

SOLUTION:

$$Q_T = \frac{V_p \cdot N}{231} = \frac{1.15\,\frac{in^3}{rev} \cdot \left(1800\,\frac{rev}{min}\right)}{231\,\frac{in^3}{gal}} = 8.96\,\frac{gal}{min}$$

The increase in flow rate in Example 3-3 will cause a corresponding increase in cylinder speed, so a decision must be made as to whether this increase in speed is tolerable. Chapters 4 and 5 discuss the effect of flow rate on actuator speed in detail.

The pumps in the previous examples were *fixed displacement,* which means they cannot vary their flow rate without changing the speed at which they are driven. Pumps may also be *variable displacement.* These pumps can vary their flow rate even while being driven at a constant speed. Several varieties of variable displacement pumps are discussed in Section 3.6.

Pumps are rated by their flow rate and the maximum pressure they can withstand. For example, a catalog may list a pump as being rated for 20 gpm and 2500 psi. Because pumps produce flow and not pressure, this rating should be interpreted as meaning that the pump produces 20 gpm and can do so at pressures up to 2500 psi.

3.3 Pump Drive Torque and Power

As stated previously, the prime mover supplies the input power to the pump by driving the pump shaft at a given rotational speed. The pressure on the pump outlet resists this rotation, therefore the prime mover must be able to generate enough *torque* to turn the pump shaft against this resistance. Torque is generated whenever a force is applied to rotate a shaft or lever arm around a pivot point, as shown in Figure 3-3. Torque is calculated by multiplying the force times the distance from the force to the pivot point:

$$T = F \cdot d \tag{3-3}$$

While Equation 3-3 is useful to explain the concept of torque, it is not easily applied to a hydraulic pump. A more useful equation would relate the required drive torque to the system pressure and the displacement of the pump, as in the following equation:

$$T_T = \frac{p \cdot V_p}{2 \cdot \pi} \tag{3-4}$$

T_T is the theoretical drive torque. The actual drive torque is greater because of friction. Typical units for this equation in the U.S. customary system are: V_p in in³/rev, p in lbs/in², and T in in · lbs. Typical units for the metric (SI) system are: V_p in m³/rev, p in N/m², and T in N · m. Displacement in the metric system, however, is usually given in cm³/rev. The conversion factor used to convert from cm³/rev to m³/rev is 1 m³ = 1,000,000 cm³. The factor 2π in

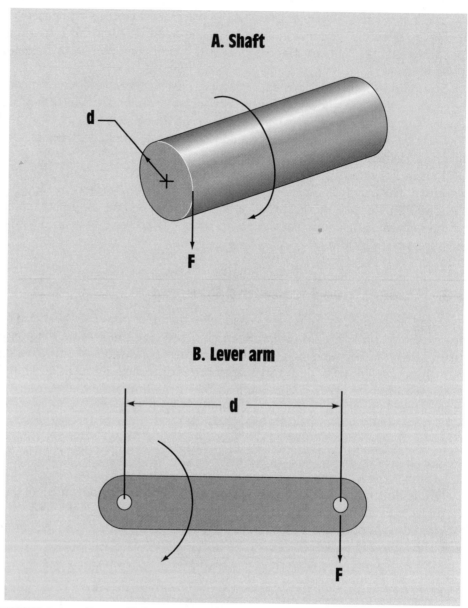

FIGURE 3-3   Torque. (A) Shaft. (B) Lever arm.

Equation 3-4 is a conversion that eliminates the unit *revolutions* (rev) from the equation (it is a conversion from revs to radians).

EXAMPLE 3-4.

A hydraulic pump with a displacement of 3 in³/rev is selected for a system that will operate at a maximum pressure of 3000 psi. What is the required drive torque? Assume 100% efficiency.

SOLUTION:

$$T_T = \frac{p \cdot V_p}{2 \cdot \pi} = \frac{3000 \frac{lb}{in^2} \cdot \left(3 \frac{in^3}{rev}\right)}{2 \cdot \pi} = 1432 \text{ in} \cdot \text{lbs}$$

EXAMPLE 3-4M.

A hydraulic pump with a displacement of 40 cm³/rev is selected for a system that will operate at a maximum pressure of 20,000 kPa. What is the required drive torque? Assume 100% efficiency.

SOLUTION:

1. Convert to m³/rev:

$$40 \frac{cm^3}{rev} \cdot \left(\frac{1 \text{ m}^3}{1,000,000 \text{ cm}^3}\right) = 0.00004 \frac{m^3}{rev}$$

2. Calculate the drive torque:

$$T_T = \frac{p \cdot V_p}{2 \cdot \pi} = \frac{20,000,000 \frac{N}{m^2} \cdot \left(0.00004 \frac{m^3}{rev}\right)}{2 \cdot \pi} = 127.3 \text{ N} \cdot \text{m}$$

We may also wish to calculate the input power provided by the prime mover. We can determine the power generated by a rotating shaft by multiplying the torque by the rotational speed, as given by the following equation:

$$HP_I = \frac{T \cdot N}{63,025} \qquad \textbf{(3-5)}$$

We use HP_I to signify that this is the input horsepower to the pump. The factor 63,025 in the denominator is a combination of conversion factors that allows us to use the most common U.S. customary units: T in $in \cdot lbs$, N in *rpms,* and HP_I in *hp.* Only these units may be used in this equation.

When using metric units, the following equation can be used to calculate the required input power in *kilowatts:*

$$kW_I = \frac{T \cdot N}{9550} \qquad \textbf{(3-5M)}$$

The factor 9550 in the denominator is a combination of conversion factors that allows us to use the most common metric units (T in $N \cdot m$, N in rpms, and kW_I in kW). Only these units may be used in this equation.

EXAMPLE 3-5.

An electric motor drives a pump at 1800 rpm with a torque of 350 in · lbs. What is the power input to this pump?

SOLUTION:

$$HP_I = \frac{T \cdot N}{63,025} = \frac{350 \text{ in} \cdot \text{lbs} \cdot (1800 \text{ rpm})}{63,025} = 10.0 \text{ hp}$$

EXAMPLE 3-5M.

An electric motor drives a pump at 1800 rpm with a torque of 40 N · m What is the power input to this pump?

SOLUTION:

$$kW_I = \frac{T \cdot N}{9550} = \frac{40 \text{ N} \cdot \text{m} \cdot (1800 \text{ rpm})}{9550} = 7.54 \text{ kW}$$

As stated in Chapter 2, the hydraulic power of a fluid flowing under pressure can be calculated with the following equations:

$$HP_H = \frac{p \cdot Q}{1714} \qquad \textbf{(3-6)}$$

$$kW_H = \frac{p \cdot Q}{60,000} \qquad \textbf{(3-6M)}$$

The output of a pump is a fluid flowing under pressure, so we can use Equations 3-6 and 3-6M to calculate the power output of a pump. We must use the following units for Equation 3-6: Q in gpm, p in psi, and HP_H in hp. The required units for Equation 3-6M are: Q in lpm, p in kPa, and kW_H in kW.

EXAMPLE 3-6.

A hydraulic pump has a flow rate of 20 gpm and is rated for a maximum pressure of 2500 psi. What is the maximum power output of this pump?

SOLUTION:

$$HP_H = \frac{p \cdot Q}{1714} = \frac{2500 \text{ psi} \cdot (20 \text{ gpm})}{1714} = 29.2 \text{ hp}$$

EXAMPLE 3-6M.

A hydraulic pump has a flow rate of 75 lpm and is rated for a maximum pressure of 20,000 kPa. What is the maximum power output of this pump?

SOLUTION:

$$kW_H = \frac{p \cdot Q}{60,000} = \frac{20,000 \text{ kPa} \cdot (75 \text{ lpm})}{60,000} = 25 \text{ kW}$$

EXAMPLE 3-7.

A hydraulic power unit (electric motor and pump) produces a flow rate of 10 gpm and has a maximum power rating of 25 hp. What is the maximum pressure at which this unit can operate? Assume 100% efficiency.

SOLUTION:

$$p = \frac{1714 \cdot HP_H}{Q} = \frac{1714 \cdot (25 \text{ hp})}{15 \text{ gpm}} = 2857 \text{ psi}$$

3.4   Pump Efficiency

The actual pump flow will be somewhat less than the theoretical flow that is calculated with Equation 3-2. The actual flow can be calculated if we know the *volumetric efficiency* of the pump. Volumetric efficiency is the ratio of the actual flow rate to the theoretical flow rate, as given by the following equation:

$$\eta_V = \frac{Q_A}{Q_T} \qquad\qquad \textbf{(3-7)}$$

where: η_V = the Greek letter eta = volumetric efficiency (no units)
Q_A = actual flow rate (gpm, lpm)
Q_T = theoretical flow rate (gpm, lpm)

A volumetric efficiency of 0.90 means that the pump will output 90% of the flow it draws in from the reservoir, or a leakage rate of 10%. Volumetric efficiencies for positive displacement pumps are typically 0.85 or higher. Plugging Equation 3-2 into Equation 3-7 and rearranging to solve for the actual flow results in the following useful equation:

$$Q_A = \eta_V \cdot \frac{V_p \cdot N}{231} \qquad \text{(3-8)}$$

The following units must be used for Equation 3-8: V_p in in³/rev, N in rpm, and Q_A in gpm. For metric units, we can plug Equation 3-2M into Equation 3-7 to obtain:

$$Q_A = \eta_V \cdot \frac{V_p \cdot N}{1000} \qquad \text{(3-8M)}$$

The following units must be used for Equation 3-8M: V_p in cm³/rev, N in rpm, and Q_A in lpm.

EXAMPLE 3-8.

A pump has a displacement of 2 in³/rev and a volumetric efficiency of 0.92. If it is driven at 1200 rpm, what will its actual flow rate be?

SOLUTION:

$$Q_A = \eta_V \cdot \frac{V_p \cdot N}{231} = 0.92 \cdot \frac{2 \, \frac{in^3}{rev} \cdot \left(1200 \, \frac{rev}{min}\right)}{231 \, \frac{in^3}{gal}} = 9.56 \, \frac{gal}{min}$$

EXAMPLE 3-8M.

A pump has a displacement of 50 cm³/rev and a volumetric efficiency of 0.90. If it is driven at 1200 rpm, what will its actual flow rate be?

SOLUTION:

$$Q_A = \eta_V \cdot \frac{V_p \cdot N}{1000} = 0.90 \cdot \frac{50\,\frac{cm^3}{rev} \cdot \left(1200\,\frac{rev}{min}\right)}{1000\,\frac{cm^3}{l}} = 54.0 \text{ lpm}$$

EXAMPLE 3-9.

A flow of 5 gpm is required in a new system. The pump type chosen has a volumetric efficiency of 0.88 and will be driven at 1200 rpm. What size (displacement) should we select?

SOLUTION:

$$V_p = \frac{231 \cdot Q_A}{\eta_V \cdot N} = \frac{231\,\frac{in^3}{gal} \cdot \left(5\,\frac{gal}{min}\right)}{0.88 \cdot \left(1200\,\frac{rev}{min}\right)} = 1.094\,\frac{in^3}{rev}$$

EXAMPLE 3-10.

For the system in the previous example, we have located a standard pump with a displacement of 1.15 in³/rev. What will the flow rate be with this pump?

SOLUTION:

$$Q_A = \eta_V \cdot \frac{V_p \times N}{231} = 0.88 \cdot \frac{1.15\,\frac{in^3}{rev} \cdot \left(1200\,\frac{rev}{min}\right)}{231\,\frac{in^3}{gal}} = 5.26\,\frac{gal}{min}$$

This is about 5% more flow than was desired. This will correspond to an increase in cylinder speed of 5%, which would be acceptable in most practical situations.

Volumetric efficiency provides us with a measure of how much power a pump loses due to leakage, but this is not the only element of inefficiency in a pump. Pumps also lose power due to mechanical losses such as friction. This is measured by mechanical efficiency (η_M), which relates the theoretical torque required to drive a pump to the actual torque, as given by the following equation:

$$\eta_M = \frac{T_T}{T_A} \qquad\qquad (3\text{-}9)$$

EXAMPLE 3-11.

A pump that has a mechanical efficiency of 0.90 and a displacement of 4 in³/rev is to be used in a system with a maximum operating pressure of 2500 psi. What is the required drive torque?

SOLUTION:

1. Calculate the theoretical drive torque:

$$T_T = \frac{p \cdot V_p}{2 \cdot \pi} = \frac{2500 \text{ psi} \cdot \left(4 \dfrac{\text{in}^3}{\text{rev}}\right)}{2 \cdot \pi} = 1592 \text{ in} \cdot \text{lbs}$$

2. Calculate the actual drive torque:

$$T_A = \frac{T_T}{\eta_M} = \frac{1592 \text{ in} \cdot \text{lbs}}{0.90} = 1769 \text{ in} \cdot \text{lbs}$$

EXAMPLE 3-11M.

A pump that has an mechanical efficiency of 0.92 and a displacement of 20 cm³/rev is to be used in a system with a maximum operating pressure of 15,000 kPa. What is the required drive torque?

SOLUTION:

1. Convert to m³/rev:

$$20 \frac{\text{cm}^3}{\text{rev}} \cdot \left(\frac{1 \text{ m}^3}{1,000,000 \text{ cm}^3}\right) = 0.00002 \frac{\text{m}^3}{\text{rev}}$$

2. Calculate T_T:

$$T_T = \frac{p \cdot V_p}{2 \cdot \pi} = \frac{15,000,000 \, \frac{N}{m^2} \cdot \left(0.00002 \, \frac{m^3}{rev}\right)}{2 \cdot \pi} = 47.75 \, N \cdot m$$

3. Calculate T_A:

$$T_A = \frac{T_T}{\eta_M} = \frac{47.75 \, N \cdot m}{0.92} = 51.90 \, N \cdot m$$

The total power loss incurred by a pump is measured by the *overall efficiency,* which is the ratio of a pump's input power to output power. This quantity includes losses due to leakage and mechanical losses due to friction. The input power is supplied by the prime mover. The output power of the pump is the fluid flowing under pressure. The overall efficiency is then given by the following equations:

$$\eta_O = \frac{HP_H}{HP_I} \tag{3-10}$$

$$\eta_O = \frac{kW_H}{kW_I} \tag{3-10M}$$

We can also express the overall efficiency as a combination of the volumetric and mechanical efficiencies:

$$\eta_O = \eta_M \cdot \eta_V \tag{3-11}$$

Equation 3-11 illustrates an important point. Whenever we have inefficiencies or losses in a system due to several factors, the individual efficiencies are multiplied together to get the overall efficiency of the system. For example, if we know that a system has a volumetric efficiency of 95% and a mechanical efficiency of 85%, the overall efficiency of the system is: $\eta_O = 0.85 \cdot 0.95 = 0.808$, or about 81%.

EXAMPLE 3-12.

A pump that has an overall efficiency of 0.85 and a flow rate of 6 gpm is to be used in a system that has a maximum operating pressure of 2500 psi. What input horsepower is required?

SOLUTION:

1. Calculate the hydraulic horsepower:

$$HP_H = \frac{p \cdot Q}{1714} = \frac{2,500 \text{ psi} \cdot (6 \text{ gpm})}{1714} = 8.75 \text{ hp}$$

2. Calculate the input horsepower:

$$HP_I = \frac{HP_P}{\eta_O} = \frac{8.75 \text{ hp}}{0.85} = 10.3 \text{ hp}$$

We will not find a standard electric motor of 10.3 *hp* from a catalog, so we would choose the next larger size to ensure that there is sufficient power.

EXAMPLE 3-12M.

A pump that has an overall efficiency of 0.87 and a flow rate of 45 lpm is to be used in a system that has a maximum operating pressure of 25,000 kPa. What input horsepower is required?

SOLUTION:

1. Calculate the hydraulic power:

$$kW_H = \frac{p \cdot Q}{60,000} = \frac{25,000 \text{ kPa} \cdot (45 \text{ lpm})}{60,000} = 18.75 \text{ kW}$$

2. Calculate the input power:

$$kW_I = \frac{kW_H}{\eta_O} = \frac{18.75 \text{ kW}}{0.87} = 2155 \text{ kW}$$

EXAMPLE 3-13.

A power unit consists of a pump with an overall efficiency of 0.82 and an electric motor with a maximum power rating of 15 hp. What is the maximum hydraulic power output for this system?

SOLUTION:

$$HP_P = \eta_O \cdot HP_I = 0.82 \cdot (15 \text{ hp}) = 12.3 \text{ hp}$$

EXAMPLE 3-14.

The pump in the previous problem produces a flow rate of 5 gpm. What is the maximum pressure this system can operate at?

SOLUTION:

$$p = \frac{1.714 \cdot HP_P}{Q} = \frac{1714 \cdot (12.3 \text{ hp})}{5 \text{ gpm}} = 4216 \text{ psi}$$

3.5   Pump Types

Three types of positive displacement pumps are commonly used in fluid power systems: piston pumps, gear pumps, and vane pumps.

3.5.1 PISTON PUMPS

The most basic of all pump designs is the single-piston pump (Figure 3-4). In this pump, the piston is drawn back quickly, creating a vacuum at its inlet. This allows atmospheric pressure to push fluid from the reservoir into the pump, as shown in Figure 3-4, part A. The piston is then driven forward to expel the fluid towards the system, as shown in part B. Note the use of *check valves* on the inlet and outlet lines. Check valves allow flow in only one direction. They are necessary to force the flow to travel from the reservoir to the system and not vice versa. Check valves are discussed in greater detail in Chapter 4.

This very simple design is used only in hand-driven pumps because it delivers flow in spurts or pulses, rather than providing a smooth flow. For power-driven pumps, multiple pistons are used, providing a smoother, more even flow. Two main types of power piston pumps are commonly used in hydraulics: radial and axial piston pumps.

The pistons in axial pumps are aligned or parallel with the pump shaft axis, as shown in the simplified cut-away drawing in Figure 3-5. The pistons are arrayed in a circular pattern around the pump shaft. Turning the pump shaft causes the pistons and piston block to rotate. The housing, end cap, and piston shoe remain stationary. The pistons ride the piston shoe, which is mounted at an angle to the shaft axis, causing the pistons to reciprocate in the piston block as the pump shaft is turned. The pistons pull back and draw in oil during 180° of their rotation and push out and expel oil during the other 180°. The oil is fed to and from the system by semicircular feed

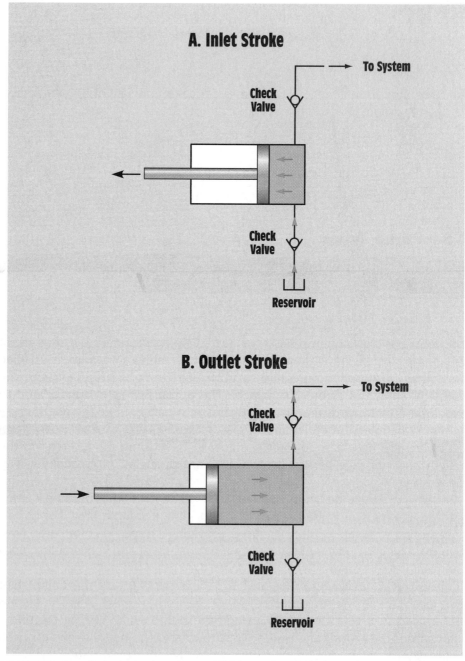

FIGURE 3-4   Basic single piston pump. (A) Inlet stroke. (B) Outlet stroke.

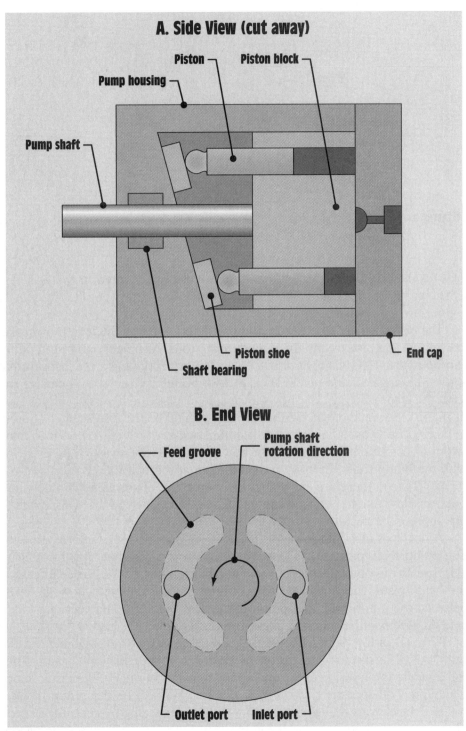

FIGURE 3-5   Axial piston pump. (A) Side view (cutaway). (B) End view.

FIGURE 3-6   Axial piston pump.
(Courtesy of Eaton Hydraulics Training)

grooves that are connected to inlet and outlet ports located in the end cap, as shown in the end view. A cutaway of an axial piston pump is shown in Figure 3-6.

The displacement of a piston pump is determined by the length of stroke, the bore of the pistons and the number of pistons. A straight-axis piston pump can be made to be variable displacement by incorporating a mechanism that allows the angle of the piston shoe to be adjusted, which would change the length of stroke. Figure 3-7 shows a variable displacement version of this type of pump. The angle of the *swash plate* is controlled by the displacement piston. In part A, the swash plate is at its maximum angle, which corresponds to maximum displacement and maximum flow. In part B, the swash plate is in-line with the pump shaft. The pistons will remain at the halfway position and not stroke. There is, therefore, no change in volume and consequently no flow output. Any flow rate in between can be obtained by adjusting the swash plate to the appropriate angle.

A bent-axis axial piston pump (Figure 3-8) operates on a similar principle. The pump shaft, pistons, and piston block all rotate together, like the straight-axis type. In this pump, however, the piston block is offset at an angle rather than the swash plate. The length of stroke of the pistons is determined by the angle between the pump shaft and piston block. This type of pump cannot be variable displacement because the angle is not adjustable.

A radial piston pump (Figure 3-9) has pistons that radiate out from the shaft axis. The pistons and piston block are keyed to the drive shaft. They rotate while the housing and cam surface remain stationary. The piston block is mounted off-center, or *eccentric*, to the cam surface. As the piston block is rotated, the pistons are kept in contact with the cam ring by centrifugal force. The eccentricity between the piston block and the cam ring causes the pistons

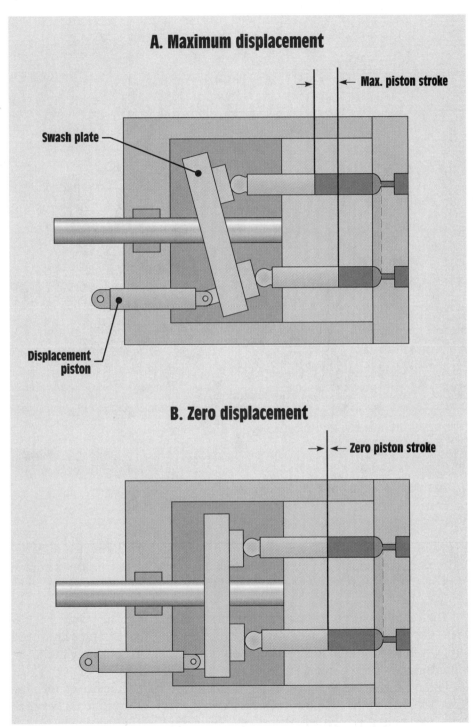

FIGURE 3-7   Variable displacement axial piston pump. (A) Maximum displacement. (B) Zero displacement.

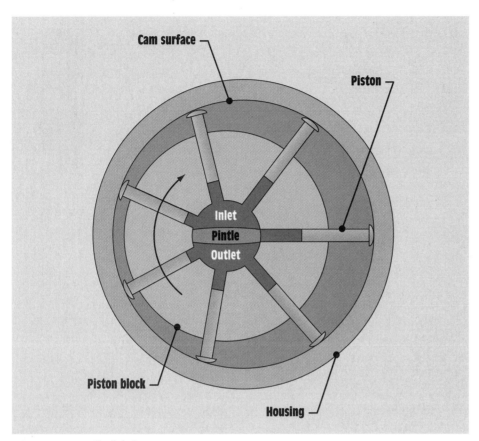

FIGURE 3-9 Radial-piston pump.

(small clearances) that can be manufactured between the piston and the bore. These tight fits, however, make them the least tolerant of contamination, making them best suited to clean environments.

3.5.2 GEAR PUMPS

Gear pumps are of two types: *external* and *internal*. A simplified cutaway of an external gear pump is shown in Figure 3-11. The pump drive shaft is keyed to the drive gear. The idler gear is driven by the drive gear. The gear teeth come into contact, or *mesh,* at the center, sealing the inlet and outlet ports. As the gears are rotated, they come out of mesh at the inlet port. This creates a vacuum due to the sudden increase in volume and fluid is pushed in by atmospheric pressure. The fluid is then carried between the gear teeth to the outlet, where the teeth mesh again and the fluid is forced out. Note that the fluid does

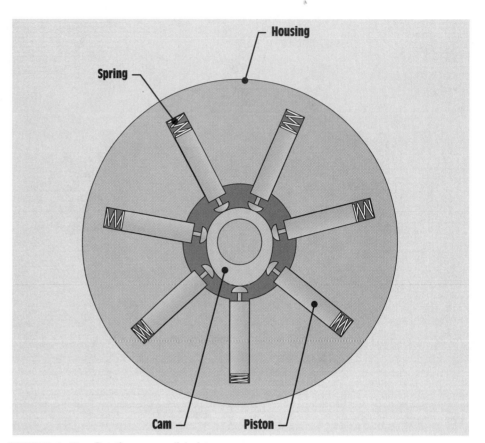

FIGURE 3-10   Rotating cam radial-piston pump.

not pass through the center, as one may assume. This type of gear pump is called an external gear pump because it consists of two pinion gears that are external to one another. An actual cutaway of an external gear pump is shown in Figure 3-12.

An internal gear pump (Figure 3-13) consists of a pinion gear and a ring gear that are eccentrically mounted. The pinion gear is driven by the pump shaft; it, in turn, drives the ring gear. As the teeth come out of mesh near the inlet port, the increasing volume creates a vacuum, allowing atmospheric pressure to push fluid in. It is then carried between the teeth of the ring gear around to the outlet, where the teeth mesh and force out the fluid. The stationary crescent seals the pumping chambers.

Gear pumps are the least efficient of the three pump types discussed in this section. They are generally the most tolerant of contamination, however, due to the greater clearances between the mating components. Pressure ratings of

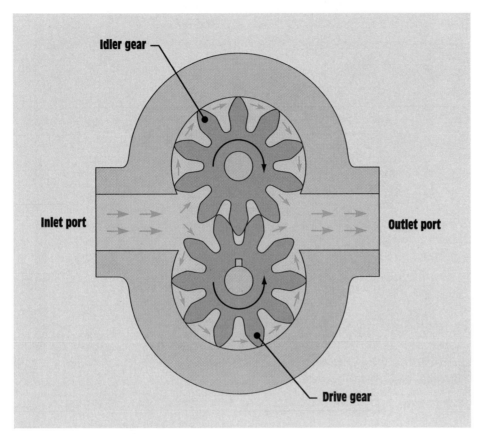

FIGURE 3-11   External gear pump.

2000 to 3000 psi are typical. Due to the basic construction of gear pumps, they are not available in variable displacement models.

3.5.3 VANE PUMPS

There are two main types of vane pumps: *unbalanced* and *balanced*. A simplified cutaway of an unbalanced vane pump is shown in Figure 3-14. The primary components of a vane pump are the cam surface and the rotor, into which vanes are slip fit. The rotor is keyed to the drive shaft and therefore rotates as the pump is driven. As the rotor spins, the vanes are kept in contact with the cam surface by centrifugal force, which may be supplemented by a spring or by fluid pressure. The cam surface and rotor are mounted eccentrically to one another, which causes the vanes to stroke as the rotor spins. As the rotor turns, pumping chambers between the vanes are opened near the inlet, creating a vacuum that

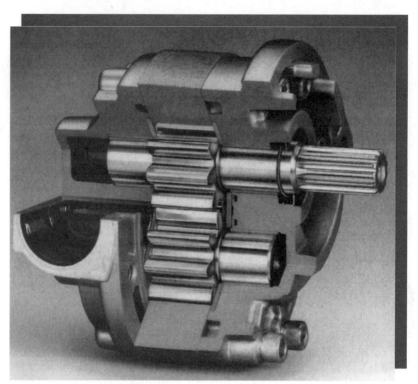

FIGURE 3-12   External gear pump.
(Courtesy of Danfoss Fluid Power. From Vockroth, Industrial Hydraulics, Albany, NY:Delmar, 1994, p. 115)

allows atmospheric pressure to push the fluid in. The fluid is then carried between the vanes to the outlet where the vanes are pushed back in, and the pumping chamber volume is reduced. The reduction in volume near the outlet causes the fluid to be forced out. A disassembled, unbalanced vane pump is shown in Figure 3-15.

An unbalanced vane pump can be made to be variable displacement by incorporating a mechanism that allows the eccentricity between the cam surface and rotor to be adjusted. As the rotor moves toward a concentric position with the rotor, the displacement is be reduced. This is easy to understand if we imagine that the rotor and cam surfaces are centered with one another, as shown in Figure 3-16. In this condition, the vanes would remain half-way out as the rotor spins and there would be no change in volume (zero displacement) and, consequently, no flow.

The design shown in Figure 3-14 is called an unbalanced vane pump because the pressurized pumping chambers are located on only one side of

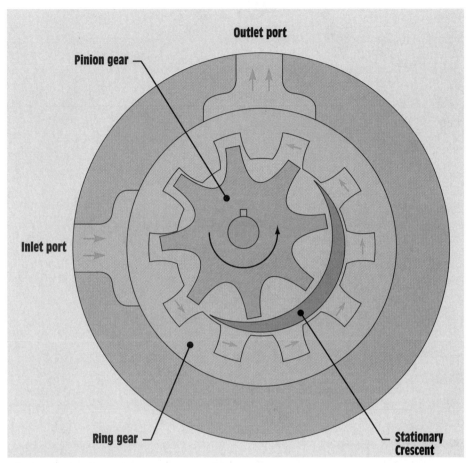

FIGURE 3-13   Internal gear pump.

the drive shaft. The outlet port at the top is under pressure, while the inlet port at the bottom is at vacuum. This unbalanced condition results in a net force on the pump shaft bearing that can cause excessive vibration and wear at high speeds or pressures. For this reason, these pumps are best suited to low-pressure applications.

A *balanced* vane pump (Figure 3-17) eliminates the high side load condition. This type of pump has an elliptical cam surface, which causes the vanes to stroke twice per revolution of the pump shaft. The pumping chambers therefore undergo an increase and decrease in volume twice per cycle, which necessitates two inlet and two outlet ports. The inlets and outlets are combined into a common inlet and outlet within the pump housing. This configuration results in

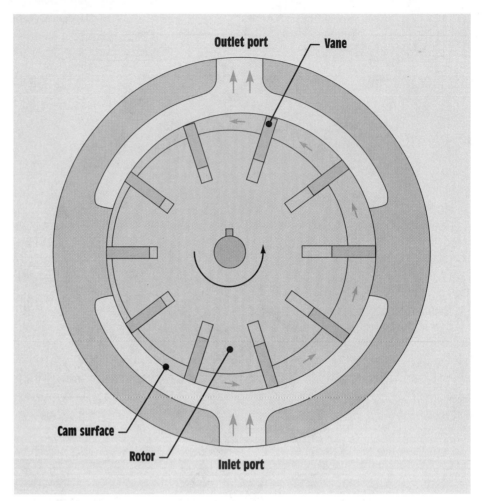

Outlet port — **Vane**

Cam surface

Rotor

Inlet port

FIGURE 3-14 Unbalanced vane pump.

FIGURE 3-15 Unbalanced vane pump.
(From Reeves, *Technology of Fluid Power,* Albany, NY:Delmar, 1996, p. 73)

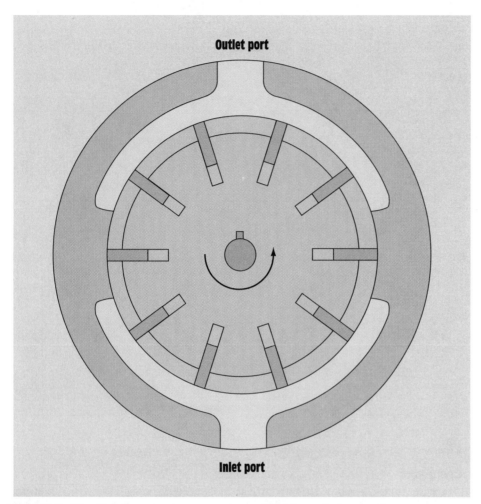

Outlet port

Inlet port

FIGURE 3-16   Unbalanced vane pump, rotor centered (zero displacement).

equal pressure on opposite sides of the pump shaft. There is therefore no net force on the shaft bearing, a condition called *balanced*. This type of pump can operate at higher speeds and pressures than the unbalanced type without excessive bearing wear. A disassembled balanced vane pump is shown in Figure 3-18.

Vane pumps are more efficient than gear pumps, but less efficient than piston pumps. They are moderately tolerant of contamination. Unbalanced vane pumps are low-pressure pumps, typically operating between 500 psi and 2000 psi. Balanced vane pumps can handle higher pressures. They can be rated for pressures as high as 4000 psi.

86

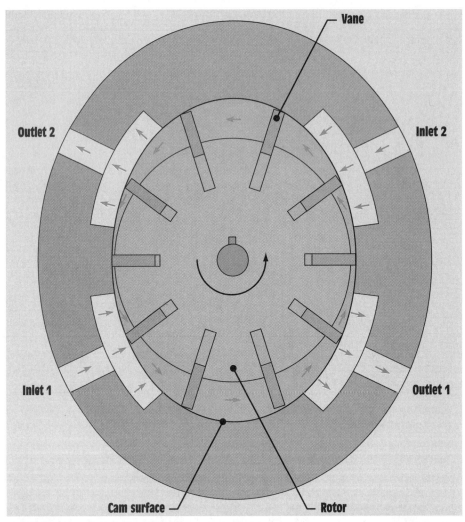

FIGURE 3-17 Balanced vane pump.

FIGURE 3-18 Balanced vane pump.

(From Reeves, *Technology of Fluid Power,* Albany, NY:Delmar, 1996, p. 73.)

3.6   Pressure-Compensated Pumps

A pressure-compensated pump (Figure 3-19) has the ability to limit the maximum pressure in a hydraulic circuit by reducing its displacement. When the pressure reaches a preset level, called the *firing pressure,* the displacement is reduced to prevent a further rise in pressure. The compensator consists of a spring that holds the cam ring in the maximum displacement position. As the pressure increases, it presses against the cam ring and eventually begins to deflect the spring, thereby reducing the displacement. The pressure level at which the compensator fires can be changed by adjusting the initial spring compression. Increasing the spring compression increases the pressure level required to deflect the spring and, therefore, increases the firing pressure. Other, more complicated mechanisms that provide pressure compensation are also common, but the basic principle is the same. Piston pumps may also be equipped with pressure compensation by incorporating a similar spring-loaded mechanism that reduces the angle of the swash plate when the pressure reaches the firing pressure. Gear pumps cannot be variable displacement, so they cannot be equipped with pressure compensation. Pressure-compensated pumps are discussed in more detail in Chapter 7, which covers hydraulic pressure control.

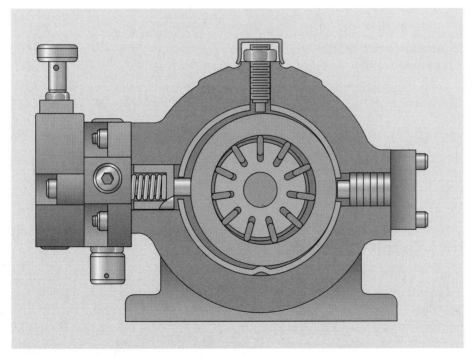

FIGURE 3-19   Pressure-compensated vane pump.
(Courtesy of Dana Fluid Power. From Norvelle, *Fluid Power Technology:* West, 1993, p. 97.)

3.7   Cavitation and Aeration

Positive displacement pumps function by creating a vacuum at their inlet, which allows atmospheric pressure to push fluid in. In some situations, the vacuum may become excessive, and a phenomenon known as *cavitation* occurs. When the pressure of a liquid reaches a low enough level, it will vaporize or boil, just as it does when the temperature becomes too high. Cavitation is the formation of oil vapor bubbles due to very low pressure (high vacuum) on the inlet side of the pump. The low pressure also causes air that is dissolved in the oil to come out of solution and form bubbles. These air and oil vapor bubbles collapse when they reach the outlet side of the pump, which is under pressure. The collapsing of these vapor bubbles causes extremely high localized pressure and fluid velocity. These pressures are so high that they actually cause pitting of the metal and, consequently, decrease the life and efficiency of the pump. Cavitation is characterized by a high-pitch, whining noise, vibration, and increased system temperature. Excessive vacuum at the inlet can be caused by several factors:

1. Undersized plumbing,
2. Clogged lines or suction filters,
3. High fluid viscosity, and
4. Too much elevation head between the reservoir and the pump inlet.

Increased fluid temperature increases the likelihood of cavitation occurring because a warmer fluid is more likely to vaporize.

Aeration is often confused with cavitation because both involve gas bubbles forming at the inlet side of the pump. With aeration, however, the source of the gas bubbles is outside air leaking into the system, rather than vapor and air bubbles forming from within the system. The air enters the system through leaky fittings and seals on the inlet side. Because the inlet side is at vacuum, any leak will cause atmospheric air to push its way in. The end result is the same as with cavitation: The bubbles collapse on the pressure side of the pump and cause increased wear, heat, vibration, and noise. Although they have the same result, cavitation and aeration have different causes and, therefore, require different remedies.

3.8   Graphic Symbols

Figure 3-20 shows the pump symbols used in hydraulic schematics. Fixed displacement pumps cannot vary their displacement and therefore cannot change their flow rate (Figure 3.20A). Variable displacement pumps have the ability to vary their displacement and consequently can vary their flow rate even while being driven at a constant speed (Figure 3-20B). Pressure-compensated variable displacement pumps can automatically reduce their displacement to prevent the pressure from rising above a preset level (Figure 3-20C).

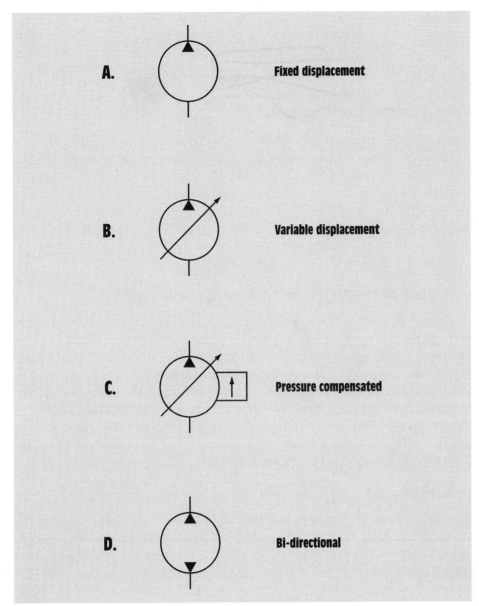

FIGURE 3-20   Pump symbols. (A) Fixed displacement. (B) Variable displacement. (C) Pressure compensated. (D) Bi-directional.

The three previous pump symbols were for *unidirectional* pumps, which are the most common. Unidirectional pumps can be rotated in only one direction. Rotating them in the wrong direction may result in a blown shaft seal because many pumps drain the leakage to the suction (inlet) side of the pump. Reversing the direction of rotation would reverse the inlet and outlet ports and, consequently, pressurizes the drain cavities. *Bi-directional,* or *reversible,* pumps are designed to be rotated in either direction (Figure 3-20D). None of the symbols in Figure 3-20 provides information on whether the pump is a gear, vane, or piston type.

3.9   Pump Specifications

This section describes the information that manufacturers commonly provide on pump performance. The format of this information can vary considerably from one manufacturer to another. The reader is encouraged to obtain several manufacturer's catalogs to compare how specifications are provided in industry. Many manufacturers provide on-line catalogs that can be accessed or downloaded via the Internet. The National Fluid Power Association's web site (www.nfpa.com) is a good place to start a search. Most manufacturers will provide a hardcopy of their catalog upon request.

3.9.1 DISPLACEMENT, FLOW RATE, AND DRIVE SPEED

Pumps are sized primarily by their displacement in in^3/rev, which is sometimes abbreviated CIR in manufacturer's catalogs. For fixed displacement pumps, a single value will be given, while a range will be given for variable displacement pumps. The theoretical flow rate can then be calculated with Equation 3-2. The actual flow rate of the pump, accounting for leakage, will usually be given at a particular drive speed. It may also be given as a graph for a variety of drive speeds and pressures. An example of such a graph is shown in Figure 3-21. Note that the 3000 psi curve is below the 1500 psi curve because there is less flow due to increased leakage at the higher pressure.

EXAMPLE 3-15.

Using Figure 3-21, determine the actual flow rate from this pump at 1800 rpm and 3000 psi. What is the volumetric efficiency at this speed?

SOLUTION:

1. Find the actual flow from the graph:

$$Q_A \approx 13.3 \text{ gpm}$$

2. Calculate the theoretical flow:

$$Q_T = \frac{V_p \cdot N}{231} = \frac{2\frac{in^3}{rev} \cdot (1800 \text{ rpm})}{231} = 15.6 \text{ gpm}$$

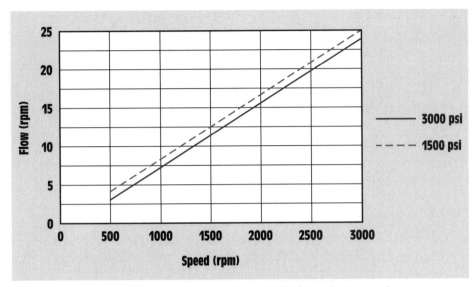

FIGURE 3-21   Actual flow versus drive speed graph (2 in³/rev displacement).

3. Calculate the volumetric efficiency:

$$\eta_V = \frac{Q_A}{Q_T} = \frac{13.3 \text{ gpm}}{15.6 \text{ gpm}} = 0.853 \quad (85\%)$$

We could also calculate the actual flow rate with Equation 3-8, if the volumetric efficiency was known. This is also read from graphs (see Section 3.9.3).

A minimum and maximum drive speed in rpm is usually specified. Operating at speeds below the minimum will result in extreme inefficiency. Operating above the maximum speed rating can cause excessive wear and decreased pump life. Maximum speeds are usually between 1800 and 3600 rpm. Reduced maximum speed ratings may be provided for fluids other than standard petroleum-based oil, such as fire-resistant synthetics or water-based fluids because these fluids are typically heavier (have a higher specific gravity) and do not lubricate as well as petroleum-based oil.

3.9.2 PRESSURE

Pressure ratings are given either as a single value or as separate values for intermittent and continuous service. Values for intermittent service are higher than those for continuous service, as the pump is subjected to these pressures for only a fraction of the duty cycle. If the pump is operated above the maximum pressure, it may become very inefficient due to increased leakage. Seal failure is also a possibility. Manufacturers may also give reduced pressure ratings for fluids other than petroleum-based oils, as is the case with the drive speed.

3.9.3 EFFICIENCY

Volumetric and overall efficiencies are usually provided in a graphical format. Figure 3-22 shows *efficiency* versus *pressure* graphs for a particular pump at two speeds: 1200 rpm (graph A) and 1800 rpm (graph B). The displacement is 1.2 in²/rev. As shown, the volumetric efficiency goes down with increasing pressure. This is as expected because leakage will increase as pressure increases. The relationship between overall efficiency and pressure is more complicated: η_o increases with increasing pressure until it reaches a peak, then begins to decrease. This is typical. These graphs are given for a particular fluid type, temperature, and viscosity (e.g., standard hydraulic oil at 120° and 100 SUS).

EXAMPLE 3-16.

Using Figure 3-22, determine the volumetric and overall efficiencies of this pump at 1200 rpm and 2000 psi. What is the actual flow rate under these conditions?

SOLUTION

1. Find the efficiencies from graph A:

$$\eta_V \approx 91\%$$

$$\eta_O \approx 82\%$$

2. Calculate the actual flow:

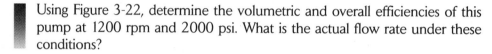

$$Q_A = \eta_V \cdot \frac{V_p \cdot N}{231} = 0.88 \cdot \frac{1.2 \frac{in^3}{rev} \cdot (1800 \text{ rpm})}{231} = 8.23 \text{ gpm}$$

It is also common for manufacturers to provide *efficiency* versus *drive speed* graphs. Figure 3-23 shows this for a particular pump at two pressures: 1500 psi (graph A) and 3000 psi (graph B). The displacement is again 1.2 in³/rev. As the graphs show, pumps are more efficient when driven faster, but only up to a point.

EXAMPLE 3-17.

Using Figure 3-23, determine the volumetric and overall efficiencies of this pump at 1800 rpm and 3000 psi. What is the actual flow rate under these conditions?

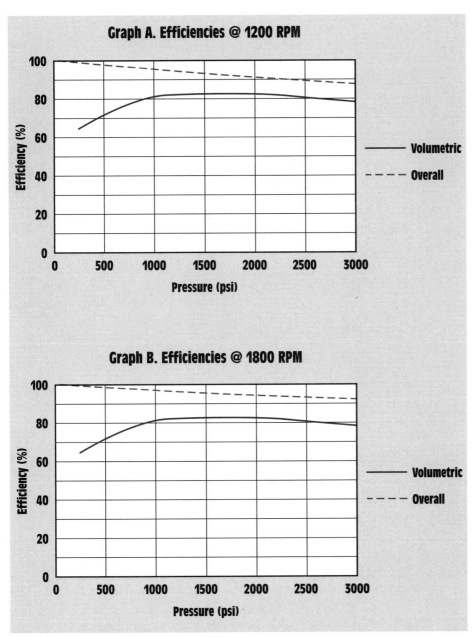

FIGURE 3-22   Efficiency versus pressure graph (1.2 in³/rev displacement). (A) Efficiencies at 1200 rpm. (B) Efficiencies at 1800 rpm.

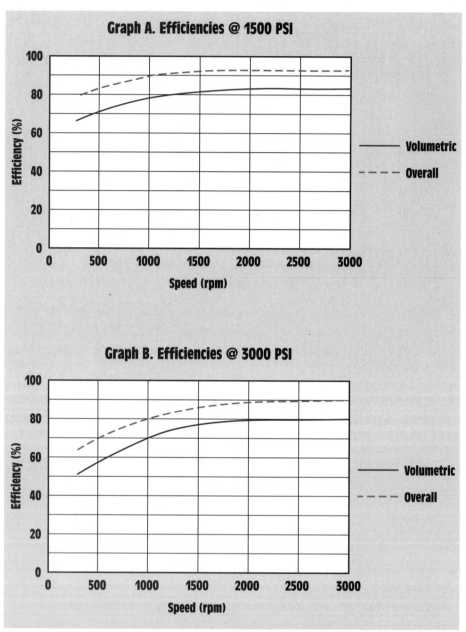

FIGURE 3-23 Efficiency versus drive speed graph (1.2 in³/rev displacement). (A) Efficiencies at 1500 psi. (B) Efficiencies at 3000 psi.

SOLUTION:

1. Find the volumetric efficiency from graph B:

$$\eta_V \approx 88\%$$

$$\eta_O \approx 78\%$$

2. Calculate the actual flow:

$$Q_A = \eta_V \cdot \frac{V_p \cdot N}{231} = 0.88 \cdot \frac{1.2 \frac{in^3}{rev} \cdot (1800 \text{ rpm})}{231} = 8.23 \text{ gpm}$$

Efficiency data can be provided in multiple formats, so care should be taken to find the graph or table that is applicable for any particular situation. Anyone working in a technical field should be able to interpret and apply data from graphs such as these.

3.9.4 FLUIDS

Viscosity is the most critical fluid property with regard to pump performance. Manufacturers usually specify a minimum and maximum fluid viscosity. Viscosity changes with temperature, so we must be sure that the fluid chosen is in this viscosity range over the temperature range at which the system will operate. If the viscosity of the fluid is too high, it becomes difficult to move through the system, causing increased power loss and heat buildup due to excessive fluid friction. High viscosity can also cause cavitation. If the viscosity is too low, excessive wear and heat buildup result because the oil no longer provides good lubrication.

Most systems and components are designed for standard petroleum-based hydraulic oil. Reduced speed and pressure ratings may be specified for other fluids. Fluid compatibility with the pump seals is another important factor. A pump designed for standard oil will often have seals that are incompatible with certain other fluids. The seals may quickly degrade and fail if used with these fluids (see Chapter 9).

3.9.5 FILTRATION

Due to the small clearances between the components in positive displacement pumps, they are very sensitive to contamination. With such tight-fitting components, a very small particle of contamination can wedge between them and cause wear. This results in increased leakage and inefficiency, possibly to the point of component failure. Manufacturers specify the level of filtration

Scale number	Particles per million	
	More than	Up to
10	5	10
11	10	20
12	20	40
13	40	80
14	80	160
15	160	320
16	320	640
17	640	1300
18	1300	2500
19	2500	5000
20	5000	10,000
21	10,000	20,000

FIGURE 3-24   ISO cleanliness code.

required for each pump model. The most common standard is ISO 4406, also known as the ISO *cleanliness code,* which assigns scale numbers to different levels of contamination. A partial list from this code is shown in Figure 3-24. Each scale number corresponds to a range of particles per milliliter (ml) present in a sample. For example, scale 18 corresponds to a range between 1300 and 2500 particles present per ml. The specification is given as two numbers: the scale number for particles larger than 5 micrometers (μm) and the scale number for particles larger than 15 μm. A micrometer is one-millionth of a meter (1 μm = 0.000001 m), which is approximately 0.000039 inches. The cleanliness code 16/13 means there are between 320 and 640 particles larger than 5 μm and between 40 and 80 particles larger than 15 μm per milliliter. When a pump manufacturer specifies the cleanliness code 16/13, the oil must be *at least* this clean. It may, of course, be cleaner.

Starting up a new system is one of the most critical times with regard to contamination. New components are often filled with particles (metal, dirt, etc.) from the manufacturing process. It is good practice to flush a system with fine filtration at startup. The filters should then be replaced with new ones before beginning regular operation under pressure. Any new components added to a system should likewise be assumed to be dirty and should be flushed before installation.

3.9.6 PHYSICAL DIMENSIONS

When selecting a pump for a particular application, the physical dimensions are an important factor to consider. The manufacturer will provide data such as:

1. Weight,
2. Overall dimensions,
3. Center-to-center distances and sizes of mounting holes,
4. Port sizes (may be threaded or flange type), and
5. Drive shaft size (may be spline or key type).

A drawing of each model with all the relevant dimensions is usually provided in the manufacturer's catalog.

3.9.7 NOISE LEVEL

Noise from a hydraulic pump is created by the vibration of the moving components, as it is with all machinery. Noise level is measured in A-weighted decibels, abbreviated dB(A). The manufacturer may provide the noise level at a particular drive speed, at several drive speeds, or may provide a graph of noise level versus drive speed. Noise levels of 60 to 80 dB(A) are common for hydraulic pumps operating at 1800 rpm. In general, vane pumps are the quietest of the three types discussed, while gear pumps are the loudest. To give a sense of scale, a vacuum cleaner has a noise level of about 70 dB(A). Most people experience pain when subjected to noise levels above 120 dB(A). The Occupational Safety and Health Administration (OSHA) requires that workers not be exposed to levels above 90 dB(A) on a continuous basis during an 8-hour shift.

3.9.8 SHAFT ROTATION

Pumps are usually designed to be rotated in one direction only. The direction is specified as either *right-hand* (clockwise) or *left-hand* (counter-clockwise) when viewing the pump from the *shaft end*. Both types are generally available as standard models.

3.9.9 INLET PRESSURE

Manufacturers will specify a maximum amount of vacuum that is permissible at the pump inlet, most commonly measured in inches of mercury (in Hg). If pumps are to be driven at high speeds (greater than 2400 rpm), it may be specified that a positive pressure be supplied at the pump inlet. This is to ensure that the pump receives sufficient flow to its inlet. The measurement of vacuum is discussed in detail in Chapter 10.

3.10 Equations

EQUATION NUMBER	EQUATION	REQUIRED UNITS
3-1	$Q_T = V_P \cdot N$	Any consistent units
3-2	$Q_T = \dfrac{V_p \cdot N}{231}$	V_P in in³/rev, N in rpm, Q_T in gpm
3-2M	$Q_T = \dfrac{V_p \cdot N}{1000}$	V_P in cm³/rev, N in rpm, Q_T in lpm
3-3	$T = F \cdot d$	Any consistent units
3-4	$T_T = \dfrac{p \cdot V_p}{2 \cdot \pi}$	Any consistent units
3-5	$HP_I = \dfrac{T \cdot N}{63,025}$	T in in · lbs, N in rpms, HP_I in hp
3-5M	$kW_I = \dfrac{T \cdot N}{9550}$	T in N · m, N in rpms, kW_I in kW
3-6	$HP_H = \dfrac{p \cdot Q}{1714}$	Q in gpm, p in psi, HP_H in hp
3-6M	$kW_H = \dfrac{p \cdot Q}{60,000}$	Q in lpm, p in kPa, kW_H in kW
3-7	$\eta_V = \dfrac{Q_A}{Q_T}$	Any consistent units

EQUATION NUMBER	EQUATION	REQUIRED UNITS
3-8	$Q_A = \eta_V \cdot \dfrac{V_P \cdot N}{231}$	V_P in in³/rev, N in rpm, Q_A in gpm
3-8M	$Q_A = \eta_V \cdot \dfrac{V_P \cdot N}{1000}$	V_P in cm³/rev, N in rpm, Q_A in lpm
3-9	$\eta_M = \dfrac{T_T}{T_A}$	Any consistent units
3-10	$\eta_O = \dfrac{HP_H}{HP_I}$	HP_H in hp, HP_I in hp, η_O is unitless
3-10M	$\eta_O = \dfrac{kW_H}{kW_I}$	kW_H in kW, kW_I in kW, η_O is unitless
3-11	$\eta_O = \eta_M \cdot \eta_V$	All quantities are unitless

3.11 Review Questions and Problems

1. Explain the basic operation of a positive displacement pump.
2. Explain the basic operation of a nonpositive displacement pump.
3. What is the purpose of a hydraulic pump?
4. What causes pressure in a hydraulic system?
5. Define *displacement.*
6. What is the difference between a fixed displacement pump and a variable displacement pump?
7. Define *volumetric efficiency.*
8. Define *mechanical efficiency.*
9. Define *overall efficiency.*
10. What is the purpose of the ISO cleanliness code?
11. Which of the three pump types (gear, vane, or piston) is most tolerant of contamination? Which is the least?
12. Which of the three pump types has the highest pressure rating?
13. Which of the three pump types is the most efficient?
14. Draw the graphic symbols for the following four pump types: fixed displacement, variable displacement, pressure compensated, and bi-directional.
15. Describe the difference between cavitation and aeration. What is the result of each?
16. A pump has a displacement of 1.15 in³/rev and is driven at 1800 rpm. What is its theoretical flow rate in gpm?

17. A pump has a displacement of 55 cm³/rev and is driven at 1800 rpm. What is its theoretical flow rate in lpm?

18. A new system requires a flow rate of at least 25 gpm. What should the minimum displacement be to achieve this flow rate if the drive speed will be 1200 rpm? What should the minimum displacement be if the drive speed is 1800 rpm?

19. A new system requires a flow rate of at least 100 lpm. What should the minimum displacement be to achieve this flow rate if the drive speed will be 1200 rpm? What should the minimum displacement be if the drive speed is 1800 rpm?

20. A hydraulic pump has a flow rate of 25 gpm and is rated for a maximum pressure of 3000 psi. What is the maximum horsepower output of this pump?

21. A hydraulic pump has a flow rate of 90 lpm and is rated for a maximum pressure of 15,000 kPa. What is the maximum power output of this pump?

22. A hydraulic pump with a displacement of 2.5 in³/rev is selected for a system that will operate at a maximum pressure of 5000 psi. What is the required drive torque? Assume 100% efficiency.

23. A hydraulic pump with a displacement of 15 cm³/rev is selected for a system that will operate at a maximum pressure of 20,000 kPa. What is the required drive torque? Assume 100% efficiency.

24. An electric motor drives a pump at 1200 rpm with a torque of 1550 in · lb. What is the horsepower input to this pump?

25. An electric motor drives a pump at 1200 rpm with a torque of 150 N · m. What is the power input to this pump?

26. A hydraulic power unit (electric motor and pump) produces a flow rate of 8 gpm and has a maximum power rating of 20 hp. What is the maximum operating pressure of this unit? Assume 100% efficiency.

27. A hydraulic power unit produces a flow rate of 85 lpm and has a maximum power rating of 40 kW. What is the maximum operating pressure of this unit? Assume 100% efficiency.

28. A pump has a displacement of 4.12 in³/rev and a volumetric efficiency of 0.90. What will its actual flow rate be if it is driven at 1200 rpm?

29. A pump has a displacement of 75 cm³/rev and a volumetric efficiency of 0.90. What will its actual flow rate be if it is driven at 1200 rpm?

30. A minimum flow of 15 gpm is required in a new system. The pump type chosen has a volumetric efficiency of 0.88 and will be driven at 1800 rpm. What is the minimum displacement required?

31. For the system in problem 30, the next larger available displacement is 2.25 in³/rev. What will the flow rate be with this pump? What is the effect of this increase in flow on actuator speed?

32. A minimum flow of 80 lpm is required in a new system. The pump type chosen has a volumetric efficiency of 0.90 and will be driven at 1800 rpm. What is the minimum displacement required?
33. A pump that has an mechanical efficiency of 0.85 and a displacement of 2.25 in³/rev is to be used in a system with a maximum operating pressure of 2000 psi. What is the required drive torque?
34. A pump that has an mechanical efficiency of 0.90 and a displacement of 35 cm³/rev is to be used in a system with a maximum operating pressure of 15,000 kPa. What is the required drive torque?
35. A pump that has an overall efficiency of 0.83 and a flow rate of 13 gpm is to be used in a system that has a maximum operating pressure of 2000 psi. What is the minimum input horsepower required?
36. A pump that has an overall efficiency of 0.87 and a flow rate of 100 lpm is to be used in a system that has a maximum operating pressure of 10,000 kPa. What is the minimum input power required?
37. A pump with an overall efficiency of 0.82 is connected to an electric motor with a maximum power rating of 25 hp. What is the maximum hydraulic horsepower output for this system?
38. A pump with an overall efficiency of 0.85 is connected to an electric motor with a maximum power rating of 30 kW. What is the maximum hydraulic power output for this system?
39. Using Figure 3-21, determine the actual flow rate from this pump at 1200 rpm and 1500 psi. What is the volumetric efficiency at this speed?
40. What is the result of operating below a pump's minimum speed rating? What is the result of operating above the maximum speed rating?
41. Using Figure 3-22, determine the volumetric efficiency of this pump at 1800 rpm and 2000 psi. What is the actual flow rate under these conditions?
42. Using Figure 3-22, determine the overall efficiency for the pump in problem 41.
43. Using Figure 3-23, determine the volumetric efficiency of this pump at 1200 rpm and 1500 psi. What is the actual flow rate under these conditions?
44. Using Figure 3-23, determine the overall efficiency for the pump in problem 43.
45. A pump manufacturer has specified that a cleanliness level of 18/15 is required. What does this translate to in terms of particle counts at 5 and 15 μm particle size?

CHAPTER 4

Hydraulic Cylinders

OUTLINE

4.1 Introduction

4.2 Cylinder Force

4.3 Cylinder Speed

4.4 Cylinder Power

4.5 Differential Flow

4.6 Cylinder Types

4.7 Graphic Symbols

4.8 Cylinder Applications

4.9 Cylinder Specifications

4.10 Equations

4.11 Review Questions and Problems

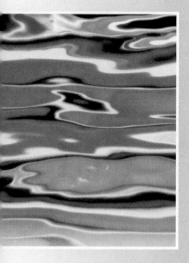

4.1 Introduction

Chapter 3 discussed pumps, which convert mechanical input into fluid power output. The *actuator,* which converts the fluid power back into mechanical power, is at the other end of a hydraulic system. The actuator is most commonly a cylinder or a motor. Cylinders are also called linear actuators because they produce straight-line motion. Motors are called rotary actuators because they produce rotational motion. Motors will be covered in Chapter 5.

If we convert mechanical power input into fluid power with the pump, and then convert fluid power back into mechanical power with the activator at the other end of the system, what is the point? As shown in Chapter 2, force can be multiplied in a hydraulic system by using a small input piston and a large output piston, for example. This increase in force is coupled with a decrease in velocity by the same factor. This is essentially what occurs in an actual hydraulic system. The pump is a smaller device that moves very quickly with a small amount of force input. The actuator is a larger device that moves more slowly with a much larger force output.

Figure 4-1 shows the construction of a typical *double-acting* hydraulic cylinder. Double-acting cylinders are able to produce force in both directions by applying pressure to either side of the piston. A cylinder consists primarily of the *barrel,* the *end caps,* the *piston,* and the *piston rod.* Motion is produced by applying pressure to either side of the piston, causing the piston to slide back and forth within the barrel. The load is attached to the piston rod through various mounting hardware, which will be discussed later in the chapter.

The type of cylinder shown in Figure 4-1 uses *tie rod* construction. Tie rods are essentially long bolts that hold the end caps onto the barrel. Cylinders with welded or threaded end caps are also common. The left end of the cylinder is called the *rod end,* the end out of which the piston rod extends. The right end is the called the *blind end.* Fluid is fed into the cylinder through the blind end and rod end ports.

Cylinders are sized primarily by their bore diameter, which is the inside diameter of the barrel, as shown. The piston requires a seal and bearing. The rod end requires a seal, a bearing, and a rod wiper. The seals are necessary to build pressure without excessive leakage. The bearings allow for a moderate amount of side loading. The rod wiper prevents contamination from entering the cylinder. The seal and wiper are made of an elastomer (synthetic rubber). The bearings are made of a soft metal such as bronze, a hard plastic, or a composite type material. The inside of the barrel and the piston rod have a very smooth finish to prevent premature wear of the bearings and seals (see Chapter 9). A photo of an actual hydraulic cylinder with tie rod construction is shown in Figure 4-2.

Figure 4-3 illustrates the operation of a double-acting hydraulic cylinder. To extend the cylinder, pump flow is sent to the blind end port, as shown in part A. Fluid from the rod end port returns to the reservoir. To retract the cylinder,

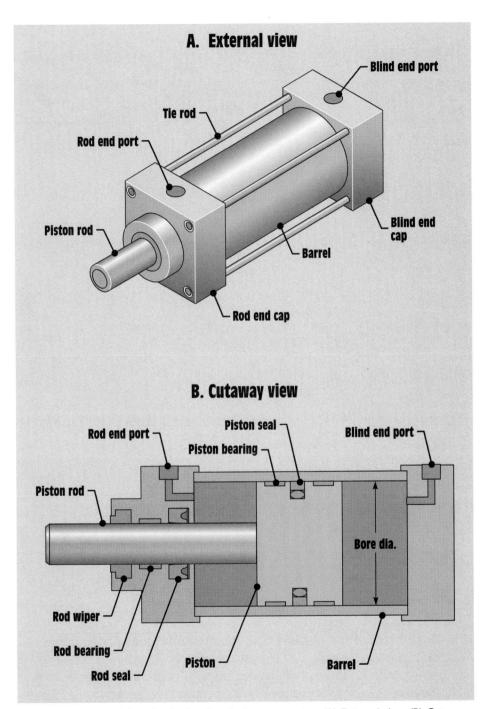

FIGURE 4-1   Double-acting hydraulic cylinder construction. (A) External view. (B) Cutaway view.

FIGURE 4-2 Double-acting hydraulic cylinder. (Courtesy of Parker Hannifin Corp.)

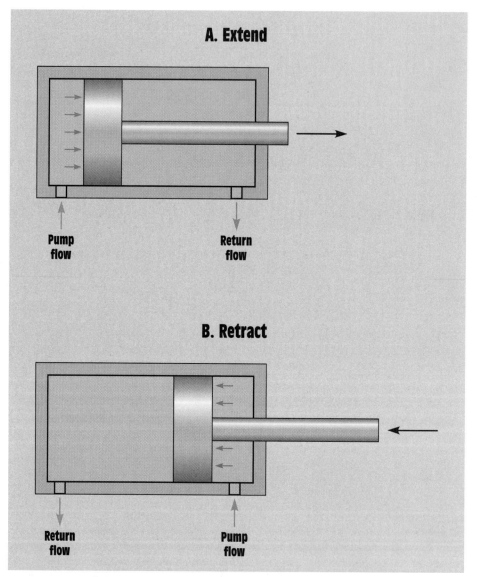

FIGURE 4-3   Cylinder operation. (A) Extend. (B) Retract.

pump flow is sent to the rod end port and fluid from the blind end returns to the reservoir, as shown in part B. The area exposed to pressure is smaller on the retract stroke than on the extend stroke due to the piston rod (Figure 4-4). The implications of this on cylinder force and speed are discussed in the following sections.

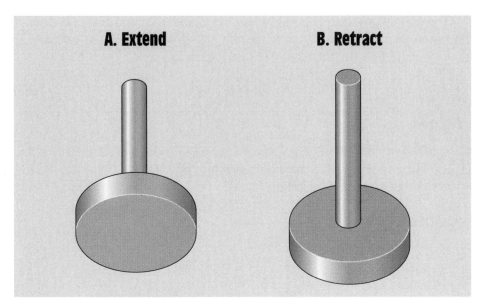

FIGURE 4-4   Piston area. (A) Extend. (B) Retract.

4.2   Cylinder Force

As explained in Chapter 2, the force generated when pressure is applied to a given area can be calculated with the following equation:

$$F = p \cdot A$$

The area exposed to pressure when the cylinder is extending is the area of the piston (Figure 4-4A). We can therefore use the following equation to calculate the force of a cylinder on the extend stroke:

$$F_E = p \cdot A_P \qquad \textbf{(4-1)}$$

where: F_E = extend force
$\quad$ p = pressure
$\quad$ A_P = area of the piston

The cylinder's bore diameter is used to calculate the area of the piston. The piston diameter is actually a few thousandths of an inch smaller than the bore diameter, but this difference is insignificant.

The area exposed to pressure when the cylinder is retracting is the area of the piston (A_P) minus the area of the rod (A_R) (Figure 4-4B). To calculate the force generated on the retract stroke, use Equation 4-2:

$$F_R = p \cdot (A_P - A_R) \qquad\qquad \textbf{(4-2)}$$

EXAMPLE 4-1.

A cylinder with a bore diameter of 2 in and a rod diameter of 1 in is to be used in a system with a maximum pressure of 2500 psi. What are the maximum extension and retraction forces?

SOLUTION:

1. Calculate the piston area:

$$A_P = \frac{\pi \cdot D_P^{\,2}}{4} = \frac{3.142 \cdot (2 \text{ in})^2}{4} = 3.142 \text{ in}^2$$

2. Calculate the rod area:

$$A_R = \frac{\pi \cdot D_R^{\,2}}{4} = \frac{3.142 \cdot (1 \text{ in})^2}{4} = 0.7854 \text{ in}^2$$

3. Calculate the extension force:

$$F_E = p \cdot A_P = 2500 \frac{\text{lbs}}{\text{in}^2} \cdot (3.142 \text{ in}^2) = 7855 \text{ lbs}$$

4. Calculate the retraction force:

$$F_R = p \cdot (A_A - A_R) = 2500 \frac{\text{lbs}}{\text{in}^2} \cdot (3.142 - 0.7854) \text{ in}^2 = 5892 \text{ lbs}$$

EXAMPLE 4-1M.

A cylinder with a bore diameter of 50 mm and a rod diameter of 20 mm is to be used in a system with a maximum pressure of 15,000 kPa. What are the maximum extension and retraction forces?

SOLUTION:

1. Calculate the piston area:

$$A_P = \frac{\pi \cdot D_P^{\,2}}{4} = \frac{3.142 \cdot (0.050 \text{ m})^2}{4} = 0.001964 \text{ m}^2$$

2. Calculate the rod area:

$$A_R = \frac{\pi \cdot D_R^{\,2}}{4} = \frac{3.142 \cdot (0.020 \text{ m})^2}{4} = 0.0003142 \text{ m}^2$$

3. Calculate the extension force:

$$F_E = p \cdot A_P = 15,000,000 \frac{N}{m^2} \cdot (0.001964 \text{ m}^2) = 29,640 \text{ N}$$

4. Calculate the retraction force:

$$F_R = p \cdot (A_P - A_R) = 15,000,000 \frac{N}{m^2} \cdot (0.001964 - 0.0003142) \text{ m}^2$$
$$= 24,750$$

EXAMPLE 4-2.

For the system in Example 4-1, what effect would doubling the bore diameter have on the output force generated on extension?

SOLUTION:

1. Calculate the piston area:

$$A_P = \frac{\pi \cdot D_P^{\,2}}{4} = \frac{3.142 \cdot (4 \text{ in})^2}{4} = 12.57 \text{ in}^2$$

2. Calculate the extension force:

$$F_E = p \cdot A_P = 2500 \frac{lbs}{in^2} \cdot (12.57 \text{ in}^2) = 31,425 \text{ lbs}$$

The previous example shows that doubling the bore diameter increases the force output by a factor of four ($31,425 \div 7855 = 4$). This is because the force output is proportional to the area ($F = p \cdot A$) and the area is proportional to the square of the diameter ($A = \frac{\pi}{4} \cdot D^2$) *The output force is therefore proportional to the square of the diameter.*

EXAMPLE 4-3.

A cylinder with a bore diameter of 3 in and a rod diameter of 1.38 in is required to extend against a load of 10,000 lbs. What pressure is required?

SOLUTION:

1. Calculate the piston area:

$$A_P = \frac{\pi \cdot D_P^{\,2}}{4} = \frac{3.142 \cdot (3 \text{ in})^2}{4} = 7.069 \text{ in}^2$$

2. Calculate the required pressure:

$$p = \frac{F_E}{A_P} = \frac{10,000 \text{ lbs}}{7.069 \text{ in}^2} = 1415 \frac{\text{lbs}}{\text{in}^2} \quad (1415 \text{ psi})$$

EXAMPLE 4-3M.

A cylinder with a bore diameter of 80 mm and a rod diameter of 25 mm is required to extend against a load of 50 kN. What pressure is required?

SOLUTION:

1. Calculate the piston area:

$$A_P = \frac{\pi \cdot D_P^{\,2}}{4} = \frac{3.142 \cdot (0.080 \text{ m})^2}{4} = 0.005027 \text{ m}^2$$

2. Calculate the required pressure:

$$P = \frac{F_E}{A_P} = \frac{50,000 \text{ N}}{0.005027 \text{ m}^2} = 9,946,000 \frac{\text{N}}{\text{m}^2} \quad (9946 \text{ kPa})$$

The pressure calculated in Examples 4.3 and 4.3M will only *balance* the load, not *move* it. Extending the cylinder requires a slightly higher pressure to overcome the inertia of the load and friction. This is difficult to determine accurately, and is usually estimated by experience. In most applications, it is adequate to simply be aware that a slightly higher pressure will be required.

EXAMPLE 4-4.

The cylinder in Example 4-3 is required to retract against a load of 500 lbs. What pressure is required?

SOLUTION:

1. Calculate the rod area:

$$A_R = \frac{\pi \cdot D_R^{\,2}}{4} = \frac{3.142 \cdot (1.38 \text{ in})^2}{4} = 1.496 \text{ in}^2$$

2. Calculate the required pressure:

$$p = \frac{F_R}{A_P - A_R} = \frac{500 \text{ lbs}}{7.069 \text{ in}^2 - 1.496 \text{ in}^2} = 89.72 \frac{\text{lbs}}{\text{in}^2}$$

The situation illustrated in Examples 4.3M and 4.4 is typical; a cylinder extends under a heavy load and retracts under a light load (often no load at all). In this situation, the extend stroke is the *working stroke* and the retract stroke is the *return stroke*. There are some applications in which the retract stroke is the working stroke, however.

EXAMPLE 4-5.

A cylinder is required to extend against a load of 25,000 lbs with a maximum pressure of 2000 psi. What cylinder size is required?

SOLUTION:

1. Calculate the piston area:

$$A_P = \frac{F}{p} = \frac{25,000 \text{ lbs}}{2000 \frac{\text{lbs}}{\text{in}^2}} = 12.5 \text{ in}^2$$

2. Calculate the bore diameter:

$$D_P = \sqrt{\frac{4 \cdot A}{\pi}} = \sqrt{\frac{4 \cdot (12.5 \text{ in}^2)}{3.142}} = 3.989 \text{ in}$$

EXAMPLE 4-5M.

A cylinder is required to extend against a load of 100 kN with a maximum pressure of 15,000 kPa. What cylinder size is required?

SOLUTION:

1. Calculate the piston area:

$$A_P = \frac{F}{p} = \frac{100,000 \text{ N}}{15,000,000 \, \dfrac{\text{N}}{\text{m}^2}} = 0.006667 \text{ m}^2$$

2. Calculate the bore diameter:

$$D_P = \sqrt{\frac{4 \cdot A}{\pi}} = \sqrt{\frac{4 \cdot (0.006667 \text{ m}^2)}{3.142}} = 0.0951 \text{ m} \quad (92.1 \text{ mm})$$

Chapter 3 discussed the concept of inefficiency with respect to hydraulic pumps. Cylinders are also not 100% efficient, primarily due to the friction of the seals and bearings, which typically consume 6% to 8% of the output force. Data regarding this, however, are usually *not* provided by the manufacturer. We must therefore always round up when sizing cylinders. For the system in Example 4-5, a cylinder with a bore of 4.50 in should be chosen. A size of 4 in is perhaps cutting it too close, unless a slightly higher pressure can be tolerated.

4.3   Cylinder Speed

Chapter 2 introduced an equation that related velocity, flow rate, and the area of a conduit through which a fluid flows:

$$Q = v \cdot A$$

Because the velocity of a cylinder is equal to the velocity of the fluid flowing through it, we can also use this equation to calculate cylinder speed. Rearranging to solve for velocity, we obtain:

$$v = \frac{Q}{A} \qquad \textbf{(4-3)}$$

The consistent U.S. customary units for this equation are: Q in in^3/min, A in in^2, and v in in/min. In hydraulic systems, however, gal/min (gpm) is the preferred unit for flow. Inserting a conversion factor of 231 in³ = 1 gal to obtain an equation that uses all of the most common units produces the following equation:

$$v = \frac{231 \cdot Q}{A} \qquad \textbf{(4-4)}$$

We must use the following units for Equation 4-4: Q in gpm, A in in², and v in in/min. The area through which the fluid flows when the cylinder is extending is the area of the piston. We can therefore use the following equation to calculate the extend velocity of a cylinder:

$$v_E = \frac{231 \cdot Q}{A_P} \qquad \textbf{(4-5)}$$

The flow area when the cylinder is retracting is the area of the piston minus the area of the rod. To calculate the speed on the retract stroke, use Equation 4-6:

$$v_R = \frac{231 \cdot Q}{A_P - A_R} \qquad \textbf{(4-6)}$$

Because we have inserted the conversion factor from in³ to gal into Equations 4-5 and 4-6, we must use the following units: Q in gpm, A in in², and v in in/min.

The preferred unit for flow in the metric system is liters per minute (lpm). We can insert the conversion factor 1 m³ = 1000 l into Equation 4-3 to create equations to calculate the extension and retraction speeds when using metric units:

$$v_E = \frac{Q}{1000 \cdot A_P} \qquad \textbf{(4-5M)}$$

$$v_R = \frac{Q}{1000 \cdot (A_P - A_R)} \qquad \textbf{(4-6M)}$$

Because we have inserted the conversion factor from m² to liters into Equations 4-5M and 4-6M, we must use the following units: Q in lpm, A in m², and v in m/min.

EXAMPLE 4-6.

A cylinder with a bore diameter of 2.5 in and a rod diameter of 1 in is to be used in a system with a 15 gpm pump. What are the extension and retraction speeds?

SOLUTION:

1. Calculate the piston area:

$$A_P = \frac{\pi \cdot D_P^2}{4} = \frac{3.142 \cdot (2.5 \text{ in})^2}{4} = 4.909 \text{ in}^2$$

2. Calculate the rod area:

$$A_R = \frac{\pi \cdot D_R^2}{4} = \frac{3.142 \cdot (1 \text{ in})^2}{4} = 0.7855 \text{ in}^2$$

3. Calculate the extension speed:

$$v_E = \frac{231 \cdot Q}{A_P} = \frac{231 \frac{\text{in}^3}{\text{gal}} \cdot \left(15 \frac{\text{gal}}{\text{min}}\right)}{4.909 \text{ in}^2} = 705.8 \frac{\text{in}}{\text{min}}$$

4. Calculate the retraction speed:

$$v_E = \frac{231 \cdot Q}{A_P - A_R} = \frac{231 \frac{\text{in}^3}{\text{gal}} \cdot \left(15 \frac{\text{gal}}{\text{min}}\right)}{(4.909 - 0.7855) \text{ in}^2} = 840.3 \frac{\text{in}}{\text{min}}$$

EXAMPLE 4-6M.

A cylinder with a bore diameter of 80 mm and a rod diameter of 25 mm is to be used in a system with a 60 lpm pump. What are the extension and retraction speeds?

SOLUTION:

1. Calculate the piston area:

$$A_P = \frac{\pi \cdot D_P^2}{4} = \frac{3.142 \cdot (0.080 \text{ m})^2}{4} = 0.005027 \text{ m}^2$$

2. Calculate the rod area:

$$A_R = \frac{\pi \cdot D_R^2}{4} = \frac{3.142 \cdot (0.024 \text{ m})^2}{4} = 0.0004909 \text{ in}^2$$

3. Calculate the extension speed:

$$v_E = \frac{Q}{1000 \cdot A_P} = \frac{60\,\frac{l}{min}}{1000\,\frac{l}{m^3} \cdot (0.005027\ m^2)} = 11.94\,\frac{m}{min}$$

4. Calculate the retraction speed:

$$v_E = \frac{Q}{1000 \cdot (A_P - A_R)} = \frac{60\,\frac{l}{min}}{100\,\frac{l}{m^3} \cdot (0.005027 - 0.0004904)m^2}$$

$$= 13.23\,\frac{m}{min}$$

EXAMPLE 4-7.

A cylinder with a bore diameter of 3 in and a rod diameter of 1.38 in is required to extend at a velocity of 0.5 ft/s. What flow rate is required to achieve this speed?

SOLUTION:

1. Calculate the piston area:

$$A_P = \frac{\pi \cdot D_P^2}{4} = \frac{3.142 \cdot (3\ in)^2}{4} = 7.069\ in^2$$

2. Convert to in/min:

$$v_E = 0.5\,\frac{ft}{s} \cdot \left(\frac{12\ in}{1\ ft}\right) \cdot \left(\frac{60\ s}{1\ min}\right) = 360\,\frac{in}{min}$$

3. Calculate the flow rate:

$$Q = \frac{v_E \cdot A_P}{231} = \frac{360\,\frac{in}{min} \cdot (7.069\ in^2)}{231\,\frac{in^3}{gal}} = 11.02\,\frac{gal}{min}$$

EXAMPLE 4-7M.

A cylinder with a bore diameter of 50 mm and a rod diameter of 20 mm is required to extend at a velocity of 1 m/s. What flow rate is required to achieve this speed?

SOLUTION:

1. Calculate the piston area:

$$A_P = \frac{\pi \cdot D_P^2}{4} = \frac{3.142 \cdot (0.050 \text{ m})^2}{4} = 0.001964 \text{ m}^2$$

2. Convert to m/min:

$$v_E = 1\frac{\text{m}}{\text{s}} \cdot \left(\frac{60 \text{ s}}{1 \text{ min}}\right) = 60 \frac{\text{m}}{\text{min}}$$

3. Calculate the flow rate:

$$Q = 1000 \cdot v_E \cdot A_P = 1000 \frac{1}{\text{m}^3} \cdot \left(60 \frac{\text{m}}{\text{min}}\right) \cdot 0.001964 \text{ m}^2 = 117.8 \frac{1}{\text{min}}$$

EXAMPLE 4-8.

A cylinder is required to extend at a minimum speed of 400 in/min in a system with a flow rate of 10 gpm. What cylinder size is required? If cylinders are available in ¼-inch increments, what size cylinder would you select? What will the extension speed be with the selected cylinder?

SOLUTION:

1. Calculate the piston area:

$$A_P = \frac{231 \cdot Q}{v_E} = \frac{231\frac{\text{in}^3}{\text{gal}} \cdot \left(10 \frac{\text{gal}}{\text{min}}\right)}{400 \frac{\text{in}}{\text{min}}} = 5.775 \text{ in}^2$$

2. Calculate the bore diameter:

$$D_P = \sqrt{\frac{4 \cdot A}{\pi}} = \sqrt{\frac{4 \cdot (5.775 \text{ in}^2)}{3.142}} = 2.711 \text{ in}$$

3. Select the bore diameter: Because the speed requirement was a *minimum*, and a smaller cylinder moves faster than a larger one for a given flow rate, we would choose the next *smaller* size (2.5 in).
4. Calculate the piston area:

$$A_P = \frac{\pi \cdot D_P^2}{4} = \frac{3.142 \cdot (2.5 \text{ in})^2}{4} = 4.909 \text{ in}^2$$

5. Calculate the actual extension speed:

$$v_E = \frac{231 \cdot Q}{A_P} = \frac{231 \frac{\text{in}^3}{\text{gal}} \cdot \left(10 \frac{\text{gal}}{\text{min}}\right)}{4.909 \text{ in}^2} = 470.6 \frac{\text{in}}{\text{min}}$$

EXAMPLE 4-8M.

A cylinder is required to extend at a minimum speed of 0.75 m/s in a system with a flow rate of 60 lpm. What cylinder size is required?

SOLUTION:

1. Convert to m/min:

$$v_E = 0.75 \frac{\text{m}}{\text{s}} \cdot \left(\frac{60 \text{ s}}{1 \text{ min}}\right) = 45 \frac{\text{m}}{\text{min}}$$

2. Calculate the piston area:

$$A_P = \frac{Q}{1000 \cdot v_E} = \frac{60 \frac{1}{\text{min}}}{1000 \frac{1}{\text{m}^3} \cdot \left(45 \frac{\text{m}}{\text{min}}\right)} = 0.001333 \text{ m}^2$$

3. Calculate the bore diameter:

$$D_P = \sqrt{\frac{4 \cdot A}{\pi}} = \sqrt{\frac{4 \cdot (0.001333 \text{ m}^{2)}}{3.142}} = 0.04119 \text{ m} \ (41.19 \text{ mm})$$

New cylinders typically have an insignificant amount of leakage, so efficiency is not usually an issue when calculating cylinder speeds. As the seals on the piston wear down, however, a phenomenon known as *blow by* may occur.

In this situation, a significant amount of the fluid entering the cylinder leaks past the piston, causing the cylinder to slow down considerably. This leakage is not visible from outside the system, so this inefficiency can sometimes go unnoticed for some time. The rod seals may, of course, also leak after being in service for some time, but this will be detected immediately because the fluid leaks outside the system.

To keep a system running efficiently, it is important to have a preventative maintenance schedule in which the seals and bearings are replaced periodically. The time between replacement depends on the number of cycles, operating pressure, system temperature, contamination level, and the type of seals used.

4.4 Cylinder Power

Cylinders convert fluid flow from a pump into the linear motion of the piston and rod, which can then be attached to a load to do work. If the movement of the piston is resisted by the load, pressure will build until sufficient force is generated to overcome the resistance of load. If the maximum system pressure is not sufficient to overcome the load, the cylinder will *stall* and generate force, but no motion. Whenever the piston is moving under load, power is being transmitted. Recall that the pump output is called *hydraulic power* and can be calculated with Equation 3-6 when using U.S. customary units:

$$HP_H = \frac{p \cdot Q}{1714}$$

The required units for this equation are: Q in gpm, p in psi, and HP_H in hp. When using metric units, Equation 3-6M can be used:

$$kW_H = \frac{p \cdot Q}{60,000}$$

The required units for this equation are: Q in lpm, p in kPa, and kW_H in kW.

Hydraulic power is the input power to the cylinder. The cylinder takes hydraulic power and converts it to mechanical power in the form of linear motion. As stated in Chapter 2, the power of a moving object can be calculated by multiplying the force times the velocity (Equation 2-23):

$$HP = \frac{F \cdot v}{550}$$

The 550 in the denominator is the conversion factor from $\frac{ft \cdot lbs}{s}$ to horse-power. Rewriting this equation, we use it to calculate the output power of a cylinder:

$$HP_O = \frac{F \cdot v}{550} \qquad \textbf{(4-7)}$$

We must use the following units for Equation 4-7. F in lbs, v in ft/s, and HP_O in hp.

A similar equation for metric units that gives the power of a moving object in kilowatts was developed in Chapter 2. Rewriting this equation, we use it to calculate the output power of a cylinder when using metric units:

$$kW_O = \frac{F \cdot v}{1000} \qquad \textbf{(4-7M)}$$

The required units for equation 4-7M are: F in N, v in m/s, and kW_O in kW.

EXAMPLE 4-10.

A cylinder is required to move a 12,000 lb load at a velocity of 2 ft/s. What is the output power?

SOLUTION:

$$HP_O = \frac{F \cdot v}{550} = \frac{12,000 \text{ lbs} \cdot \left(2\,\frac{ft}{s}\right)}{550} = 43.64 \text{ hp}$$

EXAMPLE 4-10M.

A cylinder is required to move a 70 kN load at a velocity of 0.5 m/s. What is the output power?

SOLUTION:

$$kW_O = \frac{F \cdot v}{1000} = \frac{70,000 \text{ N} \cdot \left(0.5\,\frac{m}{s}\right)}{1000} = 35 \text{ kW}$$

EXAMPLE 4-11.

A cylinder is required to move a 2400 lb load 14 in in 2 seconds. What is the output power?

SOLUTION:

1. Calculate the velocity:

$$v = \frac{d}{t} = \frac{14 \text{ in}}{2 \text{ s}} = 7 \frac{\text{in}}{\text{s}}$$

2. Convert to ft/s:

$$7 \frac{\text{in}}{\text{s}} \cdot \left(\frac{1 \text{ ft}}{12 \text{ in}}\right) = 0.5833 \frac{\text{ft}}{\text{s}}$$

3. Calculate the power:

$$HP_O = \frac{F \cdot v}{550} = \frac{2400 \text{ lbs} \cdot \left(0.5833 \frac{\text{ft}}{\text{s}}\right)}{550} = 2.545 \text{ hp}$$

EXAMPLE 4-11M.

A cylinder is required to move a 10 kN load 150 mm in 0.5 seconds. What is the output power?

SOLUTION:

1. Calculate the velocity:

$$v = \frac{d}{t} = \frac{0.150 \text{ m}}{0.5 \text{ s}} = 0.30 \frac{\text{m}}{\text{s}}$$

2. Calculate the power:

$$kW_O = \frac{F \cdot v}{1000} = \frac{10,000 \text{ N} \cdot \left(0.30 \frac{\text{m}}{\text{s}}\right)}{1000} = 3 \text{ kW}$$

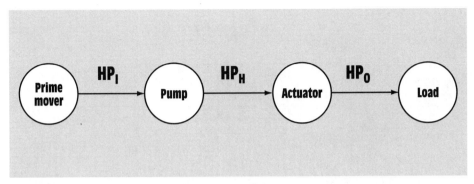

FIGURE 4-5   Power in a hydraulic system.

Figure 4-5 reviews the different forms of power in a hydraulic system. The prime mover (electric motor or engine) provides the mechanical input horsepower (HP_I) to the pump. The pump converts the mechanical input to hydraulic horsepower (HP_H). The actuator then converts the hydraulic horsepower back into mechanical output power (HP_O) and transfers it to the load.

4.5   Differential Flow

The difference in areas on the blind and rod end of the cylinder causes a phenomenon known as *differential flow,* which simply means that the flow out of the cylinder is not the same as the flow into the cylinder. When a cylinder is extending, the flow into the blind end is the pump flow (Q_{PUMP}), which pushes the piston out with a certain velocity (V_E). As the piston moves, it forces the fluid out of the rod end and back to the tank. Because the volume on the rod side is less, the flow rate out of the rod end will be less. This situation is illustrated in Figure 4-6A. We can find the return flow in this situation if we realize that the fluid velocity is the same on both sides and must be equal to the velocity of the piston (v_E). The velocity must also be equal to Q/A for each side. The following equation describes this relationship mathematically:

$$v_E = \frac{Q_{PUMP}}{A_P} = \frac{Q_{RET,E}}{A_P - A_R}$$

where: A_P = area of the piston
  A_R = area of the rod
  $Q_{RET,E}$ = return flow to tank when the cylinder is extending

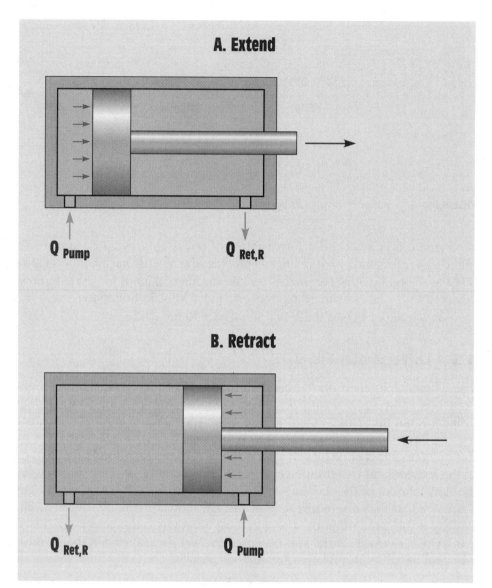

A. Extend

Q Pump Q Ret,R

B. Retract

Q Ret,R Q Pump

FIGURE 4-6   Differential flow. (A) Extend. (B) Retract.

Solving for $Q_{RET,E}$, we obtain:

$$Q_{RET,E} = \frac{Q_{PUMP} \cdot (A_P - A_R)}{A_P} \qquad \textbf{(4-8)}$$

When a cylinder is retracting, the flow into the rod end is the pump flow (Q_{PUMP}), which pushes the piston in with a certain velocity (V_R). As the piston moves, it forces the fluid out of the blind end and back to the tank. The flow rate out of the blind end will be larger because the volume on the cap side is larger. This situation is illustrated in Figure 4-6B. Again, the velocity is the same on either side (v_R), so the flow rates are then related by the following equation:

$$v_R = \frac{Q_{PUMP}}{A_P - A_R} = \frac{Q_{RET,R}}{A_P}$$

where: $Q_{RET,R}$ = return flow when the cylinder is retracting

Solving for $Q_{RET,R}$, we obtain:

$$Q_{RET,R} = \frac{Q_{PUMP} \cdot A_P}{A_P - A_R} \qquad \textbf{(4-9)}$$

EXAMPLE 4-12.

A cylinder with a bore diameter of 4 in and a rod diameter of 1.75 in is to be used in a system with a 20 gpm pump. What are the return flow rates when the cylinder is extending and retracting?

SOLUTION:

1. Calculate the piston area:

$$A_P = \frac{\pi \cdot D_P^2}{4} = \frac{3.142 \cdot (4 \text{ in})^2}{4} = 12.57 \text{ in}^2$$

2. Calculate the rod area:

$$A_r = \frac{\pi \cdot D_R^2}{4} = \frac{3.142 \cdot (1.75 \text{ in})^2}{4} = 2.406 \text{ in}^2$$

3. $Q_{RET,E}$:

$$Q_{RET,E} = \frac{Q_{PUMP} \cdot (A_P - A_R)}{A_P} = \frac{20 \text{ gpm} \cdot (12.57 \text{ in}^2 - 2.406 \text{ in})^2}{12.57 \text{ in}^2}$$
$$= 16.17 \text{ gpm}$$

4. $Q_{RET,R}$:

$$Q_{RET,R} = \frac{Q_{PUMP} \cdot A_P}{A_P - A_R} = \frac{20 \text{ gpm} \cdot (12.57 \text{ in}^2)}{12.57 \text{ in}^2 - 2.406 \text{ in}^2} = 24.73 \text{ gpm}$$

EXAMPLE 4-12M.

A cylinder with a bore diameter of 80 mm and a rod diameter of 25 mm is to be used in a system with a 80 lpm pump. What are the return flow rates when the cylinder is extending and retracting?

SOLUTION:

1. Calculate the piston area:

$$A_P = \frac{\pi \cdot D_P^{\ 2}}{4} = \frac{3.142 \cdot (0.080 \text{ m})^2}{4} = 0.005027 \text{ m}^2$$

2. Calculate the rod area:

$$A_R = \frac{\pi \cdot D_R^{\ 2}}{4} = \frac{3.142 \cdot (0.025 \text{ m})^2}{4} = 0.0004909 \text{ in}^2$$

3. $Q_{RET,E}$:

$$Q_{RET,E} = \frac{Q_{PUMP} \cdot (A_P - A_R)}{A_P} = \frac{80 \text{ lpm} \cdot (0.005027 \text{ m}^2 - 0.0004909 \text{ m}^2)}{0.005027 \text{ m}^2}$$
$$= 72.19 \text{ lpm}$$

4. $Q_{RET,R}$:

$$Q_{RET,R} = \frac{Q_{PUMP} \cdot A_P}{A_P - A_R} = \frac{80 \text{ gpm} \cdot (0.005027 \text{ m}^2)}{0.005027 \text{ m}^2 - 0.0004909 \text{ m}^2} = 88.66 \text{ lpm}$$

Why is differential flow important? Valves, filters, and other components are sized by the flow rate through the system. When sizing these components, we must be aware that the maximum flow in a hydraulic system using cylinders *is not the pump flow*. It is the return flow when the cylinder is retracting.

4.6 Cylinder Types

The standard double-acting cylinder (Figure 4-1) is by far the most commonly used type, but many other designs are also used in industry, including the *single-acting* cylinder. A single-acting cylinder acts under pressure in one direction, and returns automatically when the pressure is released. Single-acting cylinders may be either *push type* or *pull type*. The push type acts under pressure on the extend stroke and is retracted automatically when the pressure is released. A pull type acts under pressure on the retract stroke and extends automatically whenever pressure is released. Single-acting cylinders may be automatically returned by gravity or by an internal spring.

Figure 4-7 shows gravity return-type single-acting cylinders. In a push type (part A), the cylinder extends to lift a weight against the force of gravity by applying pump flow and pressure to the piston through the blind end port (labeled *pressure port*). The rod end port (labeled *vent port*) is open to the atmosphere so that air can flow freely in and out of the rod end of the cylinder. To retract the cylinder, the pressure is simply removed from the piston by connecting the pressure port to the tank. This allows the weight of the load to push the fluid out and back to tank. In a pull-type gravity return cylinder (part B), the cylinder lifts the weight from above by retracting whenever the pressure port is connected to the pump flow. Notice that the rod end port is now the pressure port and the blind end port is the vent port. This cylinder will automatically extend whenever the pressure port is connected to the tank.

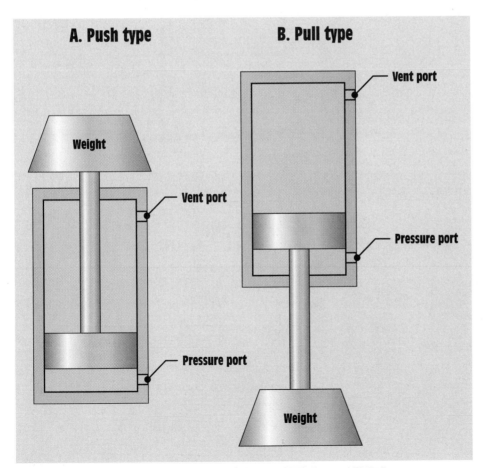

FIGURE 4-7   Gravity return single-acting cylinders. (A) Push type. (B) Pull type.

Spring return-type single-acting cylinders are shown in Figure 4-8. Part A shows a push-type spring return cylinder. To extend the cylinder, pump flow and pressure are sent to the pressure port. When pressure is released, the spring automatically returns the cylinder to the fully retracted position. The vent port performs the same function as it does with the gravity-return type. Part B shows a pull-type spring return cylinder. In this design, the cylinder retracts when the pressure port is connected to the pump flow, and automatically extends whenever the pressure port is connected to the tank.

A *telescopic* cylinder (Figure 4-9) is used when a long stroke length and a short retracted length are required. They extend in stages, each stage consisting of a

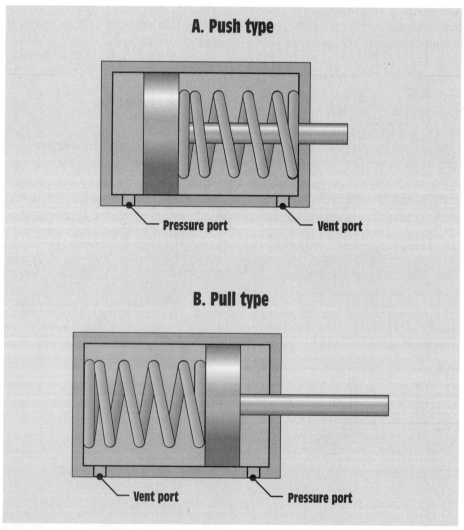

FIGURE 4-8   Spring return single-acting cylinders. (A) Push type. (B) Pull type.

sleeve that fits inside of the previous stage. The cylinder shown in Figure 4-9 is a three stage, but standard models are available with up to five stages. One application for this type of cylinder is raising a dump truck bed. In this application, the cylinder must have a long stroke to tilt the bed to a 60°–70° angle, and collapse to a very short length to fit under the bed when it is horizontal. Telescopic cylinders are available in both single-acting and double-acting models. The double-acting types are not able to create much force on the retract stroke, however, because of the small area available on the rod end. Telescopic cylinders are considerably more expensive than standard cylinders due to their more complex construction.

A *double-rod* cylinder (Figure 4-10) is a cylinder with a rod extending from both ends. This cylinder can be used in an application where work can be done by both ends of the cylinder, thereby making the cylinder more productive. Double-rod cylinders can also withstand higher side loads because they have an extra bearing (one on each rod) to withstand the loading.

A *tandem* cylinder (Figure 4-11) is used in applications where a large amount of force is required from a small-diameter cylinder. Pressure is applied to both

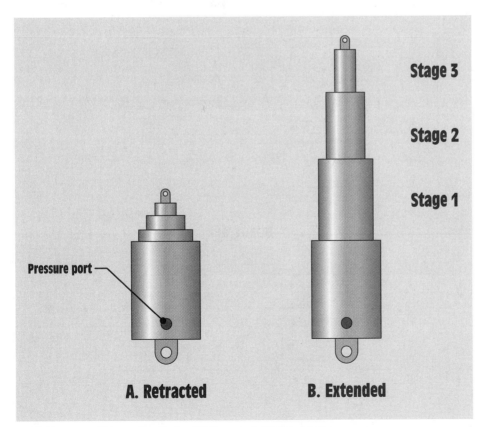

FIGURE 4-9   Telescopic cylinder. (A) Retracted. (B) Extended.

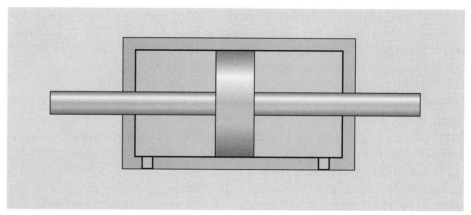

FIGURE 4-10 Double-rod cylinder.

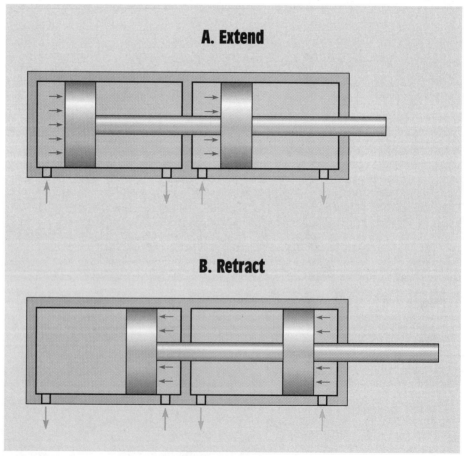

FIGURE 4-11 Tandem cylinder. (A) Extend. (B) Retract.

pistons, resulting in increased force because of the larger area. The drawback is that these cylinders must be longer than a standard cylinder of a comparable stroke length. They also require a larger flow rate than a standard cylinder to achieve an equal speed because flow must go to both pistons.

Hydraulic rams are very large single-acting cylinders capable of generating very high forces. Their force capacity is typically measured in tons (1 ton = 2,000 lbs). Figure 4-12 shows one commonly used type of hydraulic ram. The ram has a small shoulder instead of a piston. The only purpose of the shoulder is to prevent the rod from extending out of the cylinder. To extend the rod, the entire barrel is pressurized. The pressure is therefore applied to both the top and the bottom of the shoulder. The net result is that the effective area used to calculate the force is the area of the rod because the pressure on the shoulder is on both sides and cancels out. The advantage of this design is that it does not require a piston seal and bearing, making the ram much less expensive to

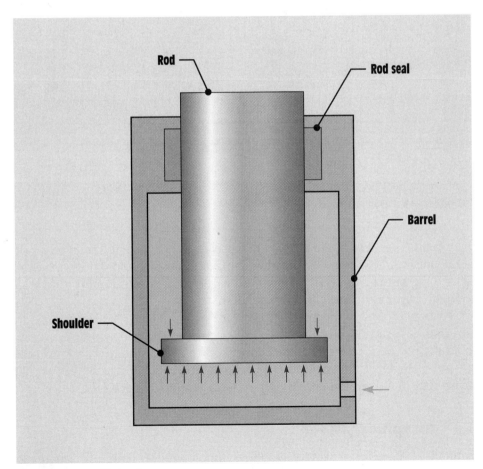

FIGURE 4-12   Hydraulic ram.

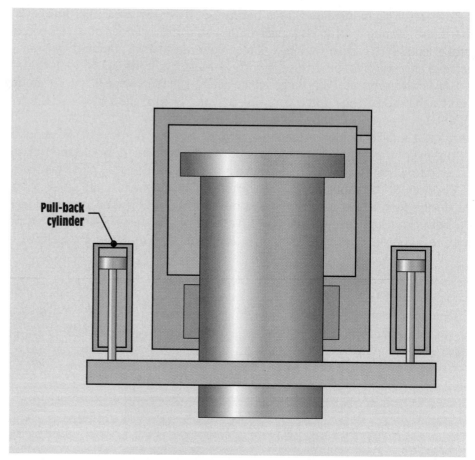

Pull-back
cylinder

FIGURE 4-13   Hydraulic ram press.

manufacture because the inside of the barrel does not have to be given a high-grade finish. It also reduces the downtime and cost required to periodically replace the piston seal. The rod does, however, require a seal and bearing and, consequently, must have a high-grade finish.

Hydraulic rams may also be downward acting (Figure 4-13). Because the ram cannot retract, it must have two pull-back cylinders whose only function is to lift the rod back to the retracted position. Notice that the pull-back cylinders are an example of an application in which the retract stroke is the working stroke. This is a common design used in large industrial presses.

4.7   Graphic Symbols

Figure 4-14 shows the cylinder graphic symbols used in hydraulic schematics. Part A shows the symbol for a double-acting cylinder. These cylinders act under

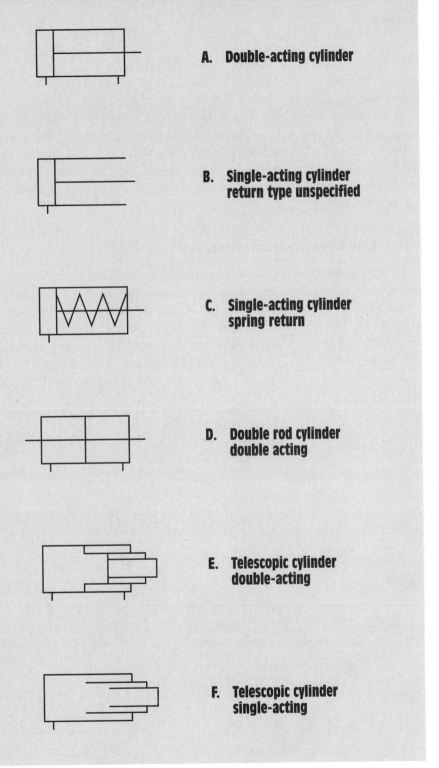

A. **Double-acting cylinder**

B. **Single-acting cylinder return type unspecified**

C. **Single-acting cylinder spring return**

D. **Double rod cylinder double acting**

E. **Telescopic cylinder double-acting**

F. **Telescopic cylinder single-acting**

FIGURE 4-14 Graphic symbols. (A) Double-acting. (B) Single-acting, return method unspecified. (C) Single-acting, spring return. (D) Double-rod, double-acting. (E) Telescopic, double-acting. (F) Telescopic, single-acting.

pressure in both the extend and retract directions. Part B shows the symbol for a single-acting cylinder. These cylinders act under pressure in one direction only, usually on the extend stroke. This symbol is the generic symbol for single-acting and does not specify the return method. It is often used for a gravity return cylinder. Part C shows the symbol for a spring return single-acting cylinder. Part D shows the symbol for a double-rod, double-acting cylinder. Parts E and F show the symbols for a double-acting and single-acting telescopic cylinder, respectively.

4.8   Cylinder Applications

A cylinder is required to work at an angle to the load in many applications. Figure 4-15 shows a cylinder mounted at an angle (a) that is required to lift a weight. The force of the load (F_{LOAD}) is due to gravity, so it will point downward (Figure 4-16A). To determine the force the cylinder must generate, we can divide the force of the weight into two components: one parallel to the axis of the cylinder (F_{CYL}) and one perpendicular to the axis of the cylinder

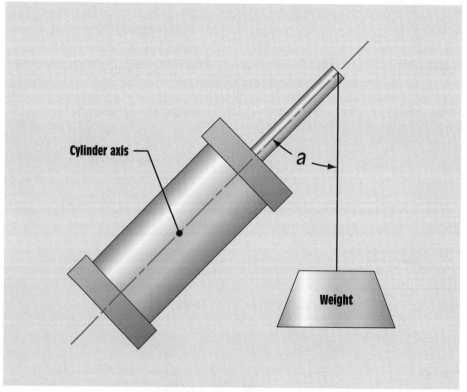

FIGURE 4-15   Cylinder working at an angle to a load.

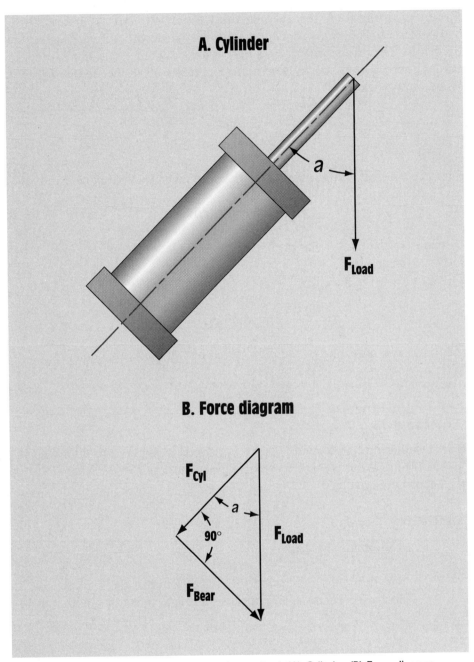

FIGURE 4-16   Cylinder working at an angle to a load. (A) Cylinder. (B) Force diagram.

(F_{BEAR}). F_{CYL} is the force the cylinder must generate, while F_{BEAR} exerts a side load on the bearings. This concept is best illustrated with a force diagram (Figure 4-16B). The three forces form a right triangle, with F_{LOAD} as the hypotenuse. We can use trigonometry to determine F_{CYL} if we recall the definition for the cosine of an angle:

$$\cos a = \frac{\text{adjacent}}{\text{hypotenuse}} = \frac{F_{CYL}}{F_{LOAD}}$$

The force the cylinder must create is therefore given by the following equation:

$$F_{CYL} = F_{LOAD} \cdot \cos a \qquad \textbf{(4-10)}$$

We can determine the force on the bearings if we recall the definition for the sine of an angle:

$$\sin a = \frac{\text{opposite}}{\text{hypotenuse}} = \frac{F_{REAR}}{F_{LOAD}}$$

The force on the bearings is therefore given by the following equation:

$$F_{BEAR} = F_{LOAD} \cdot \sin a \qquad \textbf{(4-11)}$$

EXAMPLE 4-13.

A cylinder mounted at 60° to the horizontal is required to lift a weight of 15,000 lbs. What is the force the cylinder must generate? What is the side load on the bearings?

SOLUTION:

1. Determine the angle:

    We must first determine the angle between the cylinder and the load (see Figure 4-17). Because this is a right triangle, the angle is:
    $a = 90° - 60° = 30°$

2. Calculate the cylinder force:

    $F_{CYL} = F_{LOAD} \cdot \cos a = 15,000 \text{ lbs} \cdot \cos 30 = 12,990 \text{ lbs}$

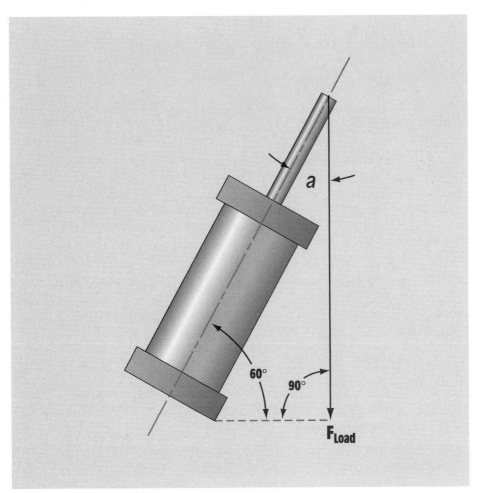

FIGURE 4-17   Example 4-13.

3. Calculate the side load:

$$F_{BEAR} = F_{LOAD} \cdot \sin a = 15,000 \text{ lbs} \cdot \sin 30 = 7500 \text{ lbs}$$

EXAMPLE 4-14.

The horizontally mounted cylinder shown in Figure 4-18 is required to extend against a load of 5000 lbs (F_{LOAD}) that is at an angle of 20° (a) to the cylinder axis. The cylinder has a bore diameter of 2 in. What is the required pressure?

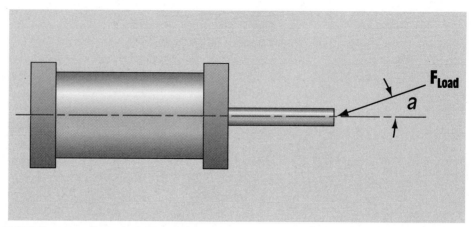

FIGURE 4-18 Example 4-14.

SOLUTION:

1. Calculate the force:

$$F_{CYL} = F_{LOAD} \cdot \cos a = 5000 \text{ lbs} \cdot \cos 20 = 4698 \text{ lbs}$$

2. Calculate the area:

$$A = \frac{\pi \cdot D^2}{4} = \frac{3.142 \cdot (2 \text{ in})^2}{4} = 3.142 \text{ in}^2$$

3. Calculate the pressure:

$$p = \frac{F}{A} = \frac{4698 \text{ lbs}}{3.142 \text{ in}^2} = 1495 \frac{\text{lbs}}{\text{in}^2}$$

EXAMPLE 4-14M.

The horizontally mounted cylinder shown in Figure 4-18 is required to extend against a load of 22,000 N (F_{LOAD}) that is at an angle of 30° (a) to the cylinder axis. The cylinder has a bore diameter of 50 mm. What is the required pressure?

SOLUTION:

1. Calculate the force:

$$F_{CYL} = F_{LOAD} \cdot \cos a = 22{,}000 \text{ N} \cdot \cos 30 = 19{,}053 \text{ N}$$

2. Calculate the area:

$$A_P = \frac{\pi \cdot D_P^{\,2}}{4} = \frac{3.142 \cdot (0.050 \text{ m})^2}{4} = 0.001964 \text{ m}^2$$

3. Calculate the pressure:

$$p = \frac{F}{A} = \frac{19{,}053 \text{ N}}{0.001964 \text{ m}^2} = 9{,}701{,}000 \frac{\text{N}}{\text{m}^2} \quad (9{,}701 \text{ kPa})$$

The friction created by the side load on the bearings, which requires an additional amount of force from the cylinder, is neglected in the preceding analysis. The magnitude of this force depends not only on the side load, but also on the bearing material and lubrication. With well-lubricated bearings, this force will normally be very small compared to the load.

Another common application for a cylinder is to rotate a lever arm (Figure 4-19). The cylinder must be allowed to swivel as the lever rotates. This necessitates the use of a *clevis mount,* which is essentially just a hinge and a pin. Figure 4-20 illustrates how the force of the lever arm can be calculated. We must first realize that only the component of the cylinder force perpendicular to the lever arm does any work. The component of the cylinder force along the lever arm simply applies a force on the pin at the base of the lever, which is not useful. The net useful force applied to the lever arm (F_{LEV}) is therefore perpendicular to the lever arm. The total output force of the cylinder (F_{CYL}) is aligned with the cylinder axis, as always. Figure 4-20B shows the force diagram. We can use the cosine function to define the following relationship:

$$\cos a = \frac{\text{adjacent}}{\text{hypotenuse}} = \frac{F_{LEV}}{F_{CYL}}$$

We can therefore use the following equation to relate the force of the lever to the force of the cylinder:

$$F_{LEV} = F_{CYL} \cdot \cos a \qquad \textbf{(4-12)}$$

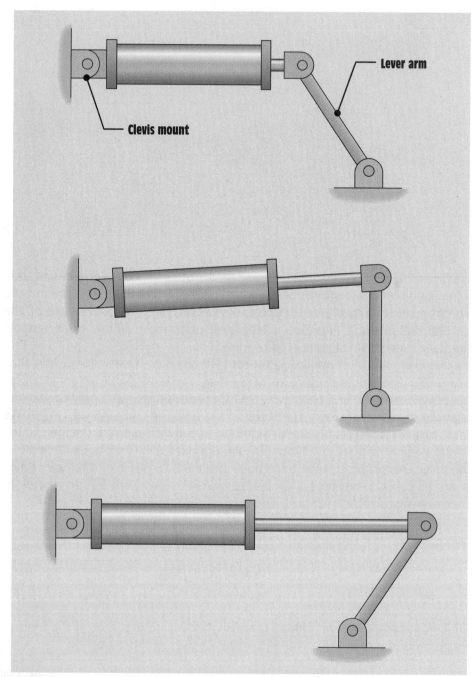

FIGURE 4-19   Cylinder rotating a lever arm.

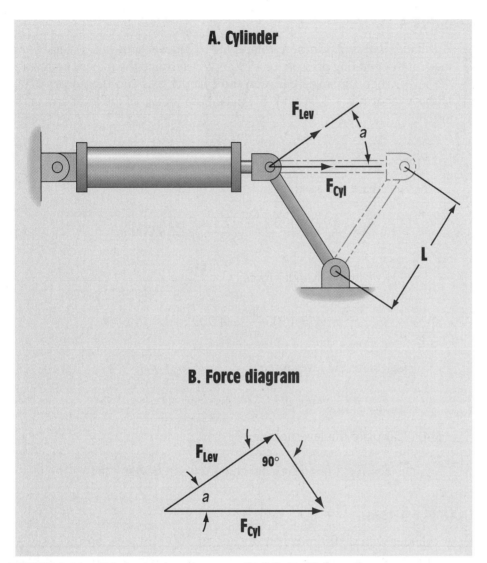

FIGURE 4–20 Cylinder rotating a lever arm. (A) Cylinder. (B) Force diagram.

Whenever a lever arm is rotated against a resisting force, torque is generated around the pivot point. If the length of the lever arm is L, the torque can be calculated with following equation:

$$T = F_{LEV} \cdot L \qquad \textbf{(4–13)}$$

EXAMPLE 4-15.

A cylinder with a 2.5 in bore rotates an 18 in lever arm in a system with a maximum operating pressure of 1500 psi. Determine the maximum force of the lever when the angle between the cylinder axis and the perpendicular of the lever arm is 40°. What is the maximum torque of the lever arm at this instant?

SOLUTION:

1. Calculate the area:

$$A = \frac{\pi \cdot D^2}{4} = \frac{3.142 \cdot (2.5 \text{ in})^2}{4} = 4.909 \text{ in}^2$$

2. Calculate the cylinder force:

$$F_{CYL} = p \cdot A = 1500 \frac{\text{lbs}}{\text{in}^2} \cdot (4.909 \text{ in}^2) = 7364 \text{ lbs}$$

3. Calculate the lever force:

$$F_{LEV} = F_{CYL} \cdot \cos a = 7364 \text{ lbs} \cdot \cos 40 = 5641 \text{ lbs}$$

4. Calculate the lever torque:

$$T = F_{LEV} \cdot L = 5{,}641 \text{ lbs} \cdot (18 \text{ in}) = 101{,}538 \text{ in} \cdot \text{lbs}$$

EXAMPLE 4-15M.

A cylinder with a 100 mm bore rotates an 300 mm lever arm in a system with a maximum operating pressure of 17,500 kPa. Determine the maximum force of the lever when the angle between the cylinder axis and the perpendicular of the lever arm is 55°. What is the maximum torque of the lever arm at this instant?

SOLUTION:

1. Calculate the area:

$$A = \frac{\pi \cdot D^2}{4} = \frac{3.142 \cdot (0.100 \text{ m})}{4} = 0.007855 \text{ m}^2$$

2. Calculate the cylinder force:

$$F_{CYL} = p \cdot A = 17,500,000 \ \frac{N}{m^2} \cdot (0.007855 \ in^2) = 137,500 \ N$$

3. Calculate the lever force:

$$F_{LEV} = F_{CYL} \cdot \cos a = 137,500 \ N \cdot \cos 55 = 78,870 \ N$$

4. Calculate the lever torque:

$$T = F_{LEV} \cdot L = 78,870 \ N \cdot (0.300 \ m) = 23,660 \ N \cdot m$$

EXAMPLE 4-16.

A cylinder with a 3 in bore is required to rotate a 12 in lever arm with a minimum torque of 10,000 ft-lbs. The maximum angle between the cylinder axis and the perpendicular of the lever arm is 30°. What is the minimum pressure required?

SOLUTION:

1. Calculate the lever force:

$$F_{LEV} = \frac{T}{L} = \frac{10,000 \ ft \cdot lbs}{12 \ in \cdot \left(\dfrac{1 \ ft}{12 \ in}\right)} = 10,000 \ lbs$$

2. Calculate the cylinder force:

$$F_{CYL} = \frac{F_{LEV}}{\cos a} = \frac{10,000 \ lbs}{\cos 30} = 11,547 \ lbs$$

3. Calculate the area:

$$A = \frac{\pi \cdot D^2}{4} = \frac{3.142 \cdot (3 \ in)^2}{4} = 7.069 \ in^2$$

4. Calculate the pressure:

$$p = \frac{F_{CYL}}{A} = \frac{11,547 \ lbs}{7.069 \ in^2} = 1633 \ \frac{lbs}{in^2}$$

When we use Equation 4-12, we are calculating the force only at a particular instant. As the lever arm rotates, the angle is constantly changing, therefore the relationship between the force output of the cylinder and the force of the lever arm is constantly changing. However, the previous problem illustrates an important point. The maximum required force output of the cylinder (and maximum pressure) will always be required when the angle between the cylinder axis and the perpendicular of the lever arm is at its maximum, so we may use this case when making design decisions.

4.9   Cylinder Specifications

As stated in Chapter 1, there are two main application areas for hydraulics: mobile and industrial. Mobile applications include excavating equipment, farm equipment, and other hydraulic systems used on vehicles. Industrial applications include machining equipment, robots, presses, and other machines used in manufacturing facilities. Cylinders used in mobile hydraulics are of lighter construction than those used in industrial hydraulic systems because weight is a significant factor and they usually see fewer cycles over a given time period. Industrial hydraulic cylinders, also called *mill-type* cylinders, have a more heavy-duty construction. They are typically required to operate on a continuous basis over long periods with as little downtime as possible. They are normally on stationary machinery, so weight is not usually a major concern.

The reader is again encouraged to obtain several manufacturers' catalogs to compare how specifications are provided in industry. Many manufacturers provide on-line catalogs that can be accessed or downloaded via the Internet. Most will provide a hardcopy of their catalog upon request.

4.9.1 SIZE

Cylinders are sized primarily by their bore diameter—this is the dimension that determines their force output capability. The stroke length, the difference between the fully retracted and fully extended length, is also important (Figure 4-21). Stroke lengths are generally available in increments of 1 in over some range. One standard rod diameter for each bore size or several rod sizes for each bore may be offered as standard models. The overall dimensions and the center-to-center dimensions of the mounting holes are also included in the manufacturer's catalog. This is important information to have when designing the physical layout of a circuit to be used in a piece of machinery. The port sizes, which are usually of the threaded type, are also given. The typical format for cylinder size information is shown in Figure 4-22.

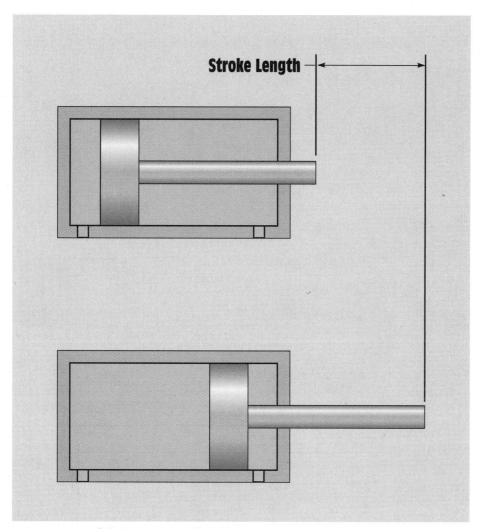

FIGURE 4-21   Cylinder stroke length.

4.9.2 MOUNTING STYLES

Several common types of mountings are used for hydraulic cylinders (Figure 4-23). In some applications, the cylinder must be allowed to pivot, as with lever applications. In others, the cylinder remains fixed in position. The clevis and trunnion styles allow the cylinder to pivot, while the others hold it fixed.

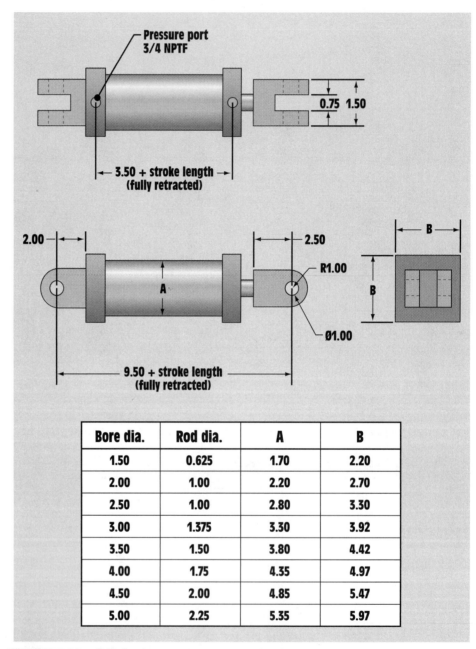

Bore dia.	Rod dia.	A	B
1.50	0.625	1.70	2.20
2.00	1.00	2.20	2.70
2.50	1.00	2.80	3.30
3.00	1.375	3.30	3.92
3.50	1.50	3.80	4.42
4.00	1.75	4.35	4.97
4.50	2.00	4.85	5.47
5.00	2.25	5.35	5.97

FIGURE 4-22   Cylinder size.

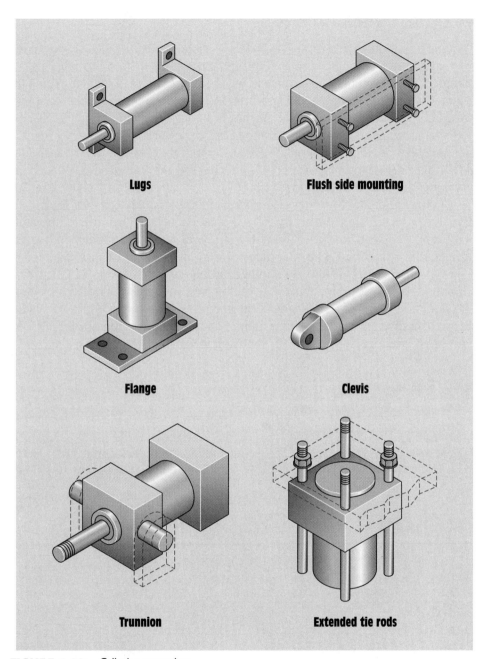

Lugs

Flush side mounting

Flange

Clevis

Trunnion

Extended tie rods

FIGURE 4-23   Cylinder mounting.

(From Vockroth, Industrial Hydraulics, Albany, NY: Delmar, 1994, p. 167.)

4.9.3 OTHER SPECIFICATIONS

The pressure rating for hydraulic cylinders is usually provided as a single value—the maximum operating pressure. Pressure ratings from 1000 psi to 5000 psi are common. The manufacturers typically use a *safety factor* of 4, which means the cylinder will not fail until reaching a pressure of four times the rated operating pressure. The rated pressure should not be exceeded, however, because all hydraulic systems experience *pressure spikes*. A pressure spike occurs when the pressure very briefly jumps well above the operating pressure due to a sudden valve shift. Exceeding the recommended pressure may result in seal failure or structural damage to the cylinder.

Manufacturers may also give an operating temperature range for their cylinders. This is largely determined by the seal material used. When the temperature is too low, the seals lose their resiliency and will not seal properly. When the temperature is too high, the seal material may degrade and cause failure. The fluid type(s) specified is also determined by the seals. As stated in Chapter 3, certain fluids are not compatible with certain seal materials. Using an incompatible fluid will result in seal degradation and, consequently, component failure due to excessive leakage. Specific materials and their compatibility with different commonly used hydraulic fluids is discussed in Chapter 9.

4.10 Equations

EQUATION NUMBER	EQUATION	REQUIRED UNITS
4-1	$F_E = p \cdot A_P$	Any consistent units
4-2	$F_R = p \cdot (A_P - A_R)$	Any consistent units
4-3	$v = \dfrac{Q}{A}$	Any consistent units
4-4	$v = \dfrac{231 \cdot Q}{A}$	Q in gpm, A in in², v in in/min
4-5	$v_E = \dfrac{231 \cdot Q}{A_P}$	Q in gpm, A_P in in², v_E in in/min
4-5M	$v_E = \dfrac{Q}{1000 \cdot A_P}$	Q in lpm, A_P in m², v_E in in/min
4-6	$v_R = \dfrac{231 \cdot Q}{A_P - A_R}$	Q in gpm, A_X in in², v_E in in/min
4-6M	$v_R = \dfrac{Q}{1000 \cdot (A_P - A_R)}$	Q in lpm, A_X in m², v_E in m/min

EQUATION NUMBER	EQUATION	REQUIRED UNITS
4-7	$HP_O = \dfrac{F \cdot v}{550}$	F in lbs, v in ft/s, HPO in hp
4-7M	$kW_O = \dfrac{F \cdot v}{1000}$	F in N, v in m/s, kW_O in kW
4-8	$Q_{RET,E} = \dfrac{Q_{PUMP} \cdot (A_P - A_R)}{A_P}$	Any consistent units
4-9	$Q_{RET,R} = \dfrac{Q_{PUMP} \cdot A_P}{A_P - A_R}$	Any consistent units
4-10	$F_{CYL} = F_{LOAD} \cdot \cos a$	Any consistent units
4-11	$F_{BEAR} = F_{LOAD} \cdot \sin a$	Any consistent units
4-12	$F_{LEV} = F_{CYL} \cdot \cos a$	Any consistent units
4-13	$T = F_{LEV} \cdot L$	Any consistent units

4.11   Review Questions and Problems

1. What is the purpose of an actuator in a hydraulic circuit?
2. Describe the difference between a single-acting cylinder and double-acting cylinder.
3. Assuming a constant flow rate, will a double-acting cylinder travel faster on the extend or the retract stroke?
4. What is the advantage of a telescopic cylinder over a standard cylinder? Are there any disadvantages?
5. Describe the construction of a double-rod cylinder. For what type of application is it best suited?
6. Describe the construction of a tandem cylinder. What is its advantage over a standard cylinder? Are there any disadvantages?
7. What is a hydraulic ram? What are the advantages of its construction? What is the function of the pull-back cylinders in a ram type press?
8. Which cylinder mounting styles allow a cylinder to pivot?
9. What can occur if a fluid type other than that specified by the manufacturer is used with a hydraulic cylinder?
10. Draw the graphic symbols for the following cylinder types: single-acting, single-acting spring return, double-acting, and double rod.
11. A cylinder with a bore diameter of 4 in and a rod diameter of 1.75 in is to be used in a system with a maximum pressure of 2000 psi. What are the maximum extension and retraction forces?

12. A cylinder with a bore diameter of 100 mm and a rod diameter of 40 mm is to be used in a system with a maximum pressure of 20,000 kPa. What are the maximum extension and retraction forces?

13. A cylinder with a bore diameter of 3 in and a rod diameter of 1.38 in is required to extend against a load of 15,000 lbs and to retract against a load of 750 lbs. What pressure is required for each stroke?

14. A cylinder with a bore diameter of 50 mm and a rod diameter of 20 mm is required to extend against a load of 60,000 N and to retract against a load of 5000 N. What pressure is required for each stroke?

15. A cylinder is required to extend against a load of 18,000 lbs with a maximum pressure of 2000 psi. What size cylinder is required?

16. A cylinder is required to extend against a load of 150 kN with a maximum pressure of 25,000 kPa. What size cylinder (in mm) is required?

17. A cylinder with a bore diameter of 3.5 in is to be used in a system with a 25 gpm pump. What is the extension speed?

18. A cylinder with a bore diameter of 1.5 in and a rod diameter of 0.625 in is to be used in a system with a 10 gpm pump. What are the extension and retraction speeds?

19. A cylinder with a bore diameter of 60 mm and a rod diameter of 20 mm is to be used in a system with a 70 lpm pump. What are the extension and retraction speeds?

20. A cylinder with a bore diameter of 4 in is required to extend at a velocity of 1.2 ft/s. What flow rate (in gpm) is required to achieve this speed?

21. A cylinder with a bore diameter of 63 mm is required to extend at a velocity of 0.75 m/s. What flow rate (in lpm) is required to achieve this speed?

22. A cylinder is required to extend at a minimum speed of 1000 in/min in a system with a flow rate of 20 gpm. What size cylinder is required?

23. If cylinders are available in ½-inch increments, what size cylinder would you select for the system in problem 22? What will the extension speed be with this cylinder?

24. A cylinder is required to extend at a minimum speed of 1.5 m/s in a system with a flow rate of 100 lpm. What cylinder size is required?

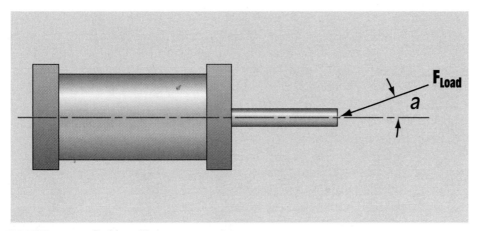

FIGURE 4-24   Problem 30.

25. A cylinder is required to move a 20,000 lb load at a velocity of 1.75 ft/s. What is the output horsepower?

26. A cylinder is required to move a 150 kN load at a velocity of 0.25 m/s. What is the output power?

27. A cylinder is required to move a 4200 lb load 20 inches in 2 seconds. What is the output horsepower?

28. A cylinder with a bore diameter of 3.5 in and a rod diameter of 1.25 in is to be used in a system with a 10 gpm pump. What are the return flow rates when the cylinder is extending and retracting?

29. A cylinder with a bore diameter of 50 mm and a rod diameter of 20 mm is to be used in a system with a 60 lpm pump. What are the return flow rates when the cylinder is extending and retracting?

30. The cylinder shown in Figure 4-24 is subjected to a 2000 lb (F_{LOAD}) load that is at a 15° (a) angle to its axis. What is the force the cylinder must generate? What is the side load on the bearings?

31. The cylinder shown in Figure 4-24 is subjected to a 17,000 N (F_{LOAD}) load that is at a 20° angle (a) to its axis. What is the force the cylinder must generate? What is the side load on the bearings?

32. The cylinder shown in Figure 4-25 has a 1.5 in bore and is mounted at 65° (a) to the horizontal, as shown. It is subjected to a 5000 lb (F_{LOAD}) vertical load. What is the required pressure?

33. The system in the previous problem has a maximum operating pressure of 3000 psi. What is the maximum load this cylinder can lift in the current configuration?

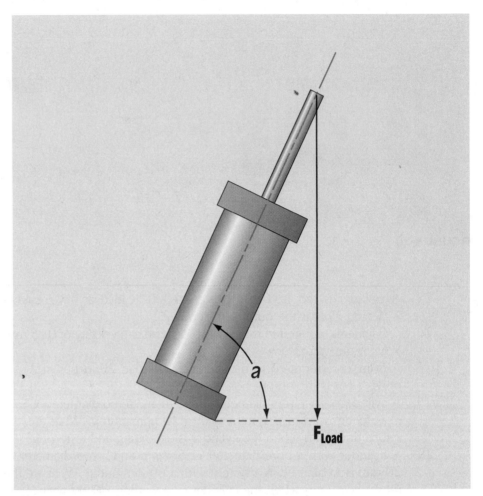

FIGURE 4-25   Problem 32.

34. The cylinder shown in Figure 4-25 has a 63 mm bore and is mounted at 70° (*a*) to the horizontal. It is subjected to a 25 kN (F_{LOAD}) vertical load. What is the required pressure?

35. The cylinder shown in Figure 4-26 has a 2 in bore and rotates a 24 in lever arm in a system with a maximum operating pressure of 1500 psi. Determine the maximum force of the lever in the position shown. What is the maximum torque of the lever arm at this instant?

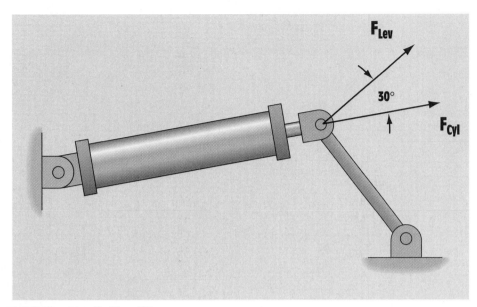

FIGURE 4-26   Problem 35.

36. A cylinder with a 80 mm bore rotates an 450 mm lever arm in a system with a maximum operating pressure of 30,000 kPa. Determine the maximum force of the lever when the angle between the cylinder axis and the perpendicular of the lever arm is 60°. What is the maximum torque of the lever arm at this instant?

37. A cylinder with a 3.5 in bore is required to rotate a 36 in lever arm with a minimum torque of 50,000 ft-lbs. The maximum angle between the cylinder axis and the perpendicular of the lever arm is 40°. What is the minimum pressure required?

CHAPTER 5

Hydraulic Motors

OUTLINE

5.1 Introduction

5.2 Motor Types

5.3 Motor Torque

5.4 Motor Speed

5.5 Motor Power

5.6 Motor Efficiency

5.7 Graphic Symbols

5.8 Motor Applications

5.9 Motor Specifications

5.10 Equations

5.11 Review Questions and Problems

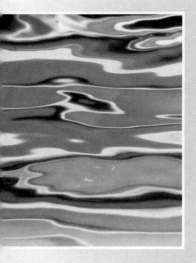

5.1   Introduction

A hydraulic motor converts fluid power into mechanical power in the form of rotational motion. Motors perform the opposite function of a pump, which converts mechanical power from an electric motor or engine into fluid power. Motors take pump flow and pressure as their input and output rotational motion and torque.

Motor *displacement* is the volume of fluid required to rotate a motor one revolution. As stated in Chapter 3, pump displacement is the volume of fluid that a pump outputs per revolution of the pump shaft, a similar concept. Like pumps, motors can be fixed or variable displacement. Increasing the displacement of a motor *decreases* its speed because it requires more fluid to turn it each revolution. Increasing displacement *increases* torque output because more area within the motor is subjected to pressure. Decreasing motor displacement increases speed and decreases torque.

5.2   Motor Types

Hydraulic motors are most commonly gear, vane, or piston type. All have a construction similar to the hydraulic pump of the same type. They also have similar properties. Gear motors are the least efficient, most dirt tolerant, and have the lowest pressure ratings of the three. Piston motors are the most efficient, least dirt tolerant, and have the highest pressure ratings. Vane and piston motors can be fixed or variable displacement, like vane and piston pumps. Gear motors, like gear pumps, are not available with variable displacement.

Figure 5-1 illustrates the operation of a gear motor. One of the gears is keyed to an output shaft, while the other is simply an idler gear. Pump flow and pressure are sent to the inlet port of the motor. The pressure is then applied to the gear teeth, causing the gears and the output shaft to rotate. The pressure builds until enough torque is generated to rotate the output shaft against the load. Most gear motors are bi-directional—the direction of rotation can be reversed by simply reversing the direction of flow.

Vane and piston types operate in a similar fashion: pump flow and pressure are applied to the pistons or vanes and the output shaft is rotated. Figure 5-2 illustrates the operation of a balanced vane motor. Recall from our discussion on vane pumps that "balanced" means that pressure is applied on both sides of the shaft; resulting in no net force on the bearings. This increases the maximum operating pressure and drive speed at which the motor can operate. The vanes extend and retract twice per revolution of the rotor, which necessitates the use of two inlet and two outlet chambers. These chambers are combined into one common inlet and one common outlet within the

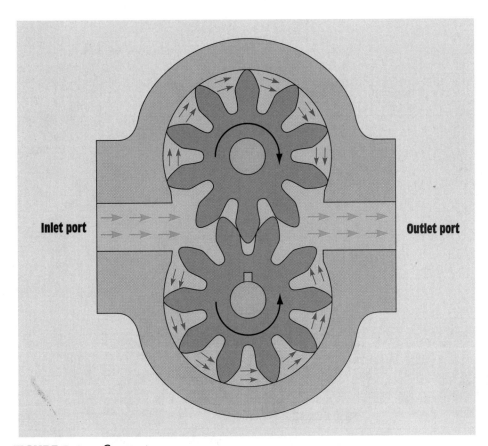

Inlet port   Outlet port

FIGURE 5-1   Gear motor.

motor housing. Figure 5-3 illustrates and explains the operation of a typical piston motor. Most vane and piston motors are bi-directional.

Some applications require a higher torque output than can be obtained with a standard hydraulic motor. In these applications, a *low-speed, high-torque* (LSHT) motor is often used. These motors have larger displacements and consequently create more torque and run at slower speeds than a standard (high-speed) hydraulic motor. Their design is usually some variant on the gear, vane, or piston type of construction. The principle of operation for LSHT motors is the same as that for high-speed motors: pressure and flow are applied to create rotational motion and torque around an output shaft.

All of these motor types provide continuous rotary motion. They can continue to rotate in the same direction indefinitely as long as pump flow is supplied to their inlet. Some applications require an actuator that allows only a partial revolution. A *rotary actuator* performs this function. In industry, the term

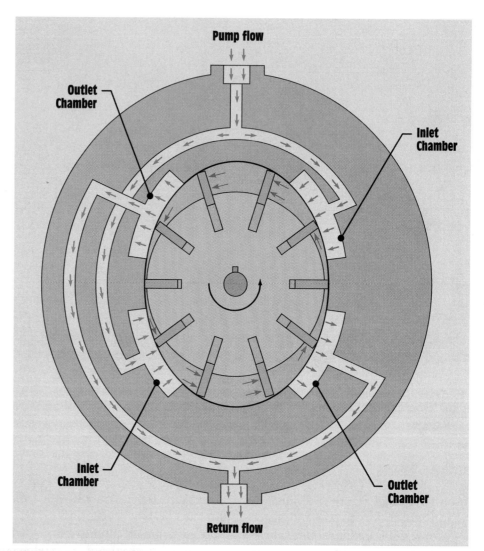

FIGURE 5-2   Balanced vane motor.

"rotary actuator" is usually reserved for a partial revolution actuator, while the term "motor" is usually used for a continuously rotating actuator. Figure 5-4 shows a simplified cutaway of a vane-type rotary actuator. In this design, a single vane is subjected to pump flow and pressure, which rotate the output shaft. A *stop* prevents the vane from rotating continuously. This model is capable of rotating 315° before hitting the stop, but models with other rotation angles are available. Models are also available with multiple vanes. A two-vane model is shown in Figure 5-5. The advantage of this design is that torque output is

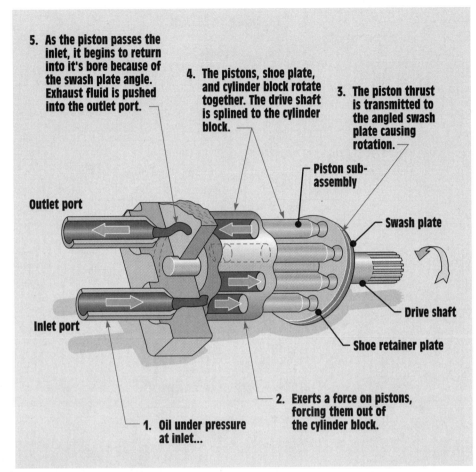

5. **As the piston passes the inlet, it begins to return into it's bore because of the swash plate angle. Exhaust fluid is pushed into the outlet port.**

4. **The pistons, shoe plate, and cylinder block rotate together. The drive shaft is splined to the cylinder block.**

3. **The piston thrust is transmitted to the angled swash plate causing rotation.**

Piston sub-assembly

Outlet port

Swash plate

Inlet port

Drive shaft

Shoe retainer plate

2. **Exerts a force on pistons, forcing them out of the cylinder block.**

1. **Oil under pressure at inlet...**

FIGURE 5-3   Piston motor.
(Courtesy of Vickers, Inc. From Norvelle, *Fluid Power Technology:* West, 1993, p. 97.)

increased because the area subjected to pressure is larger. However, two-vane models cannot rotate as many degrees as can a single-vane model. Passageways are used to connect the different chambers of the rotary actuator.

A *rack and pinion* rotary actuator is another commonly used design for obtaining partial revolution actuation. A simplified photo of this design is shown in Figure 5-6. This type is essentially a hydraulic cylinder with a rack and pinion gear mechanism. The rack gear on the piston rod turns the pinion gear, thereby converting the linear motion of the piston into rotary motion, which is transmitted to the load through the output shaft. Figure 5-7 shows a photo of an actual rack and pinion rotary actuator.

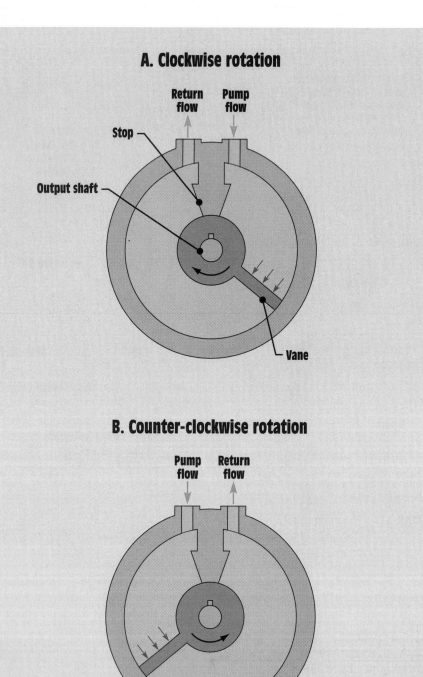

FIGURE 5-4   Vane-type rotary actuator. (A) Clockwise rotation. (B) Counter-clockwise rotation.

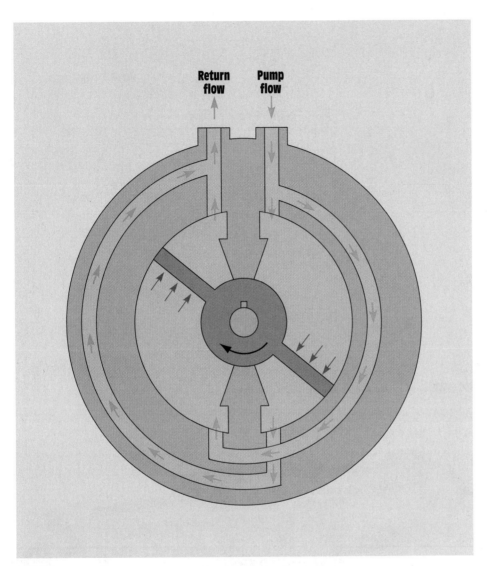

FIGURE 5-5   Two-vane rotary actuator.

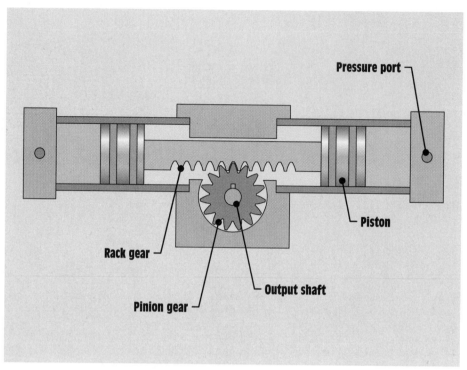

FIGURE 5-6 Rack and pinion rotary actuator.

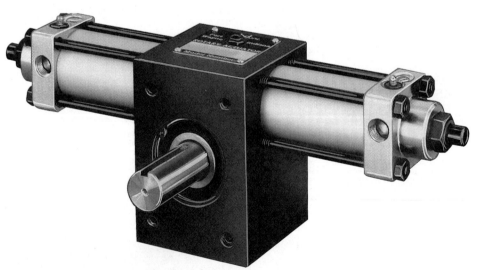

FIGURE 5-7 Rack and pinion rotary actuator.

(Photograph provided courtesy of PHD, Inc. For additional information contact us at www.phdinc.com.)

5.3   Motor Torque

Just as cylinders convert fluid pressure to force by applying the pressure to the piston area, motors convert fluid pressure to torque by applying the pressure to the vanes, gears, or pistons. The torque output of a motor will therefore depend upon the size (displacement) of the motor and the pressure of the system, as given by the following equation:

$$T_T = \frac{V_M \cdot \Delta p}{2 \cdot \pi} \qquad \textbf{(5-1)}$$

where: T_T = theoretical torque output (in · lbs, N · m)

V_M = motor displacement (volume) $\left(\dfrac{in^3}{rev}, \dfrac{m^3}{rev}\right)$

Δp = pressure differential between inlet and outlet $\left(\dfrac{lb}{in^2}, \dfrac{N}{m^2}\right)$

The Δp term, the difference in pressure between the inlet and outlet sides of the motor, is the net pressure available to create torque. The outlet pressure is usually near zero, because it is flowing back to the tank with little resistance. In that case, the Δp term is simply the system (inlet) pressure. The factor 2π is a conversion factor between revs and radians, which allows us to use the stated units. The actual torque is slightly less than the theoretical torque due to frictional losses within the motor. The actual torque is usually obtained from a graph, as we will see in the section on specifications.

EXAMPLE 5-1.

A motor has a displacement of 1.2 in³/rev and is used in a system with a maximum pressure of 2500 psi. What is its theoretical output torque?

SOLUTION:

$$T_T = \frac{V_M \cdot \Delta p}{2 \cdot \pi} = \frac{1.2 \dfrac{in^3}{rev} \cdot (2500 \text{ psi})}{2 \cdot (3.142)} = 477.5 \text{ in} \cdot \text{lbs}$$

EXAMPLE 5-1M.

A motor has a displacement of 40 cm³/rev and is used in a system with a maximum pressure of 20,000 kPa. What is its theoretical output torque?

SOLUTION:

1. Convert to m³/rev:

$$40 \frac{cm^3}{rev} \cdot \left(\frac{1 \, m^3}{1,000,000 \, cm^3} \right) = 0.0004 \frac{m^3}{rev}$$

2. Calculate the drive torque:

$$T_T = \frac{p \cdot V_M}{2 \cdot \pi} = \frac{20,000,000 \frac{N}{m^2} \cdot \left(0.00004 \frac{m^3}{rev} \right)}{2 \cdot \pi} = 127.3 \, N \cdot m$$

EXAMPLE 5-2.

A motor with a displacement of 2 in³/rev must produce a torque of 600 in · lbs. What pressure is required? Assume 100% efficiency.

SOLUTION:

$$\Delta p = \frac{2 \cdot \pi \cdot T_T}{V_M} = \frac{2 \cdot (3.142) \cdot (600 \, in \cdot lbs)}{2 \frac{in^3}{rev}} = 1885 \, psi$$

EXAMPLE 5-2M.

A motor with a displacement of 32 cm³/rev must produce a torque of 200 in · lbs. What pressure is required? Assume 100% efficiency.

SOLUTION:

1. Convert to m³/rev:

$$32 \frac{cm^3}{rev} \cdot \left(\frac{1 \, m^3}{1,000,000 \, cm^3} \right) = 0.000032 \frac{m^3}{rev}$$

2. Calculate the pressure:

$$\Delta p = \frac{2 \cdot \pi \cdot T_T}{V_M} = \frac{2 \cdot (3.142) \cdot (200 \, N \cdot m)}{0.000032 \frac{m^3}{rev}} = 39,275,000 \frac{N}{m^2} \quad (39,275 \, kPa)$$

EXAMPLE 5-3.

A motor must produce a torque of 220 ft · lbs in a system with an operating pressure of 3000 psi. What size motor should we select? Assume 100% efficiency.

SOLUTION:

1. Convert to in-lbs:

$$200 \text{ ft} \cdot \text{lbs} \cdot \left(\frac{12 \text{ in}}{1 \text{ ft}}\right) = 2640 \text{ in} \cdot \text{lbs}$$

2. Calculate the displacement:

$$V_M = \frac{2 \cdot \pi \cdot T_T}{\Delta p} = \frac{2 \cdot (3.142) \cdot (2640 \text{ in} \cdot \text{lbs})}{3000 \text{ psi}} = 5.529 \frac{\text{in}^3}{\text{rev}}$$

EXAMPLE 5-3M.

A motor must produce a torque of 350 N · m in a system with an operating pressure of 25,000 kPa. What size motor should we select? Assume 100% efficiency.

SOLUTION:

1. Calculate the displacement:

$$V_M = \frac{2 \cdot \pi \cdot T_T}{\Delta p} = \frac{2 \cdot (3.142) \cdot (350 \text{ N} \cdot \text{m})}{25,000,000 \frac{\text{N}}{\text{m}^2}} = 0.00008798 \frac{\text{m}^3}{\text{rev}}$$

2. Convert to cm³/rev:

$$0.00008798 \frac{\text{m}^3}{\text{rev}} \left(\frac{1,000,000 \text{ cm}^3}{1 \text{ m}^3}\right) = 87.98 \frac{\text{cm}^3}{\text{rev}}$$

5.4   Motor Speed

The rotational speed of a hydraulic motor depends on the flow supplied by the pump and the size (displacement) of the motor. The larger the motor, the slower it will rotate for a given flow rate. The flow rate necessary to achieve a particular rotational speed can be calculated by multiplying the displacement by the rotational speed, as given by the following equation:

$$Q_T = V_M \cdot N \qquad \text{(5-2)}$$

where: N = desired rotational speed $\left(\dfrac{\text{rev}}{\text{min}}\right)$

V_M = motor displacement $\left(\dfrac{\text{in}^3}{\text{rev}}, \dfrac{\text{m}^3}{\text{rev}}\right)$

Q_T = theoretical pump flow rate required $\left(\dfrac{\text{in}^3}{\text{min}}, \dfrac{\text{m}^3}{\text{min}}\right)$

The flow rate calculated with this equation is called the *theoretical* flow rate. The actual flow rate required will be slightly greater than the theoretical flow rate due to internal leakage in the motor. This issue will be addressed in the section on efficiency.

In most pump literature, the flow rate is expressed in gallons per minute (gpm). We can insert the conversion factor of 1 gal = 231 in³ in the previous equation so that it contains all of the most common units:

$$Q_T = \frac{V_M \cdot N}{231} \qquad \text{(5-3)}$$

The following units must be used for Equation 5-3: Q_T in gpm, N in rpm, and V_M in in³/rev.

In metric units, the displacement is typically given in cm³/rev, which will result in units of cm³/min for Q when using Equation 5-2. The preferred unit for flow in the metric system, however, is liters per minute (lpm). We can insert the conversion factor of 1 l = 1000 cm³ into Equation 3-1 so that it contains the most common metric units:

$$Q_T = \frac{V_M \cdot N}{1000} \qquad \text{(5-3M)}$$

The following units must be used for Equation 5-3M: Q_T in lpm, N in rpm, and V_M in cm³/rev.

EXAMPLE 5-4.

A motor with a 5 in³/rev displacement must rotate at 600 rpm. What flow rate is required? Assume 100% efficiency.

SOLUTION:

$$Q_T = \frac{N \cdot V_M}{231} = \frac{600\,\frac{\text{rev}}{\text{min}} \cdot \left(5\,\frac{\text{in}^3}{\text{rev}}\right)}{231\,\frac{\text{in}^3}{\text{gal}}} = 13.0\,\frac{\text{gal}}{\text{min}} \quad (13.0\text{ gpm})$$

EXAMPLE 5-4M.

A motor with a 50 cm³/rev displacement must rotate at 500 rpm. What flow rate is required? Assume 100% efficiency.

SOLUTION:

$$Q_T = \frac{N \cdot V_M}{1000} = \frac{500\,\frac{\text{rev}}{\text{min}} \cdot \left(50\,\frac{\text{cm}^3}{\text{rev}}\right)}{1000\,\frac{\text{cm}^3}{1}} = 25.5\,\frac{1}{\text{min}} \quad (25.0\text{ lpm})$$

EXAMPLE 5-5.

A system with a flow rate of 10 gpm requires a motor to turn at 300 rpm. What size motor should be selected? Assume 100% efficiency.

SOLUTION:

$$V_M = \frac{231 \cdot Q_T}{N} = \frac{231\,\frac{\text{in}^3}{\text{gal}} \cdot \left(10\,\frac{\text{gal}}{\text{min}}\right)}{300\,\frac{\text{rev}}{\text{min}}} = 7.7\,\frac{\text{in}^3}{\text{rev}}$$

EXAMPLE 5-5M.

A system with a flow rate of 45 lpm requires a motor to turn at 1000 rpm. What size motor should be selected? Assume 100% efficiency.

SOLUTION:

$$V_M = \frac{1000 \cdot Q_T}{N} = \frac{1000\,\frac{\text{cm}^3}{1} \cdot \left(45\,\frac{1}{\text{min}}\right)}{1000\,\frac{\text{rev}}{\text{min}}} = 45\,\frac{\text{cm}^3}{\text{rev}}$$

EXAMPLE 5-6.

A motor with a 2.5 in³/rev displacement is to be used in a system with a flow rate of 5 gpm. What is the rotational speed? Assume 100% efficiency.

SOLUTION:

$$N = \frac{231 \cdot Q_T}{V_M} = \frac{231 \frac{in^3}{gal} \cdot \left(5 \frac{gal}{min}\right)}{2.5 \frac{in^3}{rev}} = 462 \frac{rev}{min}$$

EXAMPLE 5-6M.

A motor with a 15 cm³/rev displacement is to be used in a system with a flow rate of 20 lpm. What is the rotational speed? Assume 100% efficiency.

SOLUTION:

$$N = \frac{1000 \cdot Q_T}{V_M} = \frac{1000 \frac{cm^3}{1} \cdot \left(20 \frac{1}{min}\right)}{15 \frac{cm^3}{rev}} = 1333 \frac{rev}{min}$$

5.5 Motor Power

Motors convert fluid flow from a pump into the rotational motion of an output shaft, which can then be attached to a load to do work. If the load resists the rotation of the shaft, pressure will build until sufficient torque is generated to overcome the resistance of the load. If the maximum system pressure is not sufficient to overcome the load, the motor will *stall* and generate torque, but no motion. Whenever the shaft is rotating under load, power is being transmitted and can be calculated with the following equation:

$$HP_O = \frac{T \cdot N}{63,025} \tag{5-4}$$

HP_O is the output power of a motor. Recall that the input horsepower to the pump (HP_I) is calculated with the same equation; this equation can be used to calculate the horsepower of any rotating shaft. The factor 63,025 in the denominator is a combination of conversion factors that allows us to use the most common U.S. customary units: T in in · lbs, N in rpms, and HP_O in hp. Only these units may be used in this equation.

When using metric units, the following equation can be used to calculate the output power in kilowatts:

$$kW_O = \frac{T \cdot N}{9550} \qquad \textbf{(5-4M)}$$

The factor 9550 in the denominator is a combination of conversion factors that allows us to use the most common metric units (T in N · m, N in rpms, and kW_O in kW). Only these units may be used in this equation.

EXAMPLE 5-7.

■ A motor is required to drive a load at 300 rpm with 200 in · lbs of torque. What is the output horsepower?

SOLUTION:

1. Convert to in-lbs:

$$200 \text{ ft} \cdot \text{lbs} \cdot \left(\frac{12 \text{ in}}{1 \text{ ft}} \right) = 2400 \text{ in} \cdot \text{lbs}$$

2. Calculate the horsepower:

$$HP_O = \frac{T \cdot N}{63,025} = \frac{2400 \text{ in} \cdot \text{lbs} \cdot \left(300 \frac{\text{rev}}{\text{min}} \right)}{63,025} = 11.42 \text{ hp}$$

EXAMPLE 5-7M.

■ A motor is required to drive a load at 500 rpm with 1000 N · m of torque. What is the output power?

SOLUTION:

$$kW_O = \frac{T \cdot N}{9550} = \frac{1000 \text{ N} \cdot \text{m} \cdot \left(500 \frac{\text{rev}}{\text{min}} \right)}{9550} = 52.36 \text{ kW}$$

5.6 Motor Efficiency

Motors, like pumps and all other power transmission devices, are not 100% efficient. Like pumps, motors have both mechanical losses (due to friction) and volumetric losses (due to leakage). The mechanical efficiency of a hydraulic motor is the ratio of the actual torque to the theoretical torque output at a given pressure, as given by the following equation:

$$\eta_M = \frac{T_A}{T_T} \tag{5-5}$$

The actual torque is always less than the theoretical torque due to friction. We can plug this equation into Equation 5-1 to obtain the following equation, which allows us to calculate the actual torque:

$$T_A = \frac{V_M \cdot \Delta p \cdot \eta_M}{2 \cdot \pi} \tag{5-6}$$

The volumetric efficiency of a hydraulic motor is the ratio of the theoretical flow to the actual flow required to achieve a particular speed, as given by the following equation:

$$\eta_V = \frac{Q_T}{Q_A} \tag{5-7}$$

The actual flow required is always greater than the theoretical flow due to leakage. We can plug this equation into Equations 5-3 and 5-3M to obtain the following equations, which allow us to calculate the actual flow required to achieve a particular speed, N:

$$Q_A = \frac{V_M \cdot N}{231 \cdot \eta_V} \tag{5-8}$$

$$Q_A = \frac{V_M \cdot N}{1000 \cdot \eta_V} \tag{5-8M}$$

The overall efficiency of a hydraulic motor is the ratio of the output power to the input power, accounting for the losses due to both friction and leakage. Recall that the input power to the motor is the pump output or hydraulic power. Overall efficiency is therefore given by the following equations:

$$\eta_O = \frac{HP_O}{HP_H} \tag{5-9}$$

$$\eta_O = \frac{kW_O}{kW_H} \qquad \textbf{(5-9M)}$$

While Equations 5-9 and 5-9M are not often used in motor calculations, the overall efficiency is an excellent way to measure a motor's performance under particular conditions. We can also express the overall efficiency as a combination of the volumetric and mechanical efficiencies, as given by the following equation:

$$\eta_O = \eta_M \cdot \eta_V \qquad \textbf{(5-10)}$$

EXAMPLE 5-8.

A motor with a displacement of 2.5 in³/rev and a mechanical efficiency of 93% is used in a system with a maximum pressure of 3000 psi. What is its output torque?

SOLUTION:

$$T_A = \frac{V_M \cdot \Delta p \cdot \eta_M}{2 \cdot \pi} = \frac{2.5 \, \frac{\text{in}^3}{\text{rev}} \cdot \left(3000 \, \frac{\text{lb}}{\text{in}^2}\right) \cdot 0.93}{2 \cdot (3.142)} = 1110 \text{ in} \cdot \text{lbs}$$

EXAMPLE 5-8M.

A motor with a displacement of 20 cm³/rev and a mechanical efficiency of 93% is used in a system with a maximum pressure of 22,000 kPa. What is its output torque?

SOLUTION:

1. Convert to m³/rev:

$$20 \, \frac{\text{cm}^3}{\text{rev}} \cdot \left(\frac{1 \, \text{m}^3}{1,000,000 \, \text{cm}^3}\right) = 0.000020 \, \frac{\text{m}^3}{\text{rev}}$$

2. Calculate T_A:

$$T_A = \frac{V_M \cdot \Delta p \cdot \eta_M}{2 \cdot \pi} = \frac{0.000020 \, \frac{\text{m}^3}{\text{rev}} \cdot \left(22,000,000 \, \frac{N}{m^2}\right) \cdot 0.93}{2 \cdot (3.142)} = 65.12 \text{ N} \cdot \text{m}$$

EXAMPLE 5-9.

A motor with a displacement of 4 in³/rev and a mechanical efficiency of 95% must produce a torque of 95 ft · lbs. What pressure is required?

SOLUTION:

1. Convert to in-lbs:

$$95 \text{ ft} \cdot \text{lbs} \cdot \left(\frac{12 \text{ in}}{1 \text{ ft}} \right) = 1140 \text{ in} \cdot \text{lbs}$$

2. Calculate the pressure:

$$\Delta p = \frac{2 \cdot \pi \cdot T_A}{V_M \cdot \eta_M} = \frac{2 \cdot (3.142) \cdot (1140 \text{ in} \cdot \text{lbs})}{4 \dfrac{\text{in}^3}{\text{rev}} \cdot (0.95)} = 1885 \text{ psi}$$

EXAMPLE 5-10.

A motor must produce a torque of 4000 in · lbs in a system with an operating pressure of 3000 psi. What size motor should we select if the mechanical efficiency is 95%?

SOLUTION:

$$V_M = \frac{2 \cdot \pi \cdot T_T}{\Delta p \cdot \eta_M} = \frac{2 \cdot (3.142) \cdot (4000 \text{ in} \cdot \text{lbs})}{3000 \dfrac{\text{lb}}{\text{in}^2} \cdot (0.95)} = 8.81 \dfrac{\text{in}^3}{\text{rev}}$$

EXAMPLE 5-11.

For the system in the previous problem, standard motors with displacements of 8.5 in³/rev and 9 in³/rev are available. What would the torque be with each of these motors?

SOLUTION:

$$T_{8.5} = \frac{V_M \cdot \Delta p \cdot \eta_M}{2 \cdot \pi} = \frac{8.5 \dfrac{\text{in}^3}{\text{rev}} \cdot \left(3000 \dfrac{\text{lb}}{\text{in}^2} \right) \cdot 0.95}{2 \cdot (3.142)} = 3856 \text{ in} \cdot \text{lbs}$$

$$T_9 = \frac{V_M \cdot \Delta p \cdot \eta_M}{2 \cdot \pi} = \frac{9 \dfrac{\text{in}^3}{\text{rev}} \cdot \left(3000 \dfrac{\text{lb}}{\text{in}^2} \right) \cdot 0.95}{2 \cdot (3.142)} = 4082 \text{ in} \cdot \text{lbs}$$

Note that the larger motor generates more torque at a particular pressure.

EXAMPLE 5-12.

A motor with a 1.75 in³/rev displacement and a volumetric efficiency of 97% must rotate at 700 rpm. What flow rate is required?

SOLUTION:

$$Q_A = \frac{N \cdot V_M}{231 \cdot \eta_V} = \frac{700\,\frac{rev}{min} \cdot \left(1.75\,\frac{in^3}{rev}\right)}{231\,\frac{in^3}{gal} \cdot (0.97)} = 5.47 \text{ gpm}$$

EXAMPLE 5-12M.

A motor with a 25 cm³/rev displacement and a volumetric efficiency of 95% must rotate at 1000 rpm. What flow rate is required?

SOLUTION:

$$Q_A = \frac{N \cdot V_M}{1000 \cdot \eta_V} = \frac{1000\,\frac{rev}{min} \cdot \left(25\,\frac{cm^3}{rev}\right)}{1000\,\frac{cm^3}{1} \cdot (0.95)} = 26.32 \text{ lpm}$$

EXAMPLE 5-13.

A motor with a 7.5 in³/rev displacement and a volumetric efficiency of 94% is used in a system with a flow rate of 20 gpm. What is the rotational speed of the motor?

SOLUTION:

$$N = \frac{231 \cdot Q_A \cdot \eta_V}{V_M} = \frac{231\,\frac{in^3}{gal} \cdot \left(20\,\frac{gal}{min}\right) \cdot 0.94}{7.5\,\frac{in^3}{rev}} = 579 \text{ rpm}$$

EXAMPLE 5-14.

A system with a flow rate of 15 gpm requires a motor to turn at 400 rpm. What size motor should be selected? The motor has a volumetric efficiency 95%.

SOLUTION:

$$V_M = \frac{321 \cdot Q_A \cdot \eta_V}{N} = \frac{231\frac{in^3}{gal} \cdot \left(15\frac{gal}{min}\right) \cdot 0.95}{400\frac{rev}{min}} = 8.23\frac{in^3}{rev}$$

EXAMPLE 5-15.

For the system in the previous problem, standard motors with displacements of 8 in³/rev and 8.3 in³/rev are available. What would the speed be with each of these motors?

SOLUTION:

$$N_8 = \frac{231 \cdot Q_A \cdot \eta_V}{V_M} = \frac{231\frac{in^3}{gal} \cdot \left(15\frac{gal}{min}\right) \cdot 0.95}{8\frac{in^3}{rev}} = 411 \text{ rpm}$$

$$N_{8.3} = \frac{231 \cdot Q_A \cdot \eta_V}{V_M} = \frac{231\frac{in^3}{gal} \cdot \left(15\frac{gal}{min}\right) \cdot 0.95}{8.3\frac{in^3}{rev}} = 397 \text{ rpm}$$

Note that the larger motor rotates more slowly at a particular flow rate.

5.7   Graphic Symbols

Figure 5-8 shows the symbols used for hydraulic motors. Part A shows the symbol for a *fixed displacement, uni-directional* motor. A fixed displacement motor cannot change its displacement and therefore its speed cannot be adjusted without adjusting the flow rate. It also cannot vary its torque output without varying the pressure. A uni-directional motor is designed to be rotated in one direction only. Notice the drain line, which is shown as a dashed line going back to tank. Many motors require a drain line to remove internal leakage from the motor. Not connecting this drain line to tank will likely result in a blown shaft seal. Figure 5-8B shows the symbol for a *fixed displacement, bi-directional* motor. This type of motor can be rotated either clockwise or counter-clockwise without damage by simply reversing the direction of flow. Bi-directional motors are by far the most common. The opposite is true with hydraulic pumps; most pumps are uni-directional. Figure 5-8C shows the symbol for a *variable displacement, bi-directional* motor. This

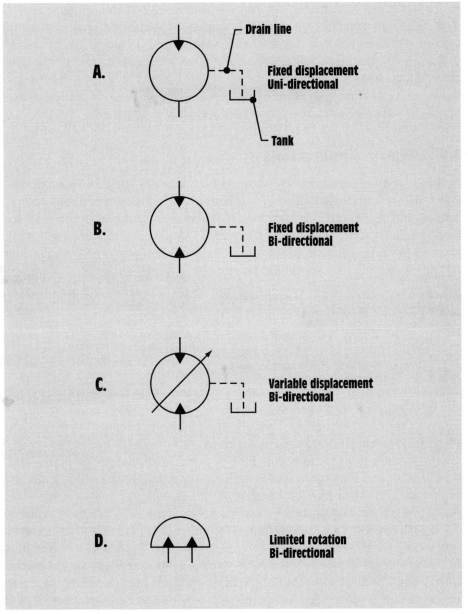

FIGURE 5-8   Hydraulic motor graphic symbols. (A) Fixed displacement, uni-directional. (B) Fixed displacement, bi-directional. (C) Variable displacement, bi-directional. (D) Limited rotation, bi-directional.

motor can adjust its displacement, so it can vary its speed and torque even if the flow rate and pressure are constant. Equation 5-1 shows that increasing a motor's displacement will increase its torque output. Equation 5-3 shows that increasing displacement will decrease the speed. Decreasing the displacement has the opposite effect: torque decreases and speed increases. Figure 5-8D shows the symbol for a *bi-directional limited rotation* motor, which is used in applications in which a partial revolution rotary motion is required.

5.8   Motor Applications

Hydraulic motors are frequently used in both industrial and mobile applications. Industrial applications include driving conveyors, winches, machine tool positioning, and many other applications that require smooth and powerful rotational motion. Some examples of mobile applications include rotating the body of an excavating machine, driving harvesting mechanisms on farm machinery, and driving boring tools in mining equipment. Two key advantages of hydraulic motors over electrical and mechanical motors include: (1) they respond quickly to speed and direction changes, and (2) they can be stalled under full load without damage to the system.

Hydraulic motors are also often used in vehicle propulsion systems. In these *hydrostatic transmission* systems, the vehicle engine drives a hydraulic pump, which provides flow to a hydraulic motor. The hydraulic motor then drives the vehicle. Hydrostatic transmissions are commonly used on excavating, mining, and farm vehicles. These systems provide very smooth power output that is well suited for use on loose terrain. They also have a high horsepower-to-weight ratio, which is obviously important for any vehicle.

Hydrostatic transmissions can have a variable displacement pump and a fixed displacement motor, a fixed displacement pump and a variable displacement motor, or both may be variable. A system with a variable displacement pump and a fixed displacement motor can vary the speed, but the torque output at a given pressure is fixed because this depends on the motor displacement. This is called a *constant torque* drive. A system with a fixed displacement pump and a variable displacement motor can control both speed and torque, because both depend on the motor displacement. With this system, however, speed and torque cannot be controlled independently. Any increase in displacement of the motor results in a speed decrease and torque increase by the same factor (at a fixed pressure). This system is called a *constant horsepower* drive because the output horsepower remains fixed at a particular pressure. The horsepower output of a motor depends on torque and speed (see Equation 5-4). A transmission with a variable displacement pump and a variable displacement motor allows the speed and torque to be varied independently, but is much more difficult to control.

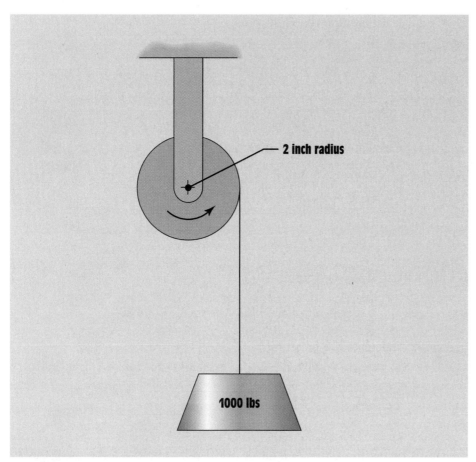

FIGURE 5-9   Example 5-16.

EXAMPLE 5-16.

A motor must be selected to operate the winch shown in Figure 5-9. The motor must be able to generate enough torque to lift the weight with a linear speed of 2 ft/s. The system flow rate will be 3 gpm and the operating pressure should not exceed 3000 psi. What size motor should be selected? Assume a volumetric efficiency of 95% and a mechanical efficiency of 93%.

SOLUTION:

1. Convert from ft/s to in/min:

$$2\,\frac{\text{ft}}{\text{s}} \cdot \left(\frac{12\ \text{in}}{1\ \text{ft}}\right) \cdot \left(\frac{60\ \text{s}}{1\ \text{min}}\right) = 1440\,\frac{\text{in}}{\text{min}}$$

2. Calculate the required rpms:

$$N = \frac{v}{2 \cdot \pi \cdot r} = \frac{1440 \frac{in}{min}}{2 \cdot (3.142) \cdot 2 \frac{in}{rev}} = 115 \frac{rev}{min}$$

3. Calculate the displacement:

$$V_M = \frac{231 \cdot Q_A \cdot \eta_V}{N} = \frac{231 \frac{in^3}{gal} \cdot \left(3 \frac{gal}{min}\right) \cdot 0.95}{115 \frac{rev}{min}} = 5.72 \frac{in^3}{rev}$$

4. Calculate the torque:

$$T = F \cdot d = 1000 \text{ lbs} \cdot (2 \text{ in}) = 2000 \text{ in} \cdot \text{lbs}$$

5. Calculate the pressure:

$$\Delta p = \frac{2 \cdot \pi \cdot T_A}{V_M \cdot \eta_M} = \frac{2 \cdot (3.142) \cdot (2000 \text{ in} \cdot \text{lbs})}{5.72 \frac{in^3}{rev} \cdot (0.93)} = 2362 \text{ psi}$$

This is below the pressure limit of 3000 psi, therefore a motor of this size will be suitable. We would now select a motor of about this size from a catalog and redo the calculations to be sure we are within acceptable limits. A larger motor will generate more torque and will rotate slower. A smaller motor will generate less torque and will rotate faster.

5.9 Motor Specifications

This section describes the information that manufacturers commonly provide on hydraulic motor performance. The format of this information varies considerably from one manufacturer to another. The reader is encouraged to obtain several manufacturer's catalogs to compare how motor specifications are provided. Many manufacturers have two main types of motors: standard (high-speed) models and LSHT (low-speed, high-torque) models. These types are usually separated in a catalog, but the format of the specifications is similar.

5.9.1 MOTOR SIZE AND PHYSICAL DIMENSIONS

Motors are sized primarily by their displacement in in^3/rev, which is sometimes abbreviated *CIR* in manufacturers' catalogs. A single value is given for fixed displacement motors, and a range is given for variable displacement models. When selecting a motor for a particular application, the physical dimensions may also be an important factor to consider. The manufacturer will provide data such as:

1. Weight,
2. Overall dimensions,
3. Center-to-center distances and sizes of mounting holes,
4. Port sizes and type, and
5. Output shaft size (may be spline or key type).

A drawing of each model with all of the relevant dimensions is usually provided in the manufacturer's catalog. These drawings are a valuable tool to the designer of the system.

5.9.2 MOTOR PERFORMANCE

The performance of a motor can be measured by its mechanical (η_M), volumetric (η_V), and overall (η_O) efficiencies under particular conditions. This information may be given in a tabular or graphical format, with the latter being more prevalent. This data will be given for a particular fluid type, temperature, and viscosity (e.g., standard hydraulic oil at 120° and 100 SUS). The performance of a particular motor may be significantly different for other fluids and conditions.

The volumetric efficiency of a motor may be given versus drive speed (Figure 5-10). Note that volumetric efficiency is lower (the leakage is higher) at higher pressures (i.e., the 3000 psi curve is below the 2000 psi curve). Also note that that volumetric efficiency is lower at lower speeds. In the example shown in Figure 5-10, the volumetric efficiency drops off rapidly at speeds lower than 1000 rpm. Another way to give the same information is to give the actual input flow (Q_A) versus drive speed (Figure 5-11). The actual input flow is the flow required to achieve a particular speed, accounting for leakage. This graph shows that at higher pressures, more input flow is required to achieve a particular speed. Again, this is due to leakage, which increases with increasing pressure.

The mechanical efficiency of a motor is not often given directly. Instead, the actual torque (T_A) is given versus the rotational speed (Figure 5-12). Note that the output torque is fairly constant as the speed varies. This is because torque

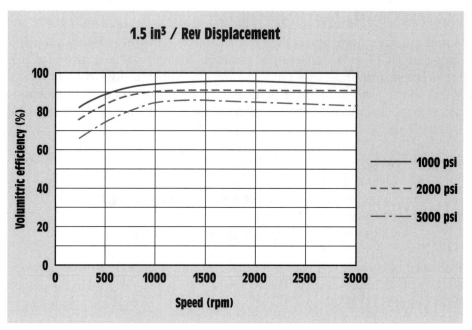

FIGURE 5-10   Volumetric efficiency versus drive speed.

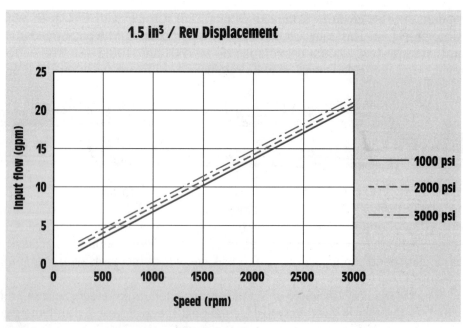

FIGURE 5-11   Actual input flow versus drive speed.

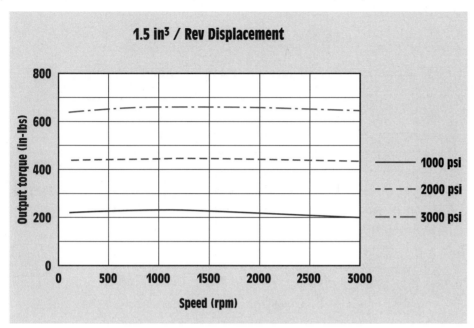

FIGURE 5-12   Actual output torque versus drive speed.

depends on pressure and displacement, not speed, as shown in Equation 5-6. It does vary slightly, however, because the frictional losses, which reduce torque output, do depend on shaft speed to some extent. The torque values given in Figure 5-12 are called the *running torque,* the torque produced when the motor is rotating. The manufacturer may also give the *starting torque,* the torque produced when the motor is started under load. This value will be less than the running torque at a given pressure.

The overall efficiency of a motor is usually given versus drive speed (Figure 5-13). As the graph shows, motors are generally more efficient at higher pressures. This is true even though the volumetric efficiency decreases. This is because the mechanical efficiency increases more rapidly due to the fact that frictional losses become a smaller percentage of the torque output. Also note that motors typically become very inefficient when driven at slow speeds.

EXAMPLE 5-17.

Use Figure 5-10 to determine the input flow required to rotate this motor at 1500 rpm. The operating pressure is 2000 psi. Look at Figure 5-11 to see if you get the same answer.

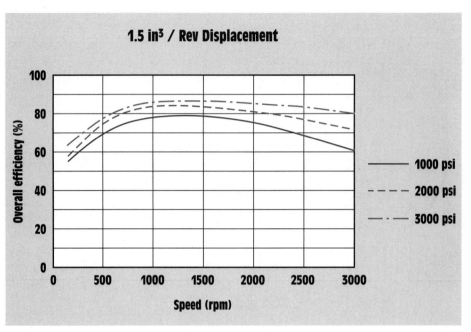

FIGURE 5-13   Overall efficiency versus drive speed.

SOLUTION:

1. Find η_V from Figure 5-10:

$$\eta_V \approx 92\%$$

2. Calculate the input flow:

$$Q_A = \frac{N \cdot V_M}{231 \cdot \eta_V} = \frac{1.5 \, \frac{in^3}{rev} \cdot \left(1500 \, \frac{rev}{min}\right)}{231 \frac{in^3}{gal} \cdot (0.92)} = 10.6 \frac{gal}{min}$$

EXAMPLE 5-18.

Use Figure 5-11 to determine the input flow required to rotate this motor at 2000 rpm if the operating pressure is 3000 psi.

SOLUTION:

Find Q_A from Figure 5-12: $Q_A \approx 15$ gpm

EXAMPLE 5-19.

Use Figure 5-12 to determine the output torque of this motor at 1000 rpm and 2000 psi. Also calculate the mechanical efficiency under these conditions.

SOLUTION:

1. Find T_A from Figure 5-13:

$$T_A \approx 440 \text{ in} \cdot \text{lbs}$$

2. Calculate the theoretical torque:

$$T_T = \frac{V_M \cdot \Delta p}{2 \cdot \pi} = \frac{1.5 \frac{\text{in}^3}{\text{rev}} \cdot \left(2000 \frac{\text{lb}}{\text{in}^2} \right)}{2 \cdot (3.142)} = 477 \text{ in} \cdot \text{lbs}$$

3. Calculate η_M:

$$\eta_M = \frac{T_A}{T_T} = \frac{440 \text{ in} \cdot \text{lbs}}{477 \text{ in} \cdot \text{lbs}} = 0.922 \quad (92.2\%)$$

EXAMPLE 5-20.

Use Figure 5-13 to determine the overall efficiency of this motor at (a.) 1200 rpm and 1000 psi and (b.) 2400 rpm and 3000 psi.

a. At 1200 rpm and 1000 psi: $\eta_O \approx 79\%$
b. At 2400 rpm and 3000 psi: $\eta_O \approx 85\%$

5.9.3 SPEED AND PRESSURE RATINGS

Manufacturers provide minimum and maximum drive speed ratings for each of their models. LSHT motors will have much lower values for both minimum and maximum drive speeds than will high-speed motors. Minimum speeds of 50 to 300 rpm and maximum speeds of 3600 to 5000 rpm are common for high-speed motors. Minimum speeds of 2 to 20 rpm and maximum speeds of 100 to 500 rpm are common for LSHT motors. Operating below the minimum or above the maximum can result in very low efficiency. Operating above the maximum speed can also result in bearing or seal damage. Manufacturers may give reduced speed ratings for fluids other than petroleum-based oils.

Pressure ratings are given either as a single value or as separate values for intermittent and continuous service. Values for intermittent service are higher than those for continuous service because the motor is subjected to these pressures for only a fraction of the duty cycle. Maximum pressures between 1000 psi and 5000 psi are common. If the motor is operated above the maximum pressure, it may become very inefficient due to increased leakage. Seal failure is also a possibility. Manufacturers may give reduced pressure ratings for fluids other than petroleum-based oils.

5.9.4 FLUIDS

Viscosity is the most important fluid property with respect to hydraulic motor performance, as is the case with hydraulic pumps. This is because the fluid is also the lubricant for the motor, and the viscosity of the fluid determines the quality of the lubrication. Manufacturers usually specify a minimum and a maximum fluid viscosity for use with their motors. Viscosity changes with temperature, so we must be sure that the fluid chosen will be in this viscosity range over the temperature range at which the system will operate.

Most systems and components are designed for standard petroleum-based hydraulic oil. A motor designed for standard oil will often have seals that are incompatible with certain other fluids. The seals may degrade quickly and fail if used with these fluids (see Chapter 9).

5.9.5 FILTRATION

Hydraulic motors, like hydraulic pumps, are sensitive to contamination. This is due to the close-fitting mating components within the motor. Because of these tight fits, a very small particle of contamination can wedge between the components and cause wear, resulting in increased leakage and inefficiency. Manufacturers will therefore specify the level of filtration required for each model. The filtration standard of choice for hydraulic components is the ISO cleanliness code (see Chapter 3).

5.10 Equations

EQUATION NUMBER	EQUATION	REQUIRED UNITS
5-1	$T_T = \dfrac{V_M \cdot \Delta p}{2 \cdot \pi}$	Any consistent units
5-2	$Q_T = V_M \cdot N$	Any consistent units
5-3	$Q_T = \dfrac{V_M \cdot N}{231}$	Q_T in gpm, N in rpm, V_M in in³/rev
5-3M	$Q_T = \dfrac{V_M \cdot N}{1000}$	Q_T in lpm, N in rpm, V_M in cm³/rev

EQUATION NUMBER	EQUATION	REQUIRED UNITS
5-4	$HP_O = \dfrac{T \cdot N}{63,025}$	T in in · lbs, N in rpms, HP_O in hp
5-4M	$kW_O = \dfrac{T \cdot N}{9550}$	T in N · m, N in rpms, kW_O in kW
5-5	$\eta_M = \dfrac{T_A}{T_T}$	Any consistent units
5-6	$T_A = \dfrac{V_M \cdot \Delta p \cdot \eta_M}{2 \cdot \pi}$	Any consistent units
5-7	$\eta_V = \dfrac{Q_T}{Q_A}$	Any consistent units
5-8	$Q_A = \dfrac{V_M \cdot N}{231 \cdot \eta_V}$	Q_A in gpm, N in rpm, V_M in in³/rev
5-8M	$Q_A = \dfrac{V_M \cdot N}{1000 \cdot \eta_V}$	Q_A in lpm, N in rpm, V_M in cm³/rev
5-9	$\eta_O = \dfrac{HP_O}{HP_H}$	HP_H in hp, HP_O in hp, η_O is unitless
5-9M	$\eta_O = \dfrac{kW_O}{kW_H}$	kW_H in kW, kW_O in kW, η_O is unitless
5-10	$\eta_O = \eta_M \cdot \eta_V$	All quantities are unitless

5.11   Review Questions and Problems

1. What is the purpose of a motor in a hydraulic circuit?
2. Describe the basic operation of a hydraulic motor.
3. Define *displacement* with respect to a hydraulic motor.
4. What is the effect of increasing a motor's displacement on speed and torque? What is the effect of decreasing the displacement on speed and torque?
5. What does LSHT stand for? What is the purpose of LSHT motors?
6. What are the two most common types of construction for a limited rotation motor?
7. Define *mechanical efficiency*.
8. Why is the actual torque of a motor output less than the theoretical torque?
9. Define *volumetric efficiency*.

10. Why is the actual speed of a motor at a given flow rate less than the theoretical speed?
11. Define *overall efficiency*. How is this related to volumetric and mechanical efficiency?
12. Draw the graphic symbols for the following four motor types: fixed displacement uni-directional, fixed displacement bi-directional, variable displacement bi-directional, and limited rotation bi-directional.
13. What is a *hydrostatic drive?* What are its two main components? In what applications is this type of system used?
14. Does volumetric efficiency increase or decrease with increasing pressure? Why?
15. Why is the viscosity of a fluid critical to a motor's performance?
16. Why is filtration important in hydraulic systems?
17. A motor has a displacement of 4 in³/rev and is used in a system with a maximum pressure of 3500 psi. What is the theoretical output torque?
18. A motor has a displacement of 75 cm³/rev and is used in a system with a maximum pressure of 25,000 kPa. What is the theoretical output torque?
19. A motor with a displacement of 1.0 in³/rev must produce a torque of 750 in · lbs. What pressure is required? Assume 100% efficiency.
20. A motor with a displacement of 45 cm³/rev must produce a torque of 300 N · m. What pressure is required? Assume 100% efficiency.
21. A motor must produce a torque of 200 ft · lbs in a system with an operating pressure of 2500 psi. What size motor should we select? Assume 100% efficiency.
22. A motor must produce a torque of 500 N · m in a system with an operating pressure of 20,000 kPa. What size motor should we select (in cm³/rev)? Assume 100% efficiency.
23. A motor with a 2.3 in³/rev displacement must rotate at 1500 rpm. What flow rate (in gpm) is required? Assume 100% efficiency.
24. A motor with a 32 cm³/rev displacement must rotate at 700 rpm. What flow rate (in lpm) is required? Assume 100% efficiency.
25. A motor with a 1.2 in³/rev displacement is to be used in a system with a flow rate of 10 gpm. What is the rotational speed? Assume 100% efficiency.
26. A motor with a 30 cm³/rev displacement is to be used in a system with a flow rate of 40 lpm. What is the rotational speed? Assume 100% efficiency.
27. A motor is required to drive a load at 1200 rpm with 200 ft-lbs of torque. What is the output horsepower?

28. A motor is required to drive a load at 900 rpm with 1200 N · m of torque. What is the output power?

29. A motor with a displacement of 1.5 in³/rev and a mechanical efficiency of 92% is used in a system with a maximum pressure of 3000 psi. What is its output torque?

30. A motor with a displacement of 35 cm³/rev and a mechanical efficiency of 95% is used in a system with a maximum pressure of 18,000 kPa. What is its output torque?

31. A motor with a displacement of 2.75 in³/rev and a mechanical efficiency of 95% must produce a torque of 90 ft · lbs. What pressure is required?

32. A motor with a displacement of 45 cm³/rev and a mechanical efficiency of 92% must produce a torque of 150 N · m. What pressure is required?

33. A motor must produce a minimum torque of 1750 in · lbs in a system with an operating pressure of 2500 psi. What size motor should we select if the mechanical efficiency is 92%?

34. If standard motors of 4.75 in³/rev and 5 in³/rev are available from the manufacturer, which size would you select for the system in problem 33? Why?

35. A motor must produce a minimum torque of 500 N · m in a system with an operating pressure of 20,000 kPa. What size motor should we select (in cm³/rev) if the mechanical efficiency is 95%?

36. A motor with a 3.75 in³/rev displacement and a volumetric efficiency of 97% must rotate at 300 rpm. What flow rate is required?

37. A motor with a 20 cm³/rev displacement and a volumetric efficiency of 95% must rotate at 800 rpm. What flow rate is required?

38. A motor with a 3.5 in³/rev displacement and a volumetric efficiency of 96% is used in a system with a flow rate of 15 gpm. What is the rotational speed of the motor?

39. A motor with a 22 cm³/rev displacement and a volumetric efficiency of 95% is used in a system with a flow rate of 40 lpm. What is the rotational speed of the motor?

40. A system with a flow rate of 20 gpm requires a motor to turn at 1400 rpm. What size motor should be selected? The motor has a volumetric efficiency of 95%.

41. For the system in problem 40, standard motors with displacements of 3 in³/rev and 3.25 in³/rev are available. What size would you select if a slightly greater speed is acceptable, but not a slower speed?

42. A system with a flow rate of 80 lpm requires a motor to turn at 1200 rpm. What size motor should be selected? The motor has a volumetric efficiency of 95%.

43. Use Figure 5-10 to determine the volumetric efficiency and input flow required to rotate this motor at 1000 rpm if the operating pressure is 2000 psi.
44. Repeat problem 43 using Figure 5-11 to see if a similar result is obtained.
45. Use Figure 5-12 to determine the output torque of this motor at 2000 rpm and 3000 psi. Calculate the mechanical efficiency under these conditions.
46. Use Figure 5-13 to determine the overall efficiency of this motor at 1500 rpm and 2000 psi.

CHAPTER 6

Hydraulic Directional Control

OUTLINE

6.1 Introduction

6.2 Check Valves

6.3 Shuttle Valves

6.4 Two-way Directional Control Valves

6.5 Three-way Directional Control Valves

6.6 Four-way Directional Control Valves

6.7 Directional Control Valve Actuation

6.8 Circuits

6.9 Directional Control Valve Mounting

6.10 Directional Control Valve Specifications

6.11 Equations

6.12 Review Questions and Problems

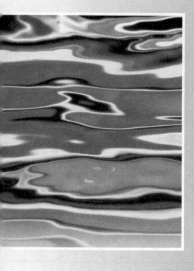

6.1   Introduction

A fluid power system can be broken down into three segments: (1) the power input segment, consisting of the prime mover and the pump; (2) the control segment, consisting of valves that control the direction, pressure, and flow rate; and (3) the power output segment, consisting of the actuator (cylinder, motor, etc.) and the load. Thus far, we have covered the power input and power output segments. Chapters 6, 7, and 8 cover the control segment of the system. A chapter is devoted to each of the following categories of control valves:

1.   Directional control valves (DCVs) (Chapter 6)
2.   Pressure control valves (PCVs) (Chapter 7)
3.   Flow control valves (FCVs) (Chapter 8)

DCVs control the direction of flow in a circuit, which, among other things, can control the direction of the actuator. PCVs control the pressure level, which controls the output force of a cylinder or the output torque of a motor. FCVs control the flow rate of the fluid, which controls the speed of the actuator.

6.2   Check Valves

The simplest directional control valve is a check valve (Figure 6-1). A check valve allows flow in one direction, but blocks flow in the opposite direction. There are several different types of construction; the most common is an inline check (Figure 6-1A), which consists of a ball with a light bias spring that holds the ball against the valve seat. Flow coming into the inlet pushes the ball off the seat against the light force of the spring and continues to the outlet. A very low pressure, usually around 15 psi, is all that is required to hold the valve open in this direction. If flow tries to enter from the opposite direction, the pressure pushes the ball against the seat and the flow cannot pass through. The graphic symbol for a check valve is shown in Figure 6-1B.

Check valves have numerous applications in hydraulic circuits. In one common application (Figure 6-2), a check valve is put in *parallel* with a component. Putting components in parallel means the fluid has the option of passing through either component. As we have discussed earlier, the fluid will always choose the path of least resistance (lowest pressure). When the flow comes in from the left, it cannot pass through the check. It is therefore forced to go through the component (Figure 6-2A). When the flow comes in from the right, however, the flow goes through the check and the component is *bypassed* (Figure 6-2B). This occurs because the check valve is designed to have

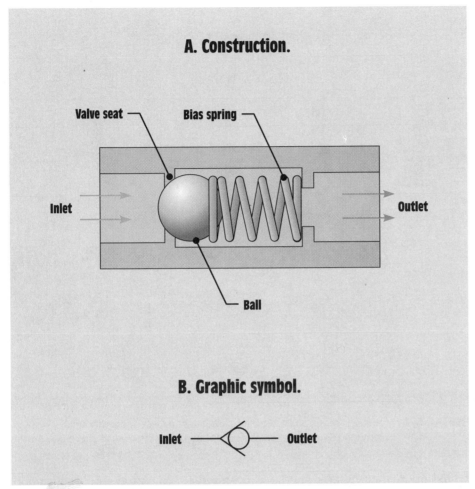

FIGURE 6-1   Inline check valve. (A) Construction. (B) Graphic symbol.

less resistance to flow than the component in this direction. Many examples of this particular application for a check valve will be seen in the next two chapters.

In addition to the simple check valve, there are many variations on the standard design. One commonly used type is the *pilot-to-open* check valve in (Figure 6-3A). *Pilot lines* are hydraulic lines that are used for control purposes. They typically send system pressure to a component, so that the component can react to pressure changes in some way. Pilot lines are shown as dashed lines on schematics. When no pressure is applied to the pilot line, pilot-to-open check acts like a standard check valve, allowing flow in one direction but not the other. However, when sufficient pressure is applied to the pilot line, the check

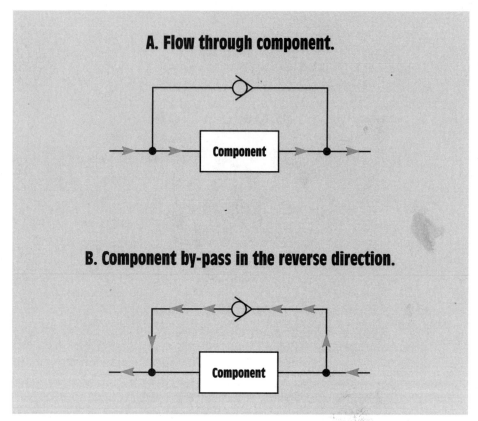

FIGURE 6-2   A check valve can be used for component bypass in one direction. (A) Flow through component. (B) Component bypass in the reverse direction.

is held open, thereby allowing flow in *both directions*. An example of an application for this valve will be shown in the circuits section later in the chapter. There are also *pilot-to-close* check valves, although they are less commonly used than the pilot-to-open variety. These valves *block* flow in both directions when they receive pressure to their pilot line. They operate as a normal check when pressure is not applied to the pilot. The graphic symbol for a pilot-to-close check valve is shown in Figure 6-3B.

6.3   Shuttle Valves

Shuttle valves allow two alternate flow sources to be connected to one branch circuit. They have two inlets (*P1* and *P2*) and one outlet (*A*). Outlet *A* receives flow from whichever inlet is at a higher pressure. Figure 6-4 illustrates the

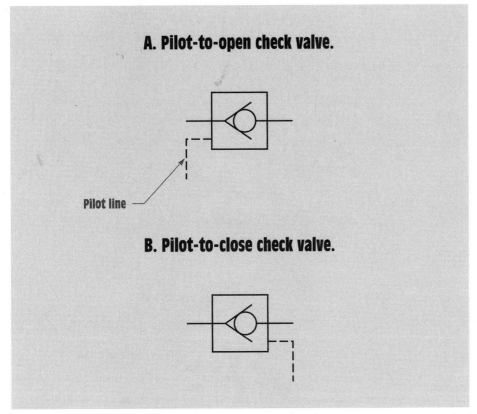

A. Pilot-to-open check valve.

Pilot line

B. Pilot-to-close check valve.

FIGURE 6-3   Pilot-operated check valves. (A) Pilot-to-open check valve. (B) Pilot-to-close check valve.

operation of a shuttle valve. If the pressure at *P1* is greater than that at *P2,* the ball slides to the right and allows *P1* to send flow to outlet *A* (Figure 6-4A). If the pressure at *P2* is greater than at *P1,* the ball slides to the left and *P2* supplies flow to outlet *A* (Figure 6-4B). Figure 6-4C shows the graphic symbol for a shuttle valve.

One application for a shuttle valve is to have a primary pump (inlet *P1*) and a secondary pump (inlet *P2*) connected to a system (outlet *A*). The secondary pump will act as a backup, supplying flow to the system if the primary pump loses pressure. A schematic for this application will be shown in the circuits section of this chapter. A shuttle valve is also called an *OR* valve because receiving a pressure input signal from either *P1* or *P2* causes a pressure output signal to be sent to *A*.

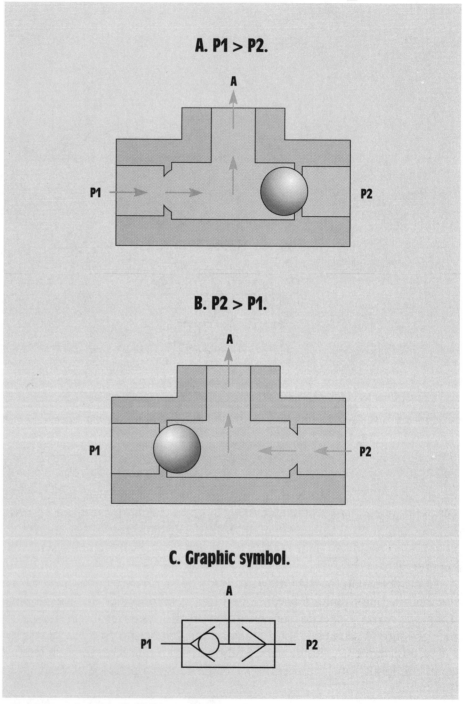

FIGURE 6-4   Shuttle valve. (A) *P1* greater than *P2*. (B) *P2* greater than *P1*. (C) Graphic symbol.

6.4   Two-way Directional Control Valves

The next three sections of this chapter deal with *two-way, three-way,* and *four-way* directional control valves (DCVs). The number of "ways" refers to the number of ports in the valve. These valves are most commonly *spool*-type valves. A two-way, spool-type DCV is shown in Figure 6-5. A spool valve consists of a cylindrical spool that slides back and forth inside the valve body to either connect or block flow between ports. The larger diameter portion of the spool, the *spool land,* blocks flow by covering a port. This particular valve has two ports, labeled *P* and *A*. *P* is connected to the pump line and *A* is the outlet to the system. Figure 6-5A shows the valve in its *normal* state and its corresponding symbol. The valve is held in this position by the force of the spring. In this position, the flow from the inlet port *P* is blocked from going to the outlet port *A*. Figure 6-5B shows the valve in its *actuated* state and its corresponding symbol. The valve is shifted into this position by applying a force to overcome the resistance of the spring. In this position, the flow is allowed to go to the outlet port around the smaller diameter portion of the spool.

The complete graphic symbol for this valve is shown in Figure 6-5C. The symbol has two blocks, one for each position of the valve. Valves may have more than two positions. The spring is on the closed position side of the symbol, which indicates that it is a *normally closed* (NC) valve. The symbol for the method of actuation (how the valve is shifted) is shown on the opposite side of the valve. In this case, the valve is push-button actuated. Thus, the graphic symbol in Figure 6-5C represents a *two-way, two-position, normally closed DCV with push-button actuation and spring return.*

Figure 6-6 shows a two-way, two-position, *normally open* (NO) DCV. The spring holds the valve in a position in which the *P* and *A* ports are connected, (part A). When the valve is actuated the flow is blocked from going to *A*, (part B). The graphic symbol (part C) has the spring on the open side of the valve and the actuation symbol on the closed side. This indicates that the valve is normally open and must be actuated to block flow. Whether a two-way valve is normally open or normally closed it has the same basic function: starting and stopping flow in a particular line.

Figure 6-7 shows an example of an application for a two-way valve. Here, a pair of two-way valves are used to fill and drain a vessel. Notice the tank (reservoir) symbols, which we have not seen previously. Although two tanks are shown in this schematic, there may in fact be only one tank in the actual system. Two tanks are drawn on the schematic to minimize the number of lines, which makes the schematic more compact and easier to read. When valve 1 and valve 2 are in the positions shown, the vessel will hold fluid because both the line from the pump and the line to the tank are blocked. This is how the schematic should be drawn, with all valves in their normal position. In Figure 6-8,

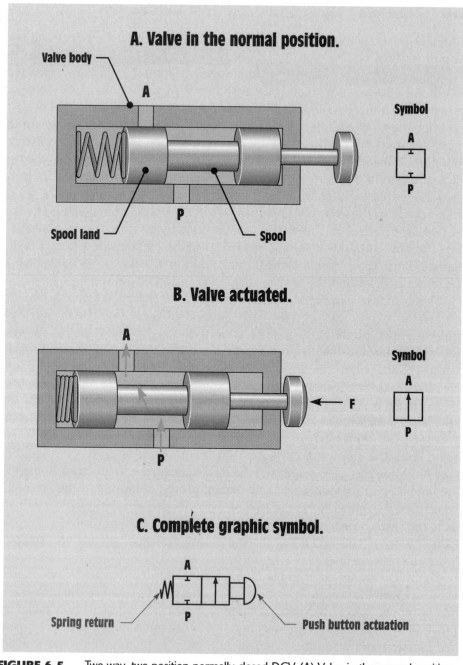

FIGURE 6-5   Two-way, two-position normally closed DCV. (A) Valve in the normal position. (B) Valve actuated. (C) Complete graphic symbol.

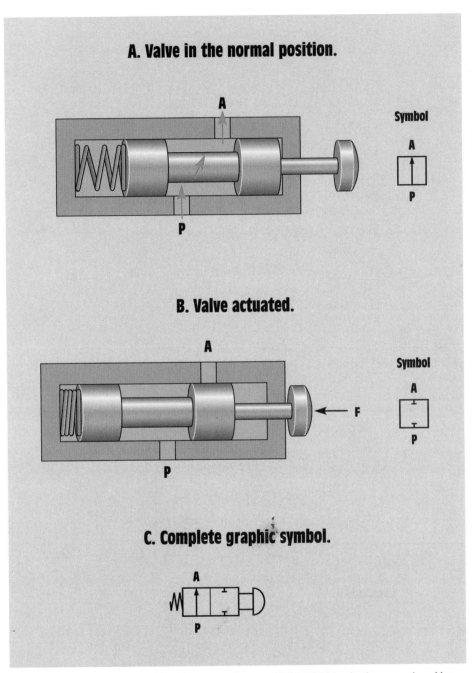

A. Valve in the normal position.

Symbol

B. Valve actuated.

Symbol

C. Complete graphic symbol.

FIGURE 6-6   Two-way, two-position normally open DCV. (A) Valve in the normal position. (B) Valve actuated. (C) Complete graphic symbol.

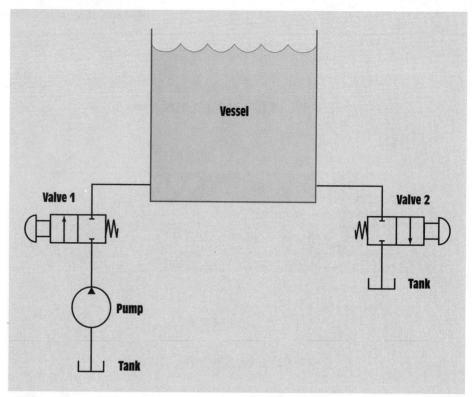

FIGURE 6-7   Two-way valves can be used to fill and drain a vessel.

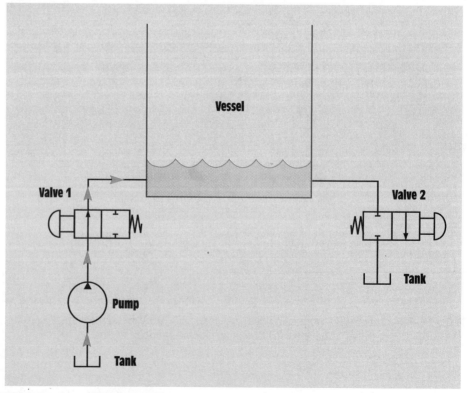

FIGURE 6-8   Filling the vessel.

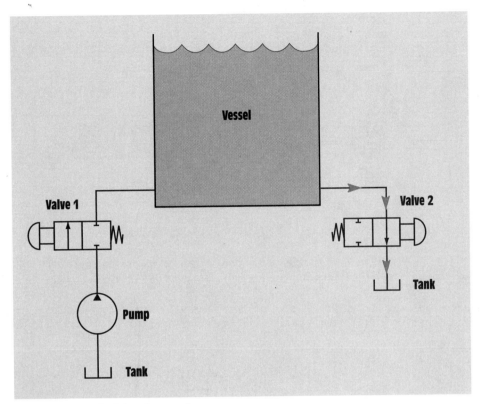

FIGURE 6-9 Draining the vessel.

valve 1 is shifted to the open position and valve 2 remains closed. This will fill the vessel. In Figure 6-9, valve 2 is shifted to the open position and valve 1 remains closed. This will drain the vessel. The next section discusses how this same function can be accomplished with one three-way valve.

There are other types of construction for two-way valves in addition to the spool type. A *ball valve* is commonly used in fluid power circuits (Figure 6-10A). This type of valve consists primarily of a ball with a hole drilled through it. That ball sits inside a valve body. Turning the handle so that the hole aligns with the flow path allows flow to the system. Turning the handle so that the hole is perpendicular to the flow path blocks the flow. Although ball valves are technically two-way valves, they are more commonly referred to as *manual shut-off valves.* Another type of manual shut-off valve, a *gate valve,* utilizes a hand wheel to raise and lower a gate into the fluid stream to start or stop flow. Gate valves are more commonly used in fluid transfer applications, rather than in fluid power. The graphic symbol for a manual shut-off valve is shown in Figure 6-10B. Note that this symbol does not tell us which type of manual shut-off valve is being used.

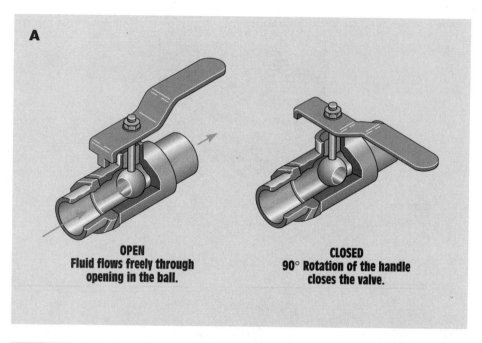

OPEN
Fluid flows freely through
opening in the ball.

CLOSED
90° Rotation of the handle
closes the valve.

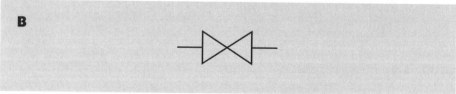

FIGURE 6-10   (A) Ball valve-type manual shutoff. (From Vockroth, Industrial Hydraulics, Albany, NY: Delmar, 1994, p. 139.) (B) Manual shut-off graphic symbol.

6.5   Three-way Directional Control Valves

As stated in the previous section, two-way valves are used to start and stop fluid flow in a particular line. They can either allow or block flow from a pump to an outlet line, for example. Three-way valves also either block or allow flow from an inlet to an outlet. They also allow the outlet to flow back to the tank when the pump flow is blocked, while a 2-way does not. A three-way valve has three ports: a pressure inlet (P), an outlet to the system (A), and a return to the tank (T). Figure 6-11 illustrates the operation of a three-way, normally closed DCV. In its normal position, just as with the two-way DCV, the valve is held in position with a spring (Figure 6-11A). In the normal position, the pressure port (P) is blocked and the outlet (A) is connected with the tank (T). This depressurizes, or *vents,* the outlet port. In the actuated position (Figure 6-11B), the pressure port is connected with the outlet and the tank port is blocked. This sends flow and pressure to the system. Figure 6-11C

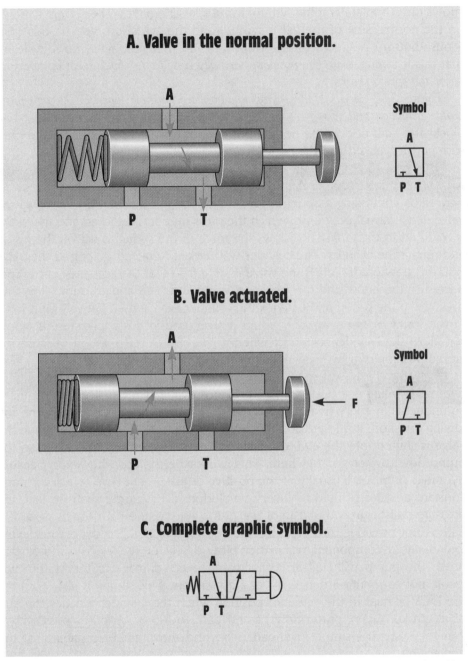

FIGURE 6-11 Three-way, two-position normally closed DCV. (A) Valve in the normal position. (B) Valve actuated. (C) Complete graphic symbol.

shows the complete graphic symbol for use in schematics. The spring is shown on the normal side of the valve symbol and the actuation type (in this case, a push-button) is shown on the opposite side. The graphic symbol indicates that this is a *three-way, two-position, normally closed DCV with push-button actuation and spring return.*

Figure 6-12 shows a *three-way, two-position, normally open DCV with push-button actuation and spring return.* This valve sends pressure to the outlet and blocks the tank port in the normal position. In the actuated position, the pressure port is blocked and the outlet is vented to the tank.

The most common application for a three-way valve in a hydraulic circuit is to control a single-acting cylinder. Figure 6-13 shows the schematic for a three-way valve controlling a spring return, single-acting cylinder. Part A shows the valve in its normal position in which the pressure port is blocked and the outlet is returned to the tank. This allows the force of the spring to act on the piston and retract the cylinder. The cylinder will remain retracted as long as the valve is in this position. In part B the valve is shifted so that the pressure port is connected to the outlet and the tank port is blocked. This applies pump flow and pressure to the piston and extends the cylinder against the relatively light force of the spring. A two-way valve could not be used in this application. It would not allow the cylinder to retract when it is in the closed position because the closed position of a two-way does not have a return to the tank.

A *pressure relief valve,* a device that limits the maximum pressure in a hydraulic circuit, is included in the previous circuit. These valves are required components in every hydraulic system. To explain why, we must ask the following question: In the previous circuit, what happens to pump flow when the valve is shifted into the closed position? The pressure port is blocked, so the pump flow cannot go anywhere. This cannot be the case, however, because hydraulic pumps are usually of the positive displacement type, which output a constant amount of fluid for every revolution of the pump shaft. In fact, the pressure will begin to build until the pump finds some path for its flow. If a path is not provided, the pressure will continue to build to dangerous levels. Eventually, a component will either blow a seal or the electric motor that drives the pump will stall. In either case, severe damage to the system will result, not to mention danger to any bystanders. This situation also occurs if the DCV is held in the actuated position when the cylinder reaches the end of its stroke. Again, pump flow has no path to follow. Another possibility is that the system is simply overloaded beyond the pressure capability of the power unit.

The pressure relief valve provides an alternate path for the pump flow by allowing it to go back to the tank when a preset maximum pressure is reached. This maximum pressure is set to a slightly higher pressure than the anticipated

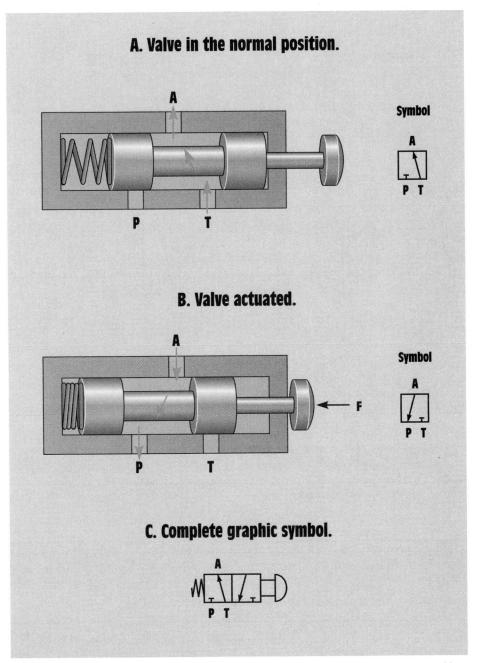

A. Valve in the normal position.

B. Valve actuated.

C. Complete graphic symbol.

FIGURE 6-12   Three-way, two-position normally open DCV. (A) Valve in the normal position. (B) Valve actuated. (C) Complete graphic symbol.

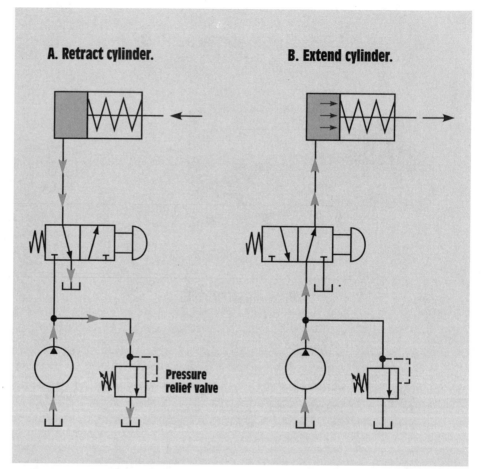

FIGURE 6-13 Three-way DCV controlling a single-acting cylinder. (A) Retract cylinder. (B) Extend cylinder.

operating pressure, but well below the maximum pressure rating of the components used in the system. The pump flow shown in Figure 6-13A is going through the pressure relief valve when the DCV is in the closed position. Pressure relief valves are discussed in great detail in Chapter 7.

Figure 6-14 shows a gravity return-type single-acting cylinder being controlled by a three-way DCV. A third valve position, called *neutral,* may be desirable for this application. This position, shown as the center position in the symbol, blocks all three ports. This holds the cylinder in a mid-stroke position because the hydraulic fluid, which is relatively incompressible, is trapped between the valve and the cylinder. Many cylinder applications require this feature. Figure 6-14 introduces another type of actuation: *manual lever* and *detent.* A detent is a mechanism that holds the valve in any position into which it is

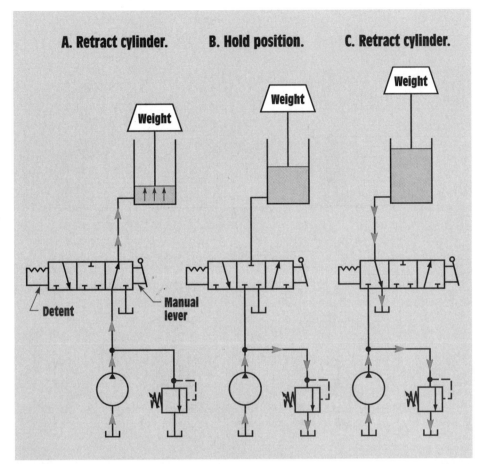

FIGURE 6-14   Three-way, three-position DCV controlling a single-acting cylinder. (A) Extend cylinder. (B) Hold position. (C) Retract cylinder.

shifted. Detented valves have no normal position because they will remain indefinitely in the last position indicated. When the valve is in the closed neutral position (Figure 6-14B) or the retract position (Figure 6-14C), pump flow goes over the pressure relief valve because the pressure port is blocked.

A three-way, three-position DCV may also be used to fill and drain a vessel (Figure 6-15). In this application, the closed neutral is required to hold the vessel at some constant fluid level.

In addition to the standard three-way DCV, there are other port configurations for a three-way valve. One type, known as a *diverter*, is shown in Figure 6-16A. Instead of having pressure inlet (*P*), outlet (*A*), and tank (*T*) ports, like the standard three-way, this valve has a pressure inlet (*P*) and two outlets (*A* and *B*). In one position, pump flow is sent to outlet *A* and outlet *B* is blocked. In the other position, pump flow is sent to outlet *B* and outlet *A* is blocked. One possible application for

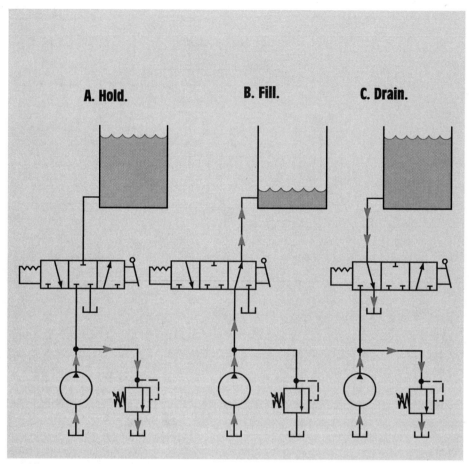

FIGURE 6-15   Three-way, three-position DCV filling & draining a vessel. (A) Hold. (B) Fill. (C) Drain.

this valve is allowing pump flow to be sent to one of two branch circuits. An example of this type of application is shown in Figure 6-16B. Here, pump flow can be diverted to one of two gravity return cylinder circuits. The major difference in the construction of the standard three-way versus the diverter is that the diverter valve is designed to withstand full pressure to all three ports. With the standard three-way, the tank port is usually rated for a lower pressure than the pressure and outlet ports.

6.6   Four-way Directional Control Valves

Four-way valves are the most commonly used directional control valves in hydraulic circuits because they are capable of controlling double-acting cylinders and bi-directional motors. Figure 6-17 illustrates the operation of a typical four-

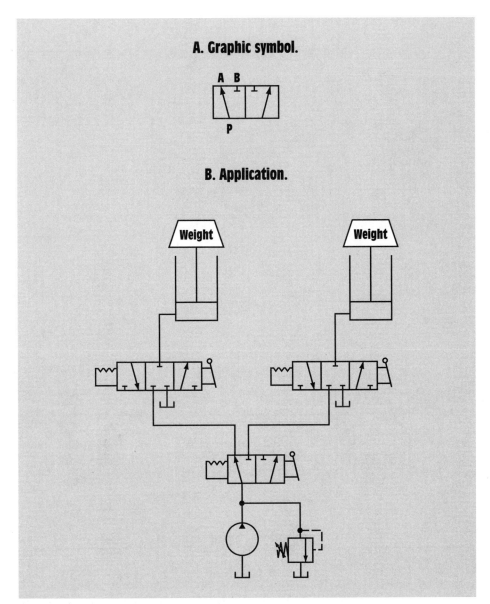

A. Graphic symbol.

B. Application.

FIGURE 6-16   Three-way diverter valve. (A) Graphic symbol. (B) Application.

way, two-position DCV. A four-way has four ports, usually labeled *P, T, A,* and *B. P* is the pressure inlet and *T* is the return to the tank. *A* and *B* are out-lets to the system. In the normal position, pump flow is sent to outlet *B* and outlet *A* is connected to the tank (Figure 6-17A). In the actuated position, pump flow is sent to port *A* and port *B* is connected to the tank (Figure 6-17B). Light-blue arrows are used to show pump flow; green arrows are used

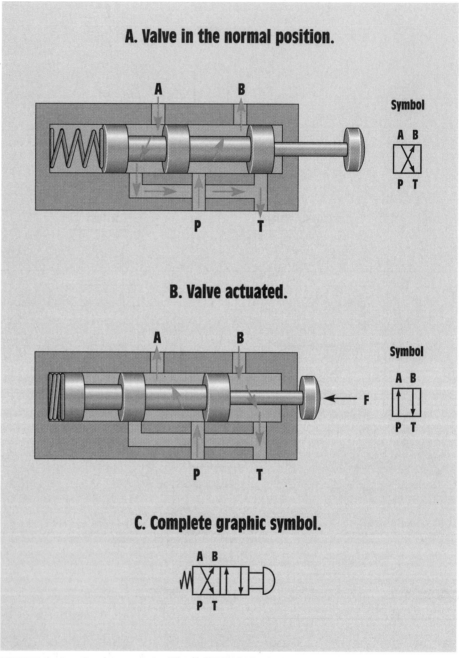

FIGURE 6-17   Four-way, two-position DCV. (A) Valve in the normal position. (B) Valve actuated. (C) Complete graphic symbol.

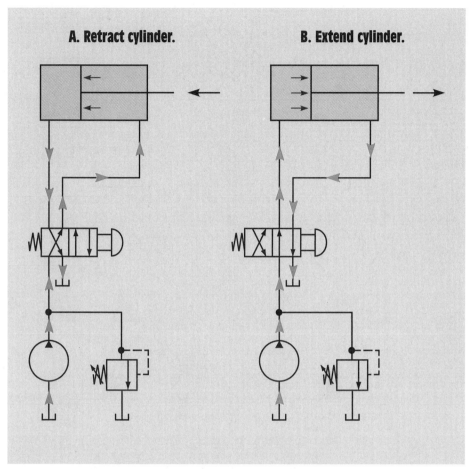

FIGURE 6-18   Four-way DCV controlling a double-acting cylinder. (A) Retract cylinder.
(B) Extend cylinder.

to show return flow to the tank. Four-way DCVs control two flows of fluid at the same time, while two-way and three-way DCVs control only one flow at a time. Figure 6-17C shows the complete graphic symbol for a *four-way, two-position* DCV.

The most common application for a four-way DCV is to control a double-acting cylinder (Figure 6-18). When the valve is in the normal position, the pump line is connected to the rod end of the cylinder and the blind end is connected to the tank (Figure 6-18A). The cylinder will therefore retract when the valve is in this position. When the cylinder is fully retracted, the pump flow will go over the pressure relief valve back to the tank, as it must whenever the pump flow cannot go to the system. In Figure 6-18B, the pump line is connected to the blind end of the cylinder and the rod end is connected to the

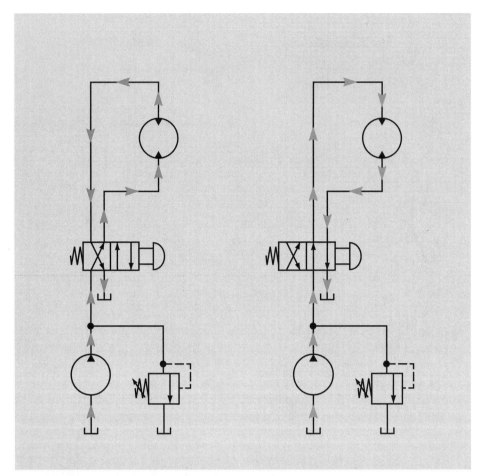

FIGURE 6-19   Four-way DCV controlling a bi-directional motor.

tank. This will cause the cylinder to extend. When the cylinder is fully extended, pump flow will again go over the pressure relief valve to the tank. A four-way is also often used to control bi-directional motors. Figure 6-19A shows schematic for this application. Unlike the cylinder, the motor rotates continuously and does not force fluid over the pressure relief valve (unless the motor is overloaded).

The four-way, two-position DCVs used in the previous two circuits are sometimes impractical because they continuously send pump flow and pressure to the actuator in one direction or the other. Many cylinder and motor applications require a third DCV position or *neutral,* in which the actuator is not subjected to pump pressure. *Four-way, three-position* DCVs are therefore used in many hydraulic circuits. Many types of neutrals are available; the most common

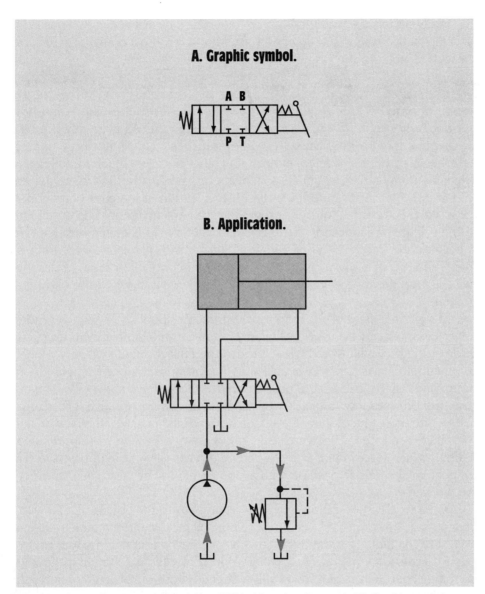

A. Graphic symbol.

B. Application.

FIGURE 6-20   Four-way, three-position DCV with a closed neutral. (A) Graphic symbol.
(B) Application.

are *closed, tandem, float, open,* and *regenerative.* Figure 6-20A shows the graphic symbol for a four-way, three position DCV with a closed neutral. Figure 6-20B shows it in a simple cylinder circuit. The valve shown here is *spring centered,* which means it will always return to the neutral position automatically when not actuated.

The behavior of any neutral can be understood by answering two questions: (1) What happens to pump flow? and (2) What is the effect on the actuator? For the closed neutral, the pump line is blocked, so the flow must go over the pressure relief valve. When the flow is going over the pressure relief valve, the pressure is at the system maximum. This is a wasteful thing to do for any period of time because it generates power in the form of pressure and flow, but does not use it. Where does all of this energy go? As is usually the case, wasted energy in a physical system goes into heat. This is not only undesirable because of the wasted energy, but also because the hydraulic, fluid becomes thinner (less viscous) as it heats up. When the fluid becomes too thin, it does not lubricate effectively. This will result in increased component wear. What is the effect of the closed neutral on the actuator? The outlet lines to the cylinder are blocked, so the cylinder will be held firmly in position. This is because the lines are full of hydraulic fluid, which is relatively incompressible. This type of neutral could also be used to control a motor. Just as with the cylinder, the motor would be held fixed in position when the valve is in neutral.

The graphic symbol for a four-way, three-position DCV with a tandem neutral is shown in Figure 6-21A. Part B shows it in a simple cylinder circuit. The pump flow is allowed to flow back to the tank through the DCV when it is in neutral. This is a very desirable situation because the only pressure in the pump line will be due to the flow resistance of the lines and DCV. This will keep the pressure low (typically 100 to 200 psi) when the valve is in neutral. In this situation, the system is said to be *unloaded* because the power consumption is reduced. This wastes much less energy than does a closed center neutral, which forces fluid over the pressure relief valve at high pressure. The cylinder will be held in position with a tandem neutral because the outlet ports are blocked.

Figure 6-22A shows the symbol for a four-way, three-position DCV with a float neutral. In this neutral, the pressure port is blocked and the outlets are connected to the tank. Figure 6-22B shows a four-way with a float neutral controlling a bi-directional motor. The pressure port is blocked, so the pump flow will be forced over the pressure relief valve. Because both outlets are connected to the tank, the motor will *float,* or spin freely, when the DCV is in neutral. This type of four-way is frequently used in motor circuits because it allows the motor to spin to a stop when the valve is shifted to neutral. This is often preferable to shifting to a closed position because motors often build up a great deal of momentum. Shifting the valve closed in this situation will cause a large pressure spike on the outlet line because the motor tends to keep spinning and will try to push the fluid to its outlet. This is known as *shifting shock.* This issue will be discussed further in the chapter on pressure control valves. Float neutrals may also be desirable for cylinder circuits in some applications.

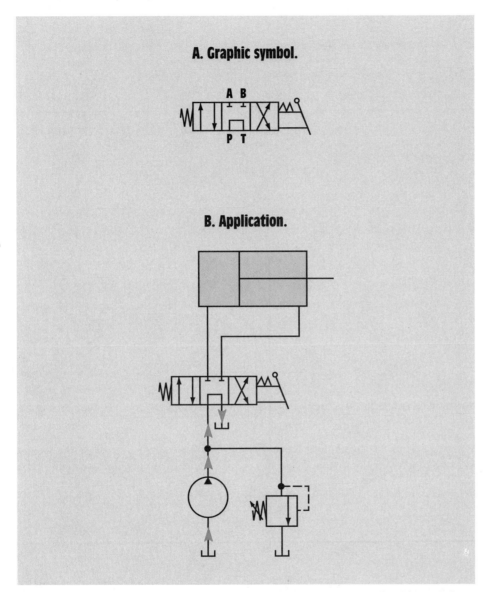

FIGURE 6-21   Four-way, three-position DCV with a tandem neutral. (A) Graphic symbol. (B) Application.

Figure 6-23A shows the symbol for a four-way, three-position DCV with an open neutral. In this neutral, the pressure port and the outlets are both connected to the tank. Figure 6-23B shows a four-way with an open neutral controlling a cylinder. Flow always follows the path of least resistance, so pump flow will go back to the tank. Because the outlets are also connected

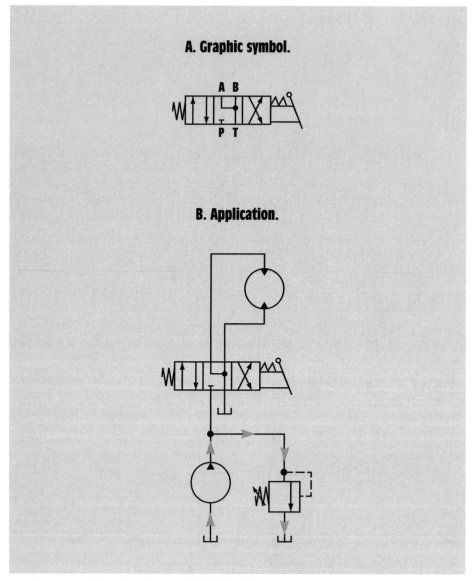

A. Graphic symbol.

B. Application.

FIGURE 6-22   Four-way, three-position DCV with a float neutral. (A) Graphic symbol. (B) Application.

to the tank, the cylinder will float when this valve is in neutral. This is desirable in a circuit in which some external force must position the cylinder when in neutral.

A *regenerative* neutral is considerably different in its function than the other types just discussed. "Regenerative" is the general term used to describe

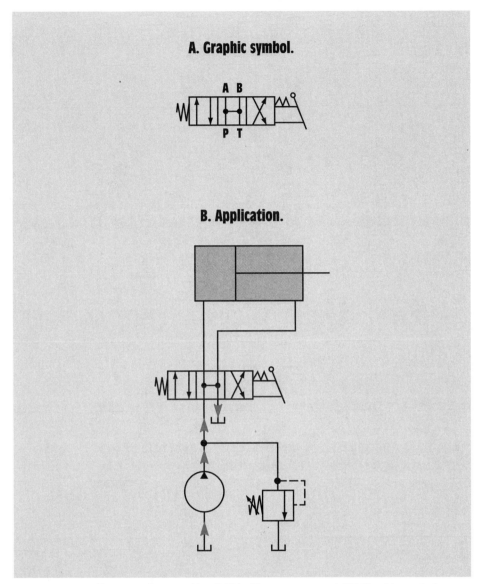

A. Graphic symbol.

B. Application.

FIGURE 6-23   Four-way, three-position DCV with a open neutral. (A) Graphic symbol.
(B) Application.

a system in which the waste is fed back into the system to supplement the input power. For example, a turbocharger takes the exhaust gas from an automobile engine and uses its energy to turn a compressor. The compressor then feeds pressurized air into the engine, giving it a boost in power. Figure 6-24A shows the symbol for a regenerative neutral. In this neutral, the pressure port

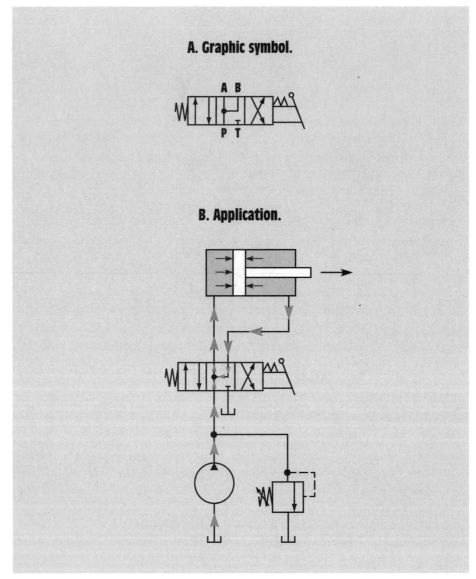

FIGURE 6-24 Four-way, three-position DCV with a regenerative neutral. (A) Graphic symbol. (B) Application.

is connected to *both* outlets and the tank port is blocked. Figure 6-24B shows a four-way with a regenerative neutral controlling a cylinder. When this valve is shifted into neutral, pump pressure is applied to both sides of the piston. Because the piston area on the rod side of the cylinder is smaller than on the blind side, there will be a net force applied to extend the piston rod (move the piston to the right in this example). As the piston moves forward, it will force the outlet flow from the rod side back to the valve, where it combines with the pump flow and goes to the blind end of the cylinder. This will cause a considerable increase in cylinder speed. This is the purpose of a regenerative neutral—instead of sending the return flow back to the tank, it sends it to the inlet side of the cylinder, thereby increasing its speed.

Chapter 4 explained how to calculate the speed of a cylinder with the following equations:

$$v_E = \frac{231 \cdot Q}{A_P} \qquad \text{U.S. customary units}$$

$$v_E = \frac{Q}{1000 \cdot A_P} \qquad \text{metric units}$$

where: Q = the flow rate from the pump
A_P = the area of the piston
v_E = the extended velocity of the cylinder

In U.S. customary units, the following units must be used: Q in gpm, A_P in in², and v_E in in/min. In metric units, the following units must be used: Q in lpm, A_P in m² and v_E in m/min.

When the return flow is being regenerated, the flow is the pump flow (Q) *plus* the flow coming from the rod end (Q_R). When we are extending in regenerative mode, the velocity is therefore given by the following equations:

$$v_{REGEN} = \frac{231 \cdot (Q + Q_R)}{A_P} \qquad \text{(6-1)}$$

$$v_{REGEN} = \frac{(Q + Q_R)}{1000 \cdot A_P} \qquad \text{(6-1M)}$$

The flow from the rod end is given by Equations 6-2 and 6-2M:

$$Q_R = \frac{v_{REGEN} \cdot (A_P - A_R)}{231} \qquad \text{(6-2)}$$

$$Q_R = 1000 \cdot v_{REGEN} \cdot (A_P - A_R) \qquad \text{(6-2M)}$$

where: A_R = the area of the rod

Substituting $\dfrac{v_{REGEN} \cdot (A_P - A_R)}{231}$ for Q_R in Equation 6-1 and solving for v_{REGEN} results in the following equation, which allows us to calculate the speed of a cylinder when in regenerative mode:

$$V_{REGEN} = \frac{231 \cdot Q}{A_R} \tag{6-3}$$

Similarly, we can substitute $1000 \cdot v_{REGEN} \cdot (A_P - A_R)$ for Q_R in Equation 6-1M and solve for v_{REGEN} to obtain an equation that allows us to calculate the regeneration speed in metric units:

$$v_{REGEN} = \frac{Q}{1000 \cdot A_R} \tag{6-3M}$$

We may also want to calculate the force available when the system is in regenerative mode. As stated previously, pressure is applied to both sides of the piston. This means that the net force will be the force due to the pressure on the blind end of the piston minus the force due to the pressure on the rod end, as given by the following equation:

$$F_{REGEN} = p \cdot A_P - p \cdot (A_P - A_R) \tag{6-4}$$

simplifying results in the following equation that can be used to calculate the force generated by a cylinder when regenerating:

$$F_{REGEN} = p \cdot A_R \tag{6-5}$$

EXAMPLE 6-1.

A cylinder with a bore diameter of 2.5 in and a rod diameter of 1.12 in is to be used in a system with a 10 gpm pump. What are the extension speeds when regenerating and when not regenerating?

SOLUTION:

1. Calculate the piston area:

$$A_P = \frac{\pi \cdot D_P^{\,2}}{4} = \frac{3.142 \cdot (2.5 \text{ in})^2}{4} = 4.909 \text{ in}^2$$

2. Calculate the rod area:

$$A_R = \frac{\pi \cdot D_R^{\,2}}{4} = \frac{3.142 \cdot (1.12 \text{ in})^2}{4} = 0.9853 \text{ in}^2$$

3. Calculate the normal extension speed:

$$v_E = \frac{231 \cdot Q}{A_P} = \frac{231 \frac{\text{in}^3}{\text{gal}} \cdot \left(10 \frac{\text{gal}}{\text{min}}\right)}{4.909 \text{ in}^2} = 471 \frac{\text{in}}{\text{min}}$$

4. Calculate the regenerative extension speed:

$$v_{REGEN} = \frac{231 \cdot Q}{A_R} = \frac{231 \frac{\text{in}^3}{\text{gal}} \cdot \left(10 \frac{\text{gal}}{\text{min}}\right)}{0.9853 \text{ in}^2} = 2344 \frac{\text{in}}{\text{min}}$$

EXAMPLE 6-1M.

A cylinder with a bore diameter of 63 mm and a rod diameter of 25 mm is to be used in a system with a 45 lpm pump. What are the extension speeds when regenerating and when not regenerating?

SOLUTION:

1. Calculate the piston area:

$$A_P = \frac{\pi \cdot D_P^{\,2}}{4} = \frac{3.142 \cdot (0.063 \text{ m})^2}{4} = 0.003118$$

2. Calculate the rod area:

$$A_R = \frac{\pi \cdot D_R^{\,2}}{4} = \frac{3.142 \cdot (0.025 \text{ m})^2}{4} = 0.0004909 \text{ m}^2$$

3. Calculate the normal extension speed:

$$v_E = \frac{Q}{1000 \cdot A_P} = \frac{\left(45 \frac{1}{\text{min}}\right)}{1000 \frac{1}{\text{m}^3} \cdot (0.003118 \text{ m}^2)} = 14.43 \frac{\text{m}}{\text{min}}$$

4. Calculate the regenerative extension speed:

$$v_{REGEN} = \frac{Q}{1000 \cdot A_R} = \frac{\left(45\,\frac{1}{min}\right)}{1000\,\frac{1}{m^3} \cdot (0.0004909\ m^2)} = 91.67\,\frac{m}{min}$$

The previous examples show that using the principle of regeneration to extend a cylinder results in a significant speed increase. The amount of speed increase is equal to the ratio of the piston area to the rod area (A_p/A_R). In Example 6-1, this ratio is 4.909/0.9853 = 4.98, resulting in a speed increase by the same factor. The next example shows that force capability is sacrificed to obtain this speed increase.

EXAMPLE 6-2.

The cylinder in Example 6-1 is used in a system with a maximum pressure of 2500 psi. What are the maximum extension forces when regenerating and when not regenerating?

SOLUTION:

1. Calculate the normal extension force:

$$F_E = p \cdot A_p = 2500\,\frac{lbs}{in^2} \cdot (4.909\ in^2) = 12,273\ lbs$$

2. Calculate the regenerative extension force:

$$F_{REGEN} = p \cdot A_R = 2500\,\frac{lbs}{in^2} \cdot (0.9853\ in^2) = 2463\ lbs$$

Regenerative neutral valves are frequently used in circuits in which the cylinder must be moved in position quickly during some portion of the machine's cycle. Regeneration is discussed further in the section on circuits later in this chapter.

Table 6-1 summarizes the five types of neutrals just discussed. As stated previously, any neutral can be understood by answering two questions: (1) What will happen to pump flow and (2) What is the effect on the actuator?

6.7   Directional Control Valve Actuation

Many other methods are used to shift a valve in addition to manual lever and push-button types of actuation. The most commonly used actuation types are

NEUTRAL TYPE	EFFECT ON THE ACTUATOR	WHERE DOES THE PUMP FLOW GO?
Closed	Holds position	Through pressure relief valve to tank
Tandem	Holds position	Through DCV to tank
Float	Floats	Through pressure relief valve to tank
Open	Floats	Through DCV to tank
Regenerative	Extends quickly with less force capability	To cylinder

TABLE 6-1   Five Types of Neutrals

shown in Figure 6-25. All are shown controlling a spring return four-way, two-position valve, but they could be used on any of the valves discussed in this chapter. Figure 6-25A and B show manual lever and push-button actuation. Manual lever is a popular method of actuation for DCVs used in mobile equipment applications such as back hoes, bulldozers, dump trucks, farm equipment, and garbage trucks, etc. Push-button actuation is more prevalent in industrial applications. Figure 6-25C shows the symbol for foot pedal actuation, which could be used in an application in which hands-free shifting of the DCV is required. Figure 6-25D shows the symbol for a mechanically actuated, or *cam* actuated, valve. These valves shift when depressed by some mechanical component of the machine (for example, a cylinder may hit a cam valve at the end of its stroke to cause the valve to shift).

Figure 6-25E shows the symbol for a pilot-operated valve. These valves are shifted with system pressure. As stated earlier, pilot-operated check valves use system pressure to hold a check valve open or closed when pressure is applied to the pilot line. Recall that a pilot line is the term used for any line that is used for control purposes. An example of an application for this valve will be shown in the next section.

Figure 6-25F shows the symbol for a solenoid-actuated DCV. These valves are shifted using electrical current, which induces a magnetic force that shifts the valve spool. Solenoid valves are widely used in industrial applications on electronically controlled machinery. These and other electronically controlled valves will be discussed in detail in Chapter 13.

Figure 6-25G shows the symbol for a pilot-operated solenoid valve. These valves are essentially two valves in one package. The solenoid is used to actuate a small pilot DCV, which in turn uses the pressure of the system to shift the main valve. This method of actuation is necessary on large valves that operate in systems at high pressures. They are necessary because the solenoid alone cannot generate enough force to shift a large valve against a high pressure. The solenoid can, however, generate enough force to shift the small pilot valve, which can then use the pressure of the system to shift the main valve. Figure 6-26

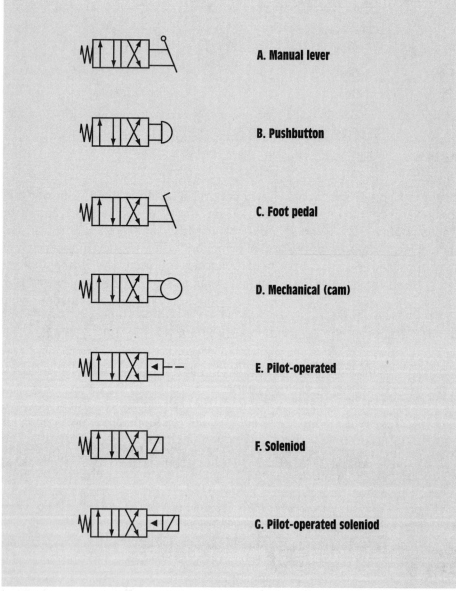

FIGURE 6-25 DCV actuation types. (A) Manual lever. (B) Push-button. (C) Foot pedal. (D) Mechanical (cam). (E) Pilot-operated. (F) Solenoid. (G) Pilot-operated solenoid.

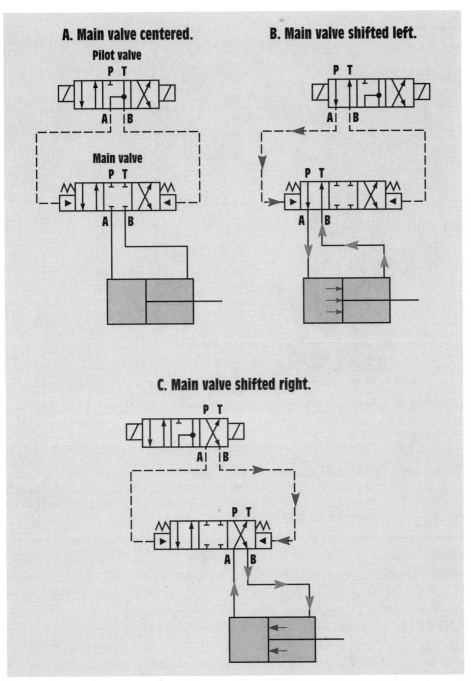

FIGURE 6-26   Complete symbol for a pilot-operated solenoid valve. (A) Main valve centered. (B) Main valve shifted left. (C) Main valve shifted right.

FIGURE 6-27   (A) Manual lever valve. (Photo courtesy of Continental Hydraulics.) (B) Cam valve. (Photo courtesy of Continental Hydraulics.) (C) Solenoid valve. (Photo courtesy of Continental Hydraulics.) (D) Pilot-operated solenoid valve. (Photo courtesy of Continental Hydraulics.)

shows a complete graphic symbol for a four-way, three-position, pilot-operated solenoid valve. The pump and tank lines have been omitted to simplify the schematic. When neither solenoid on the pilot valve is actuated, both sides of the main valve are vented to the tank and the spring pulls it into the neutral position (Figure 6-26A). When the solenoid on the left side is energized, it shifts the pilot valve and pressure is sent to the left side of main valve, causing it to shift (Figure 6-26B). Energizing the left solenoid causes the main valve to be shifted to the opposite position (Figure 6-26C). The pilot lines are only pressurized and do not receive a significant amount of flow. The complete symbol shown in Figure 6-26 is rarely used in hydraulic schematics. The simplified symbol shown in Figure 6-25G is preferred.

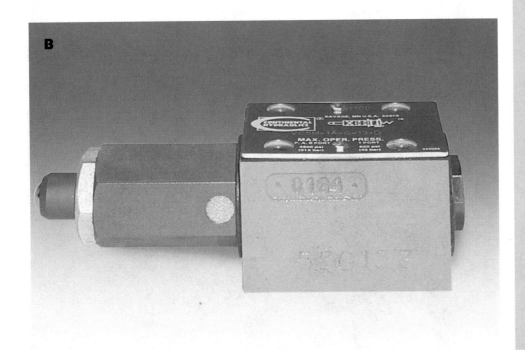

FIGURE 6-27 Continued

Figure 6-27 shows photos of some actual directional control valves with various actuation types. Figure 6-27A shows a manual lever valve, Figure 6-27B shows a cam valve, Figure 6-27C shows a solenoid valve, and Figure 6-27D shows a pilot-operated solenoid valve.

As discussed earlier, some valves are spring returned to a *normal* position and some are *detented,* which means they click into position and hold that posi-

FIGURE 6-27   Continued

tion indefinitely. Detented valves have no normal position. Figure 6-28 shows some common configurations for spring- and detent-positioned valves.

6.8   Circuits

This section examines some simple circuits that are commonly used in industry. This will help the reader to develop the ability to read hydraulic schematics and to understand the operation of basic circuits.

Figure 6-29 shows a circuit in which a cylinder is used to raise and lower a large weight from above. The cylinder is controlled by a four-way DCV with a tandem neutral. In Figure 6-29A, the DCV is in neutral, therefore the pump flow will be unloaded to the tank at low pressure. The cylinder *should* hold position

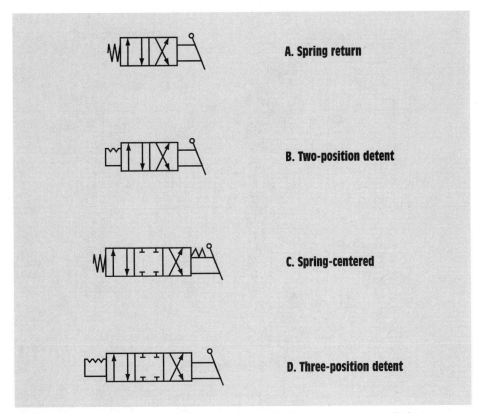

A. Spring return

B. Two-position detent

C. Spring-centered

D. Three-position detent

FIGURE 6-28 DCV positioning. (A) Spring return. (B) Two-position detent. (C) Spring centered. (D) Three-position detent.

because the outlet ports from the DCV that connect to the cylinder are blocked. It will not hold position, however, if the cylinder is in the orientation shown because the weight is pulling the cylinder down, causing pressure in the rod end line. The pressure causes a small amount of leakage within the DCV, and the cylinder will begin to creep downward. This can be remedied by placing a pilot-to-open check valve in the rod end line, as shown. The pilot-to-open check valve will not allow flow out of the rod end of the cylinder unless pressure is applied to the pilot line, thereby preventing cylinder creep. The check is acting to *counterbalance* the weight. When the DCV is shifted to extend the cylinder (lowering the weight), pump pressure from the blind end line holds open the check and allows flow to return to the tank from the blind end (Figure 6-29B). When the cylinder is retracted (the weight is raised), flow from the pump goes through the check to the rod end (Figure 6-29C). The check has no effect in this direction.

Figure 6-30 shows a circuit that utilizes a shuttle valve. This circuit allows either of two three-way buttons to operate a single-acting cylinder. Figure 6-30A shows both three-ways in their normal position. The cylinder is vented to the

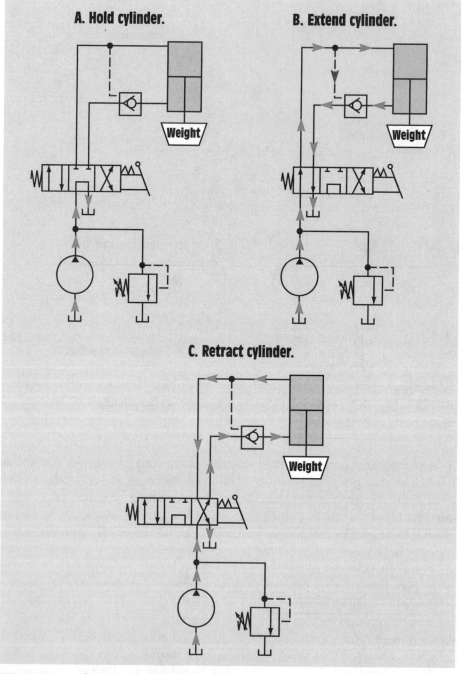

FIGURE 6-29 Counterbalance circuit. (A) Hold cylinder. (B) Extend cylinder. (C) Retract cylinder.

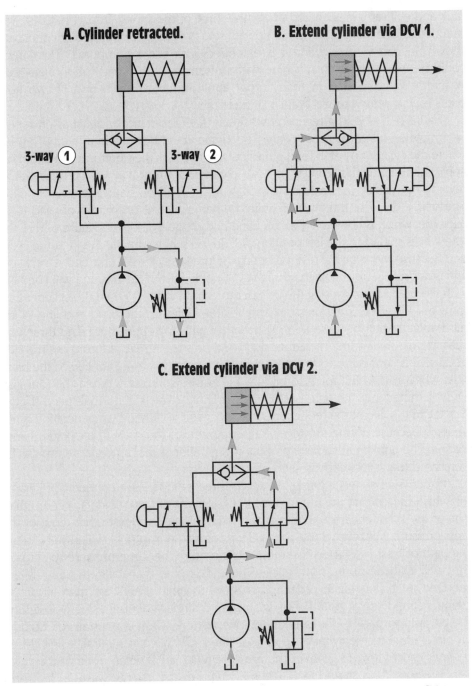

FIGURE 6-30 Shuttle valve circuit. (A) Cylinder retracted. (B) Extend cylinder via DCV 1. (C) Extend cylinder via DCV 2.

tank and will remain retracted under the force of the spring. In part B, three-way number 1 is shifted and pump flow is sent to the cylinder through the path shown. In part C, the cylinder is extended with valve number 2. This circuit could be used on a long machine with buttons on either end for convenience. A shuttle valve is used in many other applications in which one of two flow paths may supply a single branch of a circuit.

Figure 6-31 shows a regenerative circuit that automatically switches off regeneration when full force is necessary. This circuit could be used in a hydraulic press where the cylinder must extend quickly under no load, then bottoms out and must apply full force to the workpiece. Instead of using a four-way with a regenerative neutral, this circuit uses a four-way DCV in conjunction with a three-way, pilot-operated DCV. The three-way is shifted when sufficient pressure is applied to its pilot line, which is connected to the blind end of the cylinder. In Figure 6-31A, the four-way is shifted to the left position and flow is sent to the blind end of the cylinder. Because the cylinder is not loaded, the pressure in the blind end is very low and is not sufficient to shift the three-way. The flow from the rod end will combine with the pump flow, causing the cylinder to extend rapidly. When the cylinder bottoms out (Figure 6-31B), pressure immediately builds in the blind end line to the relief valve setting because there is no other path for pump flow. The three-way valve is then shifted into the left position and pressure is relieved from the rod side because it is connected to the tank port. Pressure is then only applied to the blind side, which causes full force to be applied to the workpiece. When the four-way DCV is shifted into the right position, the cylinder retracts at normal speed (Figure 6-31C). In this circuit the reduction in force capability caused by regeneration is not an issue because during the regeneration portion of the cycle, the cylinder is not loaded. The primary advantage of using regeneration is that a smaller pump can be purchased that is less expensive to buy and operate.

Figure 6-32 shows a circuit in which three cylinders are connected in parallel, which means that the pump flow can go to any of the cylinders. If more than one of the cylinders are actuated at the same time, the cylinder that requires the least pressure will receive the pump flow because pump flow will always follow the path of least resistance. If the cylinders require the same pressure to actuate, they will split the pump flow. Closed neutrals, rather than tandem neutrals, are required in this circuit in order for pressure to build when less than all of the cylinders are being actuated. This is because a tandem neutral valve would send the pump flow back to the tank under low pressure when in neutral, and no flow would ever go to the cylinders if at least one DCV were in neutral. Using closed neutrals causes flow to go over the pressure relief valve when none of the cylinders is actuated, as shown in Figure 6-32. This causes energy waste in the form of heat, which can cause increased component wear because the fluid becomes thinner and does not lubricate as well. Chapter 7 discusses ways to reduce the power consumption of circuits when full flow and pressure are not required.

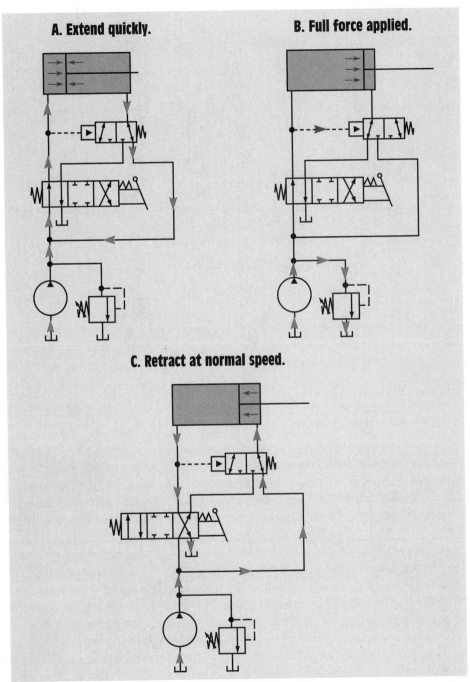

FIGURE 6-31   Regenerative circuit. (A) Extend quickly. (B) Full force applied. (C) Retract at normal speed.

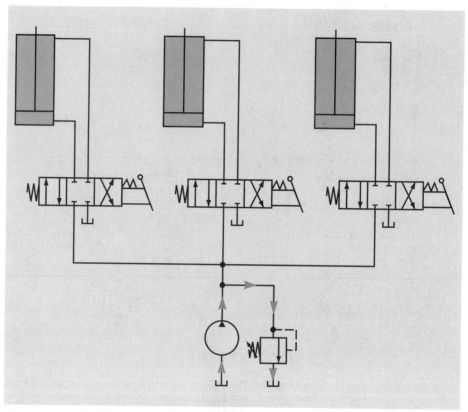

FIGURE 6-32   Parallel circuit.

Many mobile applications, such as hydraulic circuits used in excavating equipment, utilize a *multispool* directional control valve to create a parallel circuit. multispool DCVs are multiple four-way DCVs contained in one housing. A three-spool version of a multispool valve is shown operating three cylinders in Figure 6-33. Each valve has two extra ports, so flow passes through the neutral of each valve and goes back to the tank when all of the valves are in neutral. The vertical line that runs along the left in Figure 6-33 is the pressure line; the vertical line that runs along the right is the tank line. This figure introduces a new symbol, a *jump,* which is used to show that two lines are crossing on the schematic, but are not connected. Whenever two lines are connected, it should be indicated with a *junction,* which is also labeled on the schematic.

Figure 6-34 shows the flow when one of the cylinders is actuated. The path from the pump to the tank through the DCVs is blocked and flow is forced to go to the blind end of the cylinder. Flow from the rod end of the cylinder goes back to the tank. Figure 6-35 shows the flow when two of the cylinders are actuated. Because the cylinders are drawing from the same inlet, they are connected in

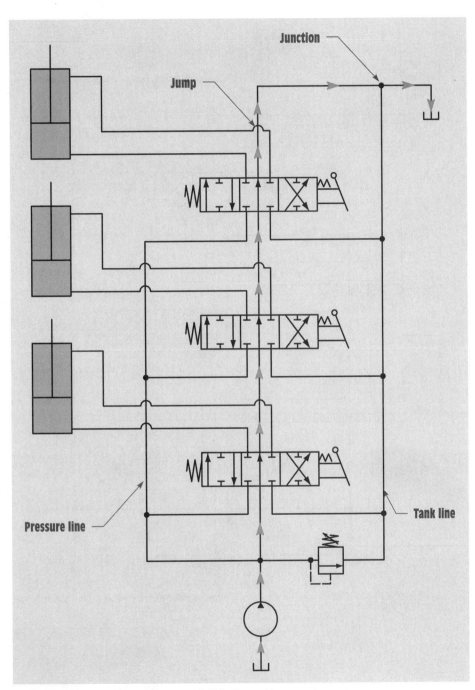

FIGURE 6-33   Multispool, four-way DCV, all spools centered.

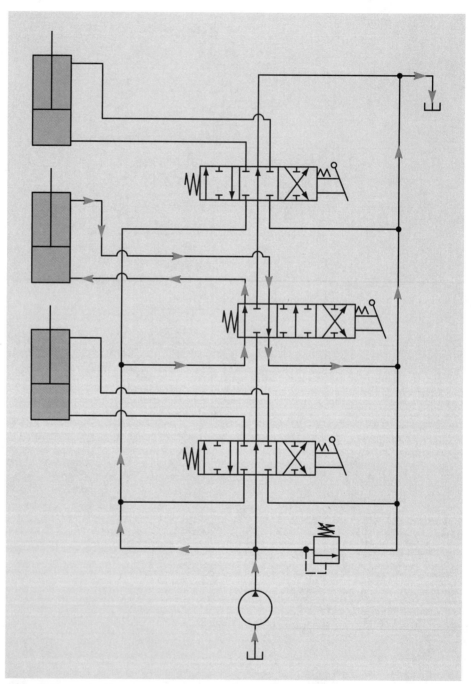

FIGURE 6-34   Multispool, four-way DCV, one cylinder extending.

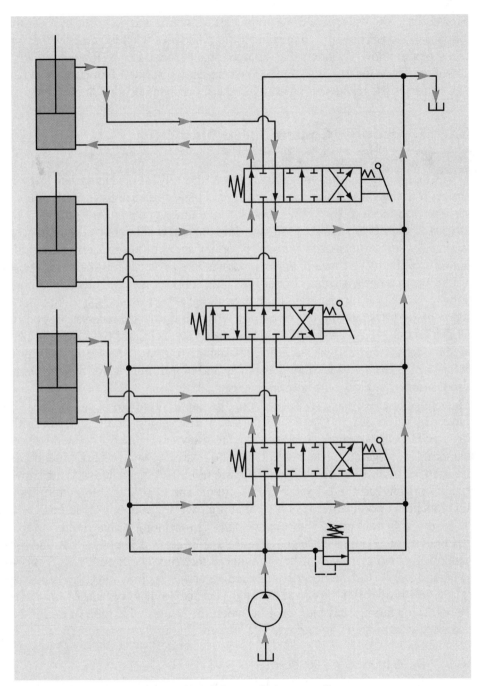

FIGURE 6-35   Multispool, four-way DCV, two cylinders extending.

parallel. The flow will follow the path of least resistance and go to the cylinder that requires the lower pressure to operate. If the cylinders are loaded equally, they will split the pump flow. Cylinders that are not equally loaded can be forced to split the flow if the DCV for the more lightly loaded cylinder is *throttled*, or partially closed. This increases the resistance (pressure) in the more lightly loaded line.

6.9   Directional Control Valve Mounting

Directional control valves can be mounted in two ways: *inline* or *subplate*. Inline means that there are threaded connections in the valve itself. Fittings are screwed directly into the valve. This method has several major disadvantages. Each time the valve is disconnected there is the possibility of damaging the valve by stripping the threads. The threads will also wear each time the unit is disconnected, causing contamination and an increased probability of leakage. In the subplate method, the bottoms of the valves have *unthreaded* connections. The valve is then attached to a subplate that has matching connections. The subplate has the threaded connections to which the fittings are attached. Sealing at the valve/subplate interface is accomplished through the use of o-rings, which fit into small recesses around the DCV ports. A typical valve and subplate are shown in Figure 6-36. The subplate method results in less leakage, less contamination, and a smaller probability of doing damage to the valve during assembly and disassembly. Valve replacement is also a simpler and less time-consuming task.

Subplates are usually available from the same distributor that supplies the valve. The valves and subplates are available with several standard patterns for the valve ports. Four commonly used patterns are shown in Figure 6-37. The fluid ports are shown in magenta, while the mounting holes are shown in blue. D03 and D05 are used for smaller valves. D07 and D08 are used for larger valves and are shown half-scale relative to the D03 and D05 patterns. The D07 and D08 patterns have additional fluid ports labeled X and Y that are used for pilot and drain lines.

Another advantage to the subplate style of mounting is that multiple valves can be mounted on a *manifold*. A manifold is basically a subplate that has connections for two or more valves. A manifold that mounts three DCVs is shown in Figure 6-38. This method can be used to create an integrated hydraulic circuit in which many of the connections are inside the manifold itself, eliminating the need for fittings and plumbing between the valves. This type of design has several key advantages, including:

1. A more compact design,
2. Actuator response time is decreased because flow paths are shortened,
3. Reduced leakage because there are fewer threaded connections,
4. Reduced assembly time,
5. Easy replacement of valves, and

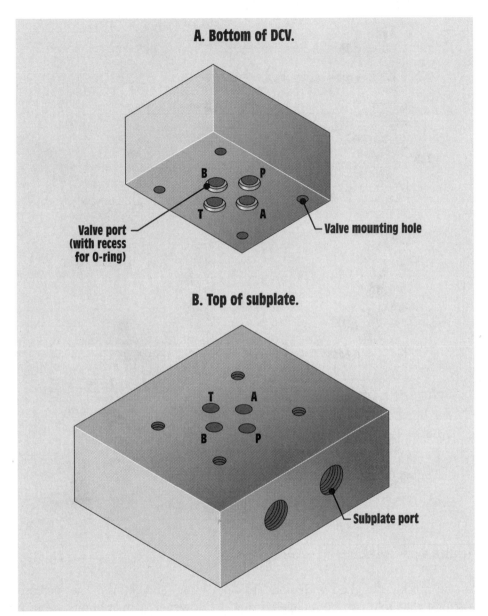

FIGURE 6-36   Valve subplate. (A) Bottom of DCV. (B) Top of subplate.

6. Less chance for contamination because there are fewer threaded connections.

The only drawback is that a manifold requires much more design and testing time, and consequently a greater initial expense. A system with inline valves

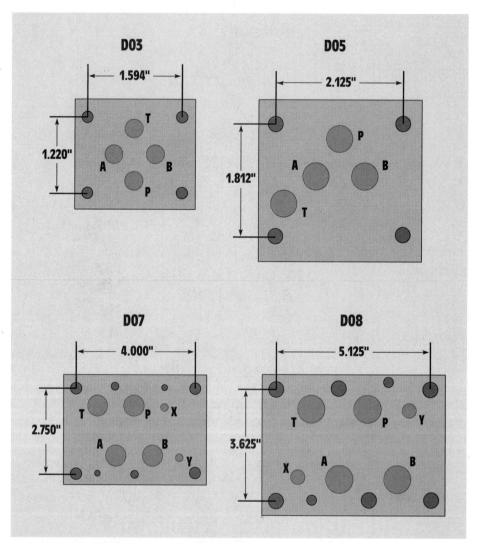

FIGURE 6-37 NFPA subplate mounting standards.

can be quickly designed with standard hose, tube, and fittings. The increased cost of a manifold can be recovered quickly, however, if many units are to be produced. Numerous companies specialize in hydraulic manifold manufacture, so outsourcing can also reduce cost.

In addition to the spool-type subplate mounted valves, *cartridge*-type valves are also commonly used in conjunction with manifolds. Cartridge valves are a very compact alternative to the spool-type design. They screw directly into a cavity in a manifold and therefore do not require separate valve port and mounting holes. Figure 6-39 shows a photo of some cartridge valves and manifolds. The six advantages of manifold circuits are further magnified when cartridge

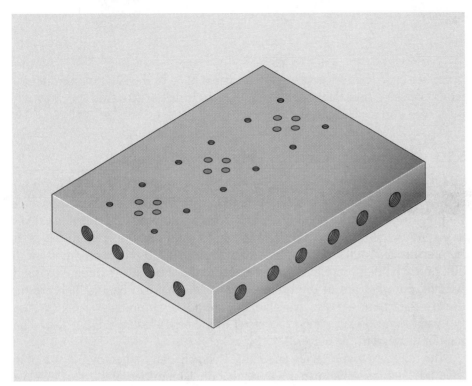

FIGURE 6-38 Valve manifold.

FIGURE 6-39 Cartridge valves and manifolds.
(Courtesy of Sun Hydraulics Corporation, Sarasota, Florida.)

valves are used. Cartridge valves are available in all of the DCV types discussed earlier. Pressure and flow control valves, which will be covered in Chapters 7 and 8, are also available as cartridge valves.

6.10 Directional Control Valve Specifications

Perhaps the most critical specifications when selecting a DCV are its maximum pressure and flow ratings. It is common for a high-pressure rating to be given for the pressure and outlet ports (*P*, *A*, and *B* on a four-way) and a lower value to be given for the tank port (*T*). Ratings of 3000 psi to 5000 psi are typical for the pressure and outlet ports; ratings of 500 psi to 1000 psi are common for the tank port. The different ratings occur because the spool seal is often exposed to the tank port, but not to the other ports. Valves that can handle high pressure to all of their ports are also available, but are not as common. Operating above the maximum pressure rating can result in increased leakage and possibly permanent damage to the valve.

The flow rate that a valve can handle is largely determined by the size of the valve itself. Larger valves can obviously handle larger flow rates, but are heavier and more expensive. Standard valves are available with flow ratings from 10 gpm to 250 gpm. Operating a DCV above its maximum flow rating will most likely result in a large pressure drop across the valve. This lost energy is converted to heat and is not only wasteful, but will also lead to increased component wear as the oil becomes thinner and does not lubricate as well. Operating above the maximum flow rating may also result in permanent damage to the valve itself.

When selecting a DCV for an application, we may also want to know what the pressure drop will be across the valve at a particular flow rate. Manufacturers typically provide graphs that relate pressure drop to flow rate through the valve for each model. A typical pressure drop-versus-flow graph for a tandem center DCV is shown in Figure 6-40. Separate curves are given for the different port-to-port connections. These curves represent data for a particular fluid and viscosity, most commonly standard hydraulic oil at 100 SSU. Manufacturers often give a correction factor for fluids at other viscosities. Fluids with a higher viscosity will have a higher pressure drop at a given flow rate because a thicker fluid is more difficult to move through the valve.

EXAMPLE 6-3.

A cylinder with a bore diameter of 2.75 in and a rod diameter of 1.25 in is to be used in a system with a 12 gpm pump. Use the graph in Figure 6-40 to determine the pressure drops across the DCV when the cylinder is retracting ($P \rightarrow B$, $A \rightarrow T$).

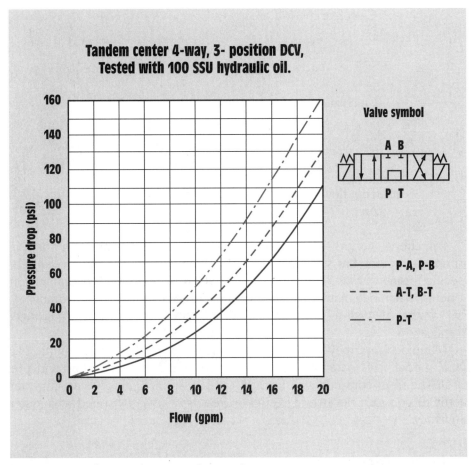

FIGURE 6-40 Pressure drop-versus-flow graph.

SOLUTION:

1. The flow from $P \rightarrow B$ is the pump flow into the rod end, so this can be read directly from the graph: $\Delta p \approx 38$ psi.
2. The flow from $A \rightarrow T$ is the return flow out of the blind end. This flow rate is greater than the pump flow and must be determined using Equation 4-9:
    a. Calculate the piston area:

$$A_P = \frac{\pi \cdot D_P^{\,2}}{4} = \frac{3.142 \cdot (2.75 \text{ in})^2}{4} = 5.940 \text{ in}^2$$

b. Calculate the rod area:

$$A_R = \frac{\pi \cdot D_R^{\,2}}{4} = \frac{3.142 \cdot (1.25 \text{ in})^2}{4} = 1.227 \text{ in}^2$$

c. Calculate the return flow:

$$Q_{RET,R} = \frac{Q_{PUMP} \cdot A_P}{A_P - A_R} = \frac{12 \text{ gpm} \cdot (5.940 \text{ in}^2)}{5.940 \text{ in}^2 - 1.227 \text{ in}^2} = 15.12 \text{ gpm}$$

d. The flow from $A \rightarrow T$ can now be read from the graph: $\Delta p \approx 77$ psi.

Directional control valves are less sensitive to contamination than are pumps or motors, but still require fine filtration. Required filtration levels are specified using the ISO cleanliness code, which is used for all hydraulic components. When deciding on an appropriate level of filtration for a hydraulic system, always be sure to consider the component that has the most stringent specification.

Manufacturers may also provide information on leakage rates for each DCV model. This data will be given for specific conditions (flow and pressure). The most common unit for leakage is in^3/min. Leakage is an important factor to consider because it results in reduced actuator speed and system efficiency.

6.11 Equations

EQUATION NUMBER	EQUATION	REQUIRED UNITS
6-3	$v_{REGEN} = \dfrac{231 \cdot Q}{A_R}$	Q in gpm, A_R in in^2, v_{REGEN} in in/min
6-3M	$v_{REGEN} = \dfrac{Q}{1000 \cdot A_R}$	Q in lpm, A_R in m^2, v_{REGEN} in m/min
6-5	$F_{REGEN} = p \cdot A_R$	Any consistent units

6.12 Review Questions and Problems

1. What are the three types of control valves used in hydraulic systems?
2. What is the purpose of a check valve? Describe its construction.
3. What is the purpose of putting a component in parallel with a check valve?
4. Describe the function of a pilot-to-open check valve.
5. Describe the function of a pilot-to-close check valve.
6. Describe the function of a shuttle valve. Give an example of an application for this valve.
7. Draw the graphic symbol and label the ports for a two-way, two-position DCV. What happens to the flow in each position?
8. Draw the graphic symbol and label the ports for a three-way, two-position DCV. What happens to the flow in each position?
9. Draw the graphic symbol and label the ports for a four-way, two-position DCV. Describe the path of the pump flow and return flow in each position.
10. Draw a four-way valve with the following neutrals: closed, tandem, open, float, and regenerative. Describe what happens to pump flow and the effect on the actuator for each.
11. Describe the difference between a spring return and a detented DCV.
12. Draw the graphic symbol and label the ports for the following directional control valves:
    a. Two-way, two-position, normally closed with manual lever actuation
    b. Three-way, two-position, normally closed with pilot actuation
    c. Three-way, two-position, detented with pushbutton actuation
    d. Four-way, three-position, spring-centered with solenoid actuation
    e. Four-way, three-position (closed neutral), detented with manual lever actuation
13. How is a pilot-operated DCV shifted?
14. Describe the operation of a pilot-operated solenoid DCV. What is the advantage of these valves over the simple solenoid type?
15. Describe the difference between inline and subplate mounting of hydraulic valves.
16. What are the advantages of subplate mounting? Are there any disadvantages?
17. What are cartridge valves?
18. What can occur if a DCV is operated above its maximum flow rating?
19. What can occur if a DCV is operated above its maximum pressure rating?

20. What is the effect of regeneration on cylinder speed and force output? Explain why regeneration has this effect.

21. A cylinder with a bore diameter of 3 in and a rod diameter of 1.375 in is to be used in a system with a 20 gpm pump. What are the extension speeds when regenerating and when not regenerating?

22. A cylinder with a bore diameter of 50 mm and a rod diameter of 20 mm is to be used in a system with a 50 lpm pump. What are the extension speeds when regenerating and when not regenerating?

23. The cylinder in Problem 21 is used in a system with a maximum pressure of 2000 psi. What are the maximum extension forces when regenerating and when not regenerating?

24. The cylinder in Problem 22 is used in a system with a maximum pressure of 15,000 kPa. What are the maximum extension forces when regenerating and when not regenerating?

25. Determine the pressure drops across the DCV in Example 6-3 when the cylinder is extending ($P{\rightarrow}A, B{\rightarrow}T$).

26. Determine the pressure drop across the DCV in Example 6-3 when it is in neutral ($P{\rightarrow}T$, A and B are blocked).

CHAPTER 7

Hydraulic Pressure Control

OUTLINE

7.1 Introduction

7.2 Pressure Relief Valves

7.3 Unloading Valves

7.4 Pressure-reducing Valves

7.5 Sequence Valves

7.6 Counterbalance Valves

7.7 Brake Valves

7.8 Pressure-compensated Pumps

7.9 Pressure Control Valve Mounting

7.10 Pressure Control Valve Specifications

7.11 Review Questions and Problems

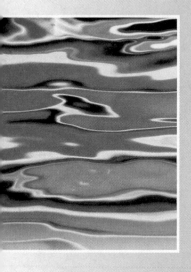

7.1   Introduction

The force of a cylinder is proportional to the pressure in a system and the area over which the pressure is applied ($F = p \cdot A$). Controlling the pressure level in a circuit will therefore allow us to control the output force of a cylinder. Controlling the pressure level in a motor circuit allows us to control the output torque of the motor. Pressure control valves control the maximum pressure level in a hydraulic circuit, which provides control of the maximum output force or torque. They also protect the circuit from excessive pressure, which could damage components and possibly cause serious injury. Pressure control valves may control the maximum pressure of the entire circuit, or simply one branch of the circuit. Some types of pressure control valves simply react to pressure changes rather than control the pressure.

7.2   Pressure Relief Valves

Pressure relief valves limit the maximum pressure in a hydraulic circuit by providing an alternate path for fluid flow when the pressure reaches a preset level. The most basic type of relief valve is a *direct-acting* type (Figure 7-1). All relief valves have a pressure port that is connected to the pump line and a tank port that is connected to the tank. In the direct-acting design, a ball or poppet is subjected to pump pressure on one side and the force of a spring on the other. When the pressure in the system creates a force on the ball that is less than the spring force, it remains on its seat and the pump flow will go to the system (Figure 7-1A). When the pressure is high enough to create a force greater than the spring force, the ball will move off its seat and allow pump flow to go back to the tank through the relief (Figure 7-1B). The pressure at which the relief valve opens can be adjusted by changing the amount of spring compression, which changes the amount of force applied to the ball on the spring side. This is accomplished with an adjustment screw or knob. This type of relief valve is called direct-acting because the ball is directly exposed to pump pressure. The graphic symbol for an adjustable pressure relief valve, along with a pump, is shown in Figure 7-1C. The symbol shows that the valve is normally closed (the arrow is offline). On one side of the valve, pressure is fed in (the dashed line) to try to open the valve, while on the other side the spring is trying to keep it closed. The arrow through the spring signifies that it is adjustable, allowing adjustment of the pressure level at which the relief valve opens.

In most hydraulic circuits a *pilot-operated relief valve,* rather than a direct-acting relief valve, is used to control the maximum pressure (Figure 7-2). A pilot-operated relief valve consists of a small pilot relief valve and a main relief valve. It operates in a two-stage process: First, the pilot relief valve opens when

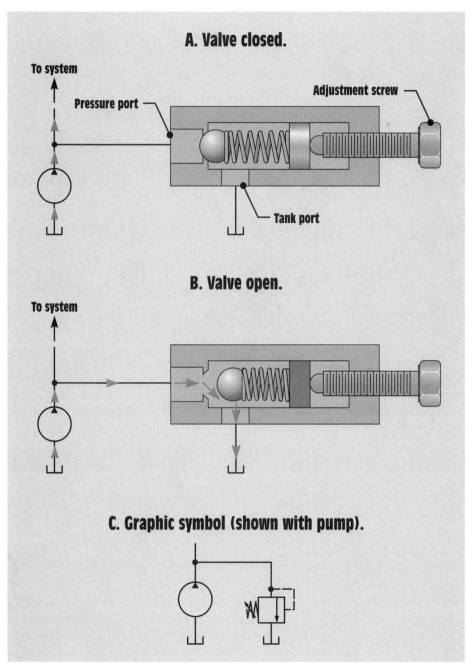

A. Valve closed.

To system

Pressure port

Adjustment screw

Tank port

B. Valve open.

To system

C. Graphic symbol (shown with pump).

FIGURE 7-1   Direct-acting pressure relief valve. (A) Valve closed. (B) Valve open. (C) Graphic symbol (shown with pump).

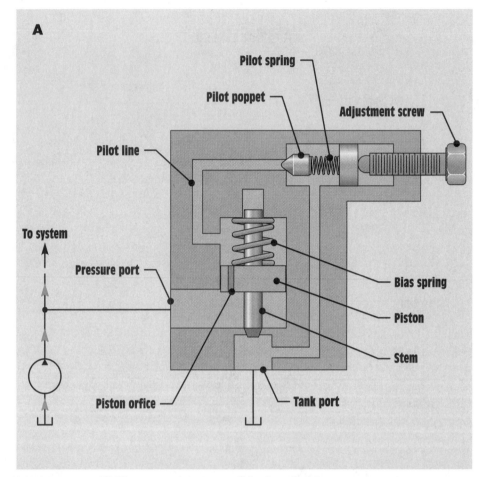

FIGURE 7-2   (A) Pilot-operated pressure relief valve—closed.

a preset maximum pressure is reached, which then causes the main relief valve to open. Just like the direct-acting type, the pilot-operated type has a pressure port that is connected into the pump line and a tank port that is connected to tank (Figure 7-2A). The pilot relief is usually a poppet type. The main relief consists of a piston and stem. The main relief piston has a hole (called the *orifice*) drilled through it. This allows pressure to be applied to the top side of the piston, as well as the bottom side. The piston has equal areas exposed to pressure on the top and bottom and is therefore *balanced*—it will have equal force on each side. It will remain stationary in the closed position. The piston has a light bias spring to ensure that it will stay closed. When the pressure is less that the relief valve setting, the pump flow goes to the system.

B

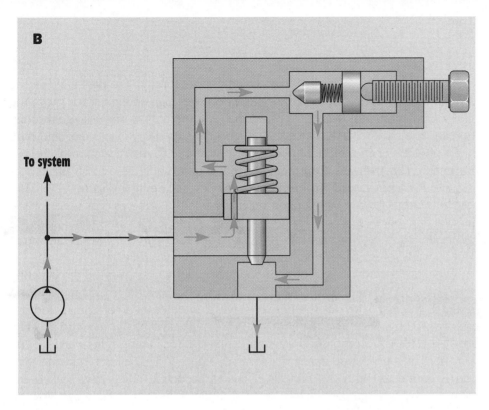

To system

C

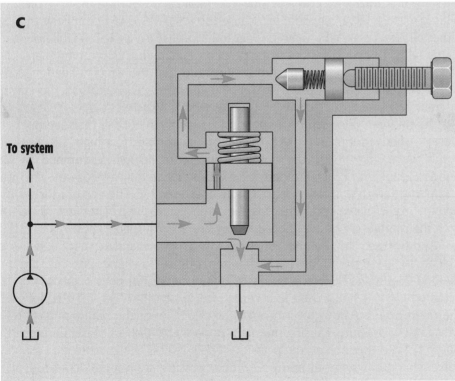

To system

FIGURE 7-2 (B) Pilot-operated pressure relief valve—pilot valve open. (C) Pilot-operated pressure relief valve—pilot and main valve open.

FIGURE 7-3 Cutaway of an actual pilot-operated pressure relief valve.

(Courtesy of Eaton Hydraulics Training.)

The pressure is also applied to the pilot poppet through the pilot line (via the piston orifice). If the pressure in the system becomes high enough, it will move the pilot poppet off its seat (Figure 7-2B). A small amount of flow begins to go through the pilot line back to tank. Once flow begins through the piston orifice and pilot line, a pressure drop is induced across the piston due to the restriction of the piston orifice. This pressure drop then causes the piston and stem to lift off its seat and the flow goes directly from the pressure port to the tank (Figure 7-2C). The graphic symbol for a pilot-operated relief valve is the same as that used for the direct-acting relief valve (see Figure 7-1). Pilot-operated pressure relief valves are also called *compound* or *two-stage* pressure relief valves. A cutaway of an actual pilot-operated pressure relief valve of this type is shown in Figure 7-3.

What are the advantages of pilot-operated relief valves over the direct-acting type? They are usually smaller than a direct-acting type for the same flow and pressure ratings. They also generally have a wider range for the maximum pressure setting. Another advantage is that they can be operated remotely (Figure 7-4). This is accomplished by connecting a direct-acting relief valve to the vent port of the pilot-operated relief valve (Figure 7-4A). Notice that the vent port is connected into the pilot line. The direct-acting relief valve, called the *remote* in this arrangement, acts as a second pilot relief valve. Flow can now go back to tank through either the onboard pilot or the remote pilot. Whichever pilot is set to a lower pressure will cause the relief valve to open. Flow through either pilot will cause the main poppet to lift off its seat and allow full flow back to the tank. The advantage of this type of arrangement is that the onboard pilot can be set to the absolute maximum pressure that the circuit is designed for, while the remote pilot can be set for a lower pressure dictated by the current operating parameters. This method of pressure control has two key advantages: (1) The onboard pilot can be made inaccessible so that if the machine operator were to inadvertently set the pressure of the remote too high, the pressure would never rise above the absolute maximum setting determined by the onboard pilot, and (2) the remote pilot can be located away from the circuit in a safe location that is easily accessible to the operator. The symbol for a pilot-operated relief valve and remote is shown in Figure 7-4B. The lines associated with the remote are dashed because they are control (pilot) lines.

Another way to use the vent port on a pilot-operated relief valve is to unload the system during periods when pump flow and pressure are not needed (Figure 7-5). Here, a two-way, two-position, normally closed, solenoid-actuated DCV is being used as a vent valve. When this vent valve is shifted to the open position, flow goes through the pilot line of the pilot-operated relief valve at low pressure, causing the main poppet to lift off its seat and return the pump flow to the tank at low pressure. The circuit is designed so that the vent valve will open whenever pump flow and pressure are not needed. Most pilot-operated relief valves have a vent port. The vent port can be plugged if the valve will not be operated remotely or vented.

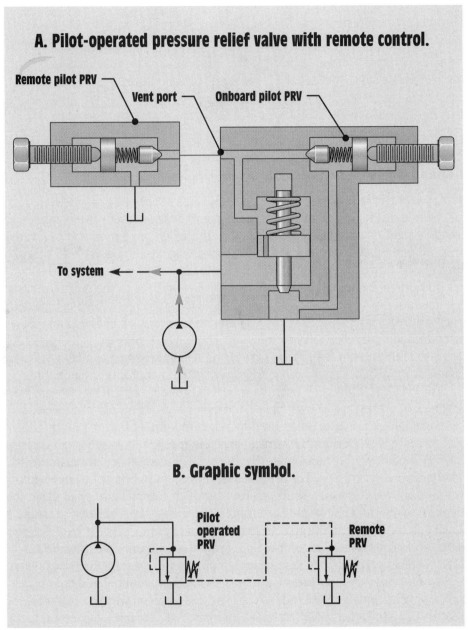

A. Pilot-operated pressure relief valve with remote control.

Remote pilot PRV

Vent port   Onboard pilot PRV

To system

B. Graphic symbol.

Pilot operated PRV

Remote PRV

FIGURE 7-4   (A) Pilot-operated pressure relief valve with remote control. (B) Graphic symbol.

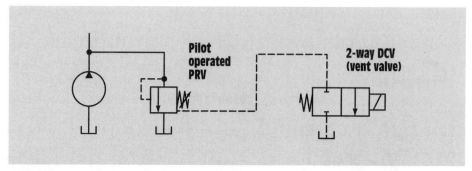

FIGURE 7-5   Pilot-operated pressure relief valve with a vent valve.

Our discussion of relief valves thus far assumes that a relief valve is either completely closed or completely open. In fact, the relief valve may also be partially open. The pressure at which the relief valve begins to open is known as the *cracking pressure*. At this pressure, the poppet just begins to lift off its seat and some of the pump flow begins to go through the relief valve back to the tank. The rest of the flow goes to the system. The pressure at which the relief valve is completely open is known as the *full-flow pressure*. At this pressure, all of the pump flow is going through the relief valve to the tank. The full-flow pressure is the maximum pressure of the system. The difference between the cracking pressure and the full-flow pressure is often called the *pressure override* in manufacturers' literature.

The concept of pressure override is best illustrated with a flow-pressure graph (Figure 7-6). This graph shows typical curves for a direct-acting relief valve and a pilot-operated relief valve when used in a system with a 50 gpm pump. The system pressure is plotted on the vertical axis and the flow going through the relief valve is shown on the horizontal axis. Both valves are set for a maximum (full-flow) pressure of 2500 psi. For the direct-acting relief valve, flow begins to go over the pressure relief valve at 1800 psi. Flow through the pressure relief valve then increases with increasing pressure until it reaches the full-flow pressure of 2500 psi. At this point, the full 50 gpm from the pump is going over the relief valve. The pilot-operated type does not crack, however, until the pressure reaches 2150 psi. This is typical; for the same maximum pressure setting, the pilot-operated type has a higher cracking pressure. In other words, a pilot-operated type typically has a smaller pressure override. A higher cracking pressure is advantageous because once the relief valve cracks, we lose flow over the relief valve even though we are not yet at the system maximum. Lost flow results in reduced actuator speed and lost power. The following example illustrates this concept.

EXAMPLE 7-1.

Compare the two types of relief valves shown in Figure 7-6 for a system operating at 2000 psi and 2250 psi.

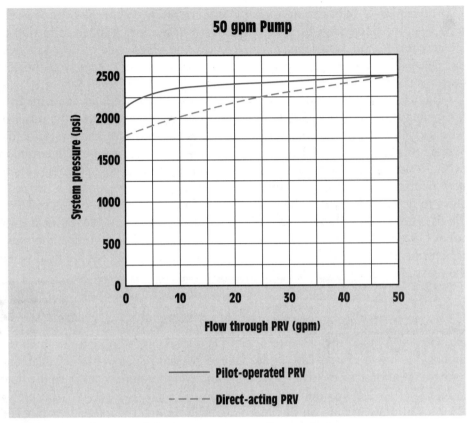

FIGURE 7-6   Pressure relief valve pressure versus flow graph.

SOLUTION:

1. At 2000 psi:
   a. Direct-acting: 9 gpm is lost across the relief valve, 41 gpm goes to the system
                       $9 \div 50 \times 100 = 18\%$ of the system flow is lost
   b. Pilot-operated: 0 gpm is lost across relief valve, 50 gpm goes to the system
                       $0 \div 50 \times 100 = 0\%$ of the system flow is lost
2. At 2250 psi:
   a. Direct-acting: 25 gpm is lost across relief valve, 25 gpm goes to the system
                       $25 \div 50 \times 100 = 50\%$ of the system flow is lost
   b. Pilot-operated: 3 gpm is lost across relief valve, 50 gpm goes to the system
                       $3 \div 50 \times 100 = 6\%$ of the system flow is lost

The results of the previous example are important to consider for two reasons. First, the percentage of system flow lost will result in a speed reduction of the actuator by the same percentage. Second, the percentage of flow lost over the relief valve represents a power loss of the same percentage and the lost power is converted to heat.

This is an opportune time to again discuss the concept of pressure in a hydraulic system. Chapter 3 stated that pressure in a fluid system is the result of resistance to flow. What does this mean about the pressure level in an actual hydraulic circuit? It means that the pressure at any particular time is determined by the resistance to flow at that time. If the pump flow is going through a tandem neutral back to the tank, the pressure will be very low because the resistance to flow is very low. If the pump flow is forced to go to a loaded cylinder, the pressure will be proportional to the load. Thus, pressure relief valves do not control the pressure level in a circuit at all times. Rather, they control the *maximum* pressure in a circuit by providing an alternate flow path for the fluid when the pressure reaches the relief setting.

The *cross-over* pressure relief valve is a variant of the pressure relief valve. These valves are used to prevent shifting shock in motor circuits. Shifting shock can occur in hydraulic motor circuits when a directional control valve is shifted into the closed position. If the motor has a large amount of momentum built up, shifting the valve closed can cause a large pressure spike on the outlet line because the motor tends to keep spinning and will therefore try to push the fluid to its outlet. A high vacuum also occurs on the inlet side because the motor attempts to draw in fluid on the inlet side, even though the inlet line is blocked at the DCV.

A cross-over pressure relief is essentially two pressure relief valves combined into one package. They are mounted in parallel and at 180° to each other. Figure 7-7 shows a cross-over relief in a motor circuit. When the DCV is shifted into neutral in this circuit, some oil is allowed to flow from the outlet to the inlet through the cross-over relief. This provides a shock-absorbing or cushioning effect on the circuit. Two pressure relief valves are required because the motor is bi-directional and the motor inlet and outlet will switch depending on the direction of rotation. The dashed line around the two relief valves is called a *component enclosure,* indicating that the relief valves are integrated into one component. The pressure setting on the cross-over relief valves must be low enough to provide sufficient cushion, but high enough to ensure that they do not open during times when the motor is running under load. Direct-acting valves are used in cross-over pressure relief valves because they have a faster response time and provide more cushion because of the larger pressure override. Cross-over pressure relief valves should be installed as close as possible to the actuator to maximize their effectiveness.

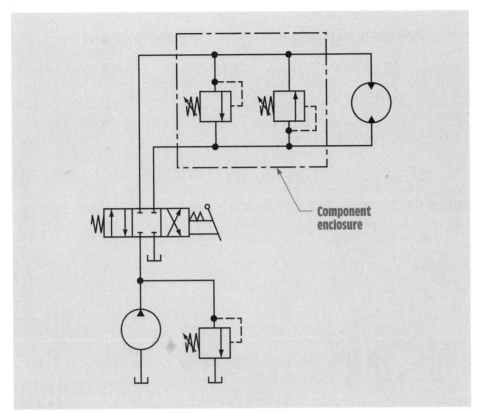

Component enclosure

FIGURE 7-7   Cross-over pressure relief valve application.

7.3   Unloading Valves

Unloading valves are very similar in construction to pressure relief valves. The symbol for each is shown in Figure 7-8 for comparison. Both send flow back to the tank when a preset pressure is reached. However, an unloading valve reads the pressure in an external line, rather than in its own line, as indicated by the dashed pilot lines.

Figure 7-9 shows an application for an unloading valve. This circuit can be used in an application in which high flow (speed) and low pressure (force) are required for a part of the cylinder's stroke, while low flow and high pressure are required for the rest, for example, a metal stamping machine. In this machine it may be desirable for the cylinder to move into position very quickly, then slow down when it reaches the workpiece. The first part of the cycle requires only minimal pressure because the only resistance is the flow resistance of the components and the friction of the cylinder. The second part of the cycle requires high pressure because the cylinder is deforming the metal.

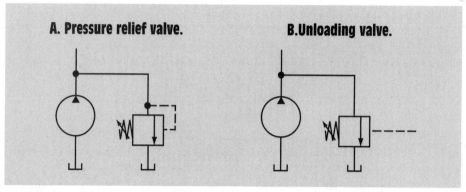

FIGURE 7-8   Unloading valve. (A) Pressure relief valve. (B) Unloading valve.

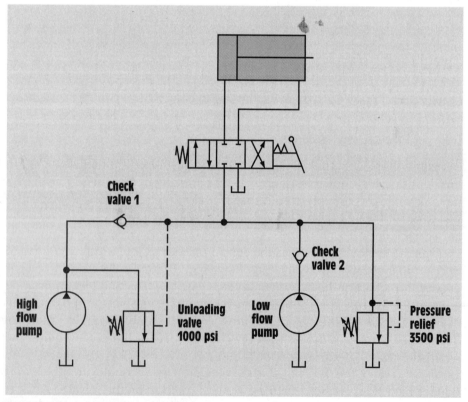

FIGURE 7-9   Unloading valve circuit.

The circuit shown in Figure 7-9 supplies the cylinder with flow from both the high-flow pump and the low-flow pump when the pressure is below 1000 psi. When the pressure reaches 1000 psi, the unloading valve opens and *unloads* the high-flow pump back to tank at low pressure. Only the low-flow pump supplies the cylinder with flow at pressures from 1000 psi to 3500 psi. If the pressure reaches 3500 psi, flow from the low-flow pump is forced over the relief valve at this pressure. Check valve 1 isolates the high-flow pump from the system pressure while it is being unloaded. Check valve 2 prevents the flow from the high-flow pump from flowing into the low-flow pump line. This would reverse the low-flow pump, which would cause damage to the power unit.

Figure 7-10 shows the operation of the previous circuit when the cylinder is extending. Part A shows the cylinder is extending under no load and the pressure is therefore very low. Both pumps are supplying flow and the cylinder is extending rapidly. Part B shows that the cylinder has encountered the workpiece and the pressure is therefore greater than 1000 psi. The high-flow pump is unloaded to the tank and only the low-flow pump is supplying the cylinder, causing it to extend much more slowly. The pressure in the high-flow branch is very low because the pressure from the circuit to the right of the check valve is used to hold the unloading valve open. Again, the check valve isolates the high-flow pump from the pressure in the rest of the circuit when the unloading valve is open. This circuit saves energy (and money) by only supplying the high flow at low pressure when it is needed. It also saves in initial cost because only the small, low-flow pump needs to be able to withstand the high pressure. The larger, high-flow pump can be a much less expensive low-pressure model.

7.4   Pressure Reducing Valves

Pressure reducing valves maintain a reduced pressure level in a branch circuit of a hydraulic system. Figure 7-11 compares the symbols for a relief valve and a reducing valve. The reducing valve is normally open, while relief valve is normally closed. The reducing valve reads the pressure *downstream* (after the valve), while the relief valve reads the pressure *upstream* (before the valve). The reducing valve has an external drain line, while a relief valve does not. When a valve has an external drain, a line must be connected from the valve's drain port to the tank. Drain lines, like pilot lines, are shown as dashed lines on a schematic.

Figure 7-12 shows an application for a pressure reducing valve. Here, two cylinders are connected in parallel. The circuit is designed to operate at a maximum pressure of 3500 psi, which is determined by the relief valve setting. This is the maximum pressure that cylinder 1 will see. For a reason determined by the function of the machine, cylinder 2 is limited to a maximum pressure of 2500 psi. This is accomplished by placing a pressure-reducing valve in the circuit in the location shown in Figure 7-12. If the pressure in the circuit rises above

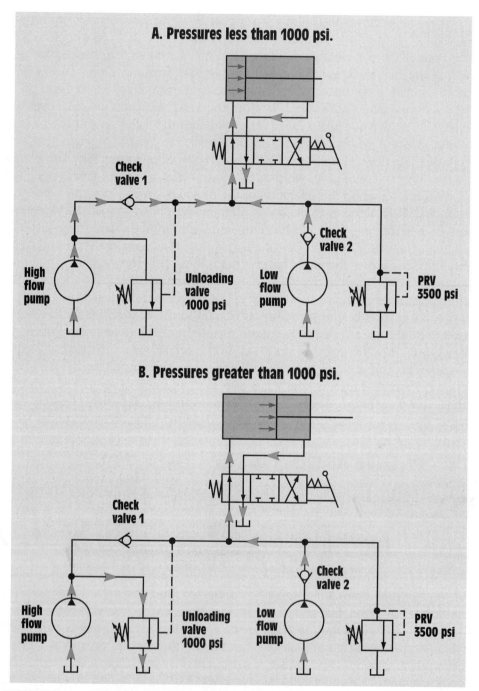

FIGURE 7-10 Unloading valve circuit with cylinder extending. (A) Pressures less than 1000 psi. (B) Pressures greater than 1000 psi.

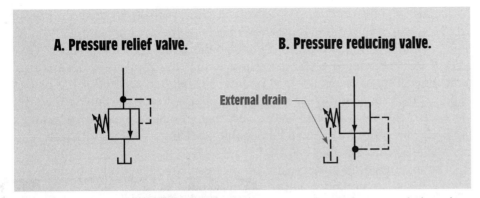

FIGURE 7-11   Pressure-reducing valve. (A) Pressure relief valve. (B) Pressure-reducing valve.

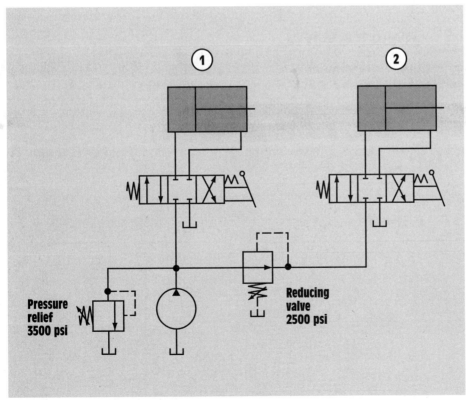

FIGURE 7-12   Pressure reducing valve application.

2500 psi, the pressure-reducing valve will close partially to create a pressure drop across the valve. The valve then maintains the pressure drop so that the outlet pressure is not allowed to rise above the 2500 psi setting. The disadvantage of this method of pressure control is that the pressure drop across the reducing valve represents lost energy that is converted to heat. If the pressure setting of the reducing valve is set very low relative to the pressure in the rest of the system, the pressure drop will be very high, resulting in excessive heating of the fluid. When the hydraulic fluid becomes too hot, its viscosity reduces, causing increased component wear.

The external drain is necessary on a pressure-reducing valve because both sides of the valve are under pressure. A relief valve does not require an external drain because one side of the valve is connected to the tank. This allows the drain cavities to be connected to the tank line internally, allowing the leakage of the valve to pass to the tank port. This is true in general for hydraulic components; those that have a low-pressure tank port can be drained internally, while those that do not have a tank port must be drained externally. Not connecting an external drain line on a component that requires one will likely result in a blown seal.

7.5   Sequence Valves

A sequence valve is a pressure control valve that is used to force two actuators to be operated in sequence. They are very similar in construction to pressure relief valves. The symbol for each is shown in Figure 7-13 for comparison. Instead of sending flow back to the tank, however, a sequence valve allows flow to a branch circuit when a preset pressure is reached. The check valve allows the sequence valve to be bypassed in the reverse direction (from *OUT* to *IN*). The

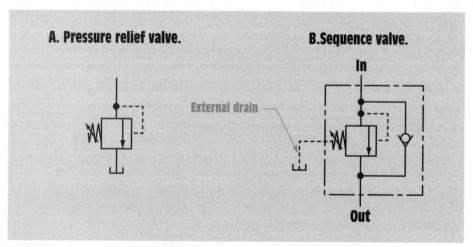

FIGURE 7-13   Sequence valve. (A) Pressure relief valve. (B) Sequence valve.

component enclosure line indicates that the check is an integral part of the component. The sequence valve has an external drain line, therefore a line must be connected from the sequence valve's drain port to the tank.

An example of an application for a sequence valve is a circuit in which a clamp cylinder extends first to hold a workpiece, then a second cylinder extends to bend the workpiece into a desired shape. This can be accomplished with the circuit shown in Figure 7-14. In this circuit, two cylinders are connected in parallel. Without the sequence valve, these cylinders would extend together, as

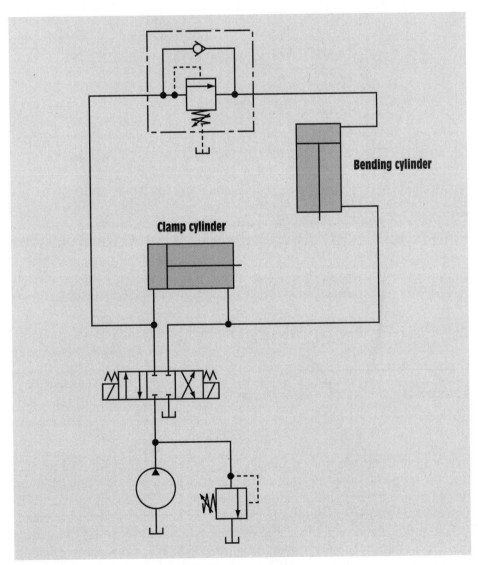

FIGURE 7-14   Sequence valve application.

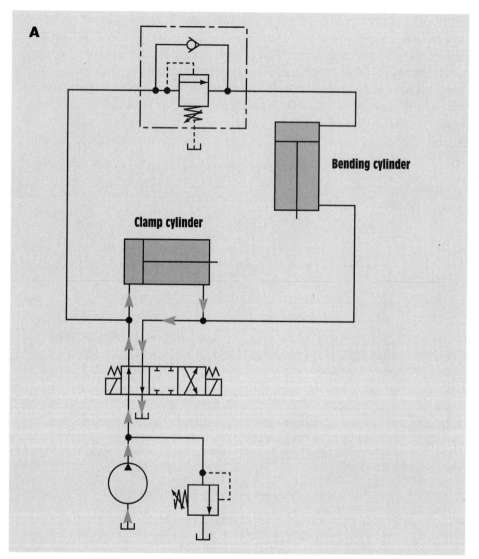

FIGURE 7-15   (A) Sequence valve application, clamp cylinder extending.

they are both unloaded when the DCV is shifted. In order for this circuit to function properly, however, the clamp cylinder must extend completely and bottom out on the workpiece before the bending cylinder begins to extend. The sequence valve accomplishes this by not allowing flow into the bending cylinder branch of the circuit until the clamp cylinder has reached the end of its stroke. The flow in this situation is shown in Figure 7-15A. When the clamp cylinder is fully extended, the pressure will rise and trip the sequence valve, allowing the

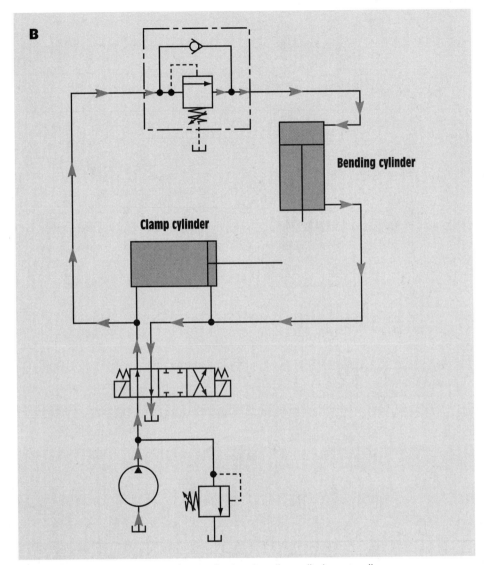

FIGURE 7-15   (B) Sequence valve application, bending cylinder extending.

bending cylinder to extend (Figure 7-15B). The sequence valve must be set high enough so that it opens only after the clamp cylinder has bottomed out. This setting should be as low as possible, however, because the pressure drop across the sequence valve is wasted energy that is converted to heat. The setting for any particular circuit is determined by trial and error.

The check valve allows the sequence valve to be bypassed when the cylinders are retracted. The sequence valve has no effect on the circuit in this situation. Both

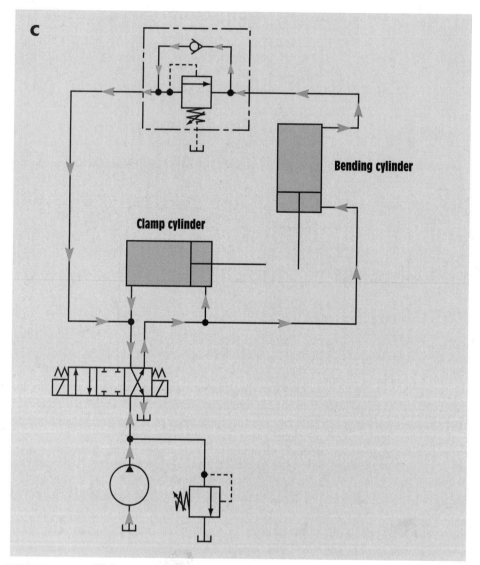

FIGURE 7-15   (C) Sequence valve application, both cylinders retracting.

cylinders will retract together because both are unloaded and will split the pump flow (Figure 7-15C). This would probably *not* be desirable in the case of a clamp and bend circuit, however. In most cases, we would want the clamp cylinder to remain extended until the bending cylinder extends *and* retracts so that the part is held in a fixed position during the entire operation. The clamp cylinder should extend first and retract last, after the bending cylinder has cleared the work area. This can be accomplished by placing a second sequence valve in the rod end line of the clamp cylinder (Figure 7-16). In this circuit, the bending cylinder will extend

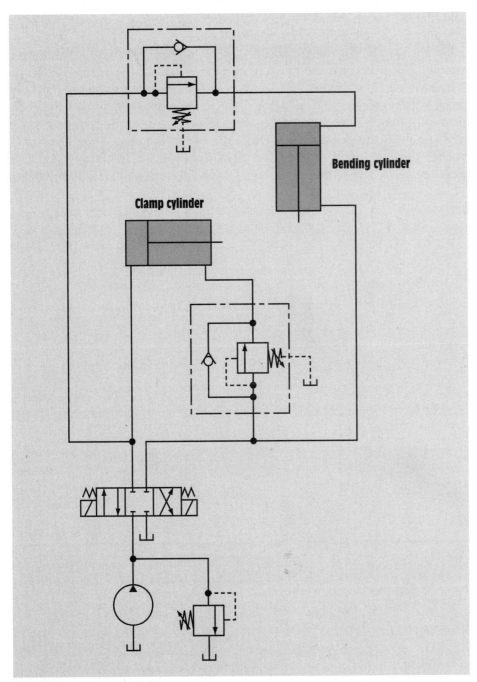

FIGURE 7-16   Clamp and bend circuit.

only after the clamp cylinder is fully extended and the clamp cylinder will retract only after the bending cylinder is fully retracted.

Sequence valves are also available with *remote sensing*. These valves have an external pilot that reads the pressure in another part of the circuit, rather than at its own inlet line. Figure 7-17 shows remotely operated sequence valve in a circuit with a cylinder and a motor. This circuit is designed so that the motor turns only after the cylinder reaches the end of its stroke. The sequence valve's pilot line reads the pressure in the cap end of the cylinder. When the cylinder reaches the end of its stroke, the pressure will build, thereby tripping the sequence valve. The sequence valve in this circuit is between the pump and the

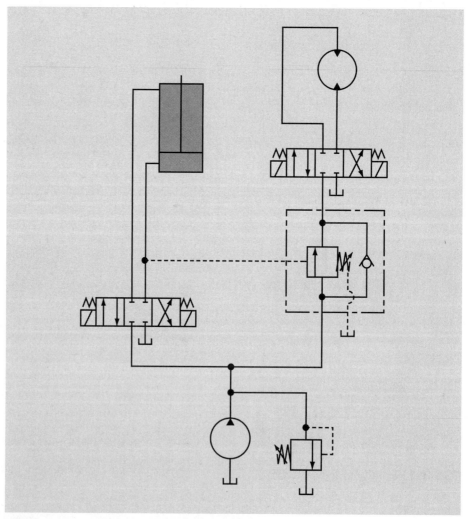

FIGURE 7-17   Sequence valve with remote sensing.

DCV that controls the motor, rather than between the DCV and the motor, as in the circuits discussed previously. This prohibits the motor from being operated in *either direction* until the cylinder reaches the end of its stroke. Operating a sequence valve remotely in this manner is desirable because it reads the pressure right at the cylinder, the device whose movement we are trying to detect.

7.6   Counterbalance Valves

Counterbalance valves are pressure control valves that are used to prevent a load from accelerating uncontrollably. This situation can occur in vertically mounted cylinders in which the load is a weight (Figure 7-18). When the DCV is shifted to extend the cylinder (lower the weight) in this circuit, the weight may cause the cylinder to accelerate too quickly. When this occurs, the load is driving the cylinder, as opposed to the more controllable situation of the cylinder driving the load. This can cause damage to the load, or even to the cylinder itself, when the load is stopped quickly at the end of its travel. This can be remedied by placing a counterbalance valve on the rod end of the cylinder (Figure 7-19). When the DCV is shifted to lower the weight, the cylinder will not extend until

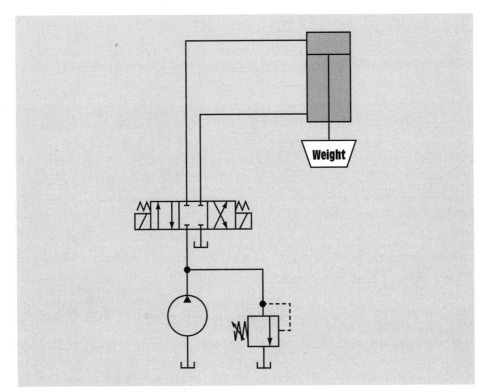

FIGURE 7-18   Vertically mounted cylinder lifting a weight.

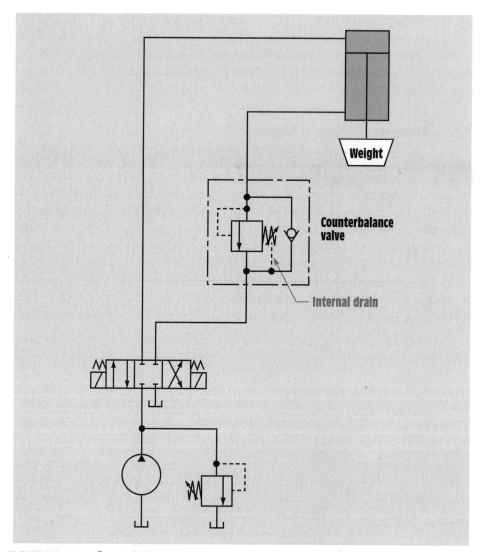

FIGURE 7-19   Counterbalance valve.

a preset pressure is reached in the rod end. This provides a backpressure against the rod end of the piston, which acts to stabilize the downward movement of the cylinder. The check valve allows the counterbalance valve to be bypassed when the cylinder is retracted. A counterbalance valve has an internal drain, unlike the sequence valve, which has an external drain.

Counterbalance valves can also be operated remotely (Figure 7-20). In this circuit, the pressure is being sensed in the cap end line. The setting of a counterbalance valve will depend on the magnitude of the weight being lowered. Just as with a sequence valve, this setting should be as low as possible because the

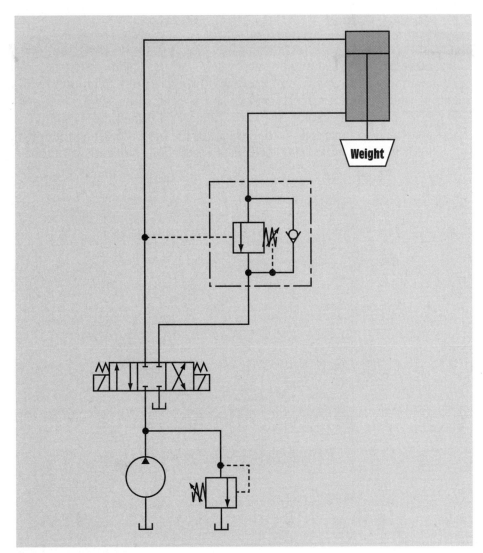

FIGURE 7-20   Counterbalance valve with remote sensing.

pressure drop across the counterbalance valve is wasted energy that is converted to heat. The setting for any particular circuit is determined by trial and error and must be adjusted if the weight is changed.

The symbols for counterbalance and sequence valves are similar. The only difference is that a sequence valve requires an external drain, while a counter-balance valve does not because it can be drained internally. Some manufacturers sell one model that can be used as either valve by plugging the external drain port if it is to be used as a counterbalance valve or by connecting the external drain port to tank if it will be used as a sequence valve.

7.7   Brake Valves

Brake valves, like counterbalance valves, are used to prevent loads from accelerating uncontrollably. Counterbalance valves are used with cylinders; brake valves are used with hydraulic motors. Brake valves are most commonly used in circuits in which the motor must lower a large weight, such as in a winch application. A simple winch circuit is shown in Figure 7-21. When the weight is lowered, it may tend to drive the motor, instead of the motor lowering the weight. This is known as an *overrunning* load. In this situation, the load will probably

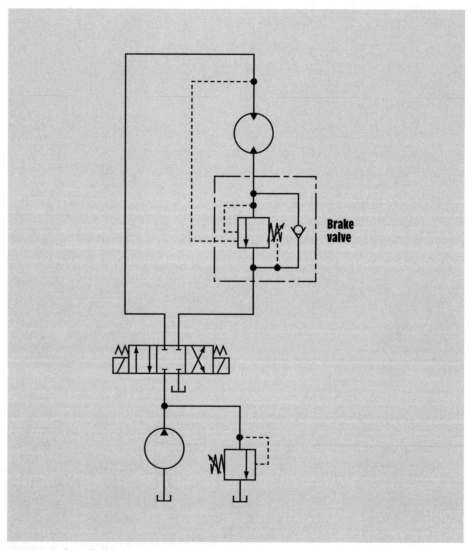

FIGURE 7-21   Brake valve circuit.

accelerate too quickly. The motor is being driven by the load and is basically acting as a pump. When this occurs, the pressure at the outlet will be higher than
the pressure at the inlet. The brake valve senses the pressure in both the inlet
and outlet lines of the motor, just as it is with a pump. Whenever the pressure
at the outlet is lower than the pressure at the inlet, the motor is functioning normally and the brake valve allows nearly unrestricted flow out of the motor.
When the pressure at the outlet is higher than at the inlet, however, the brake
valve closes partially to provide enough of a backpressure on the outlet of the
motor to keep the load in control. The check valve allows the valve to be
bypassed when the weight is being raised.

7.8   Pressure-compensated Pumps

Pressure-compensated pumps, as discussed in Chapter 3, are pumps that have
the ability to limit the maximum pressure in a hydraulic circuit by reducing their
displacement. When the pressure reaches a preset level, called the *firing pressure,*
the displacement is reduced to prevent a further rise in pressure. Pressure compensation is therefore a type of pressure control.

Figure 7-22 shows a typical flow-pressure curve for a pressure-compensated
pump. This particular pump has a flow of 20 gpm and a compensator firing pressure setting of 3000 psi. Notice that as the pressure increases, the flow from the
pump gradually decreases due to increased leakage. This is true whether or not
a pump is equipped with pressure-compensation. In a pressure-compensated
pump, however, the flow decreases rapidly when the firing pressure of 3000 psi

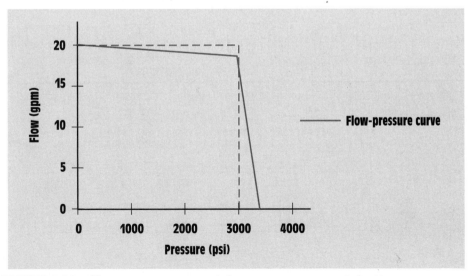

FIGURE 7-22   Flow versus pressure graph for a pressure-compensated pump.

is reached. The flow decreases rapidly until the flow is decreased to zero (at about 3400 psi in this case). In a real circuit, this pressure-compensated pump would still produce a small amount of flow at 3400 psi to make up for leakage that is occurring at various points within the system.

In the circuits discussed thus far, the maximum pressure in the system was determined by the relief valve setting. When flow is sent over the relief valve at high pressure, a great deal of energy is wasted and is transformed into heat. The hydraulic press circuit shown in Figure 7-23 explains the benefit obtained from the use of a pressure-compensated pump. In this circuit, the cylinder must extend, bottom out, and hold pressure on the platens for some fixed period of time. During this time, pressure must be maintained in order to maintain the contact pressure between the platens. A machine such as this can be used to bond two pieces of material together with an adhesive or hold a mold closed while the material is setting. This situation occurs in many other applications as well. If pressure is to be maintained, the DCV must be left in the extend position even after the cylinder is fully extended (Figure 7-24).

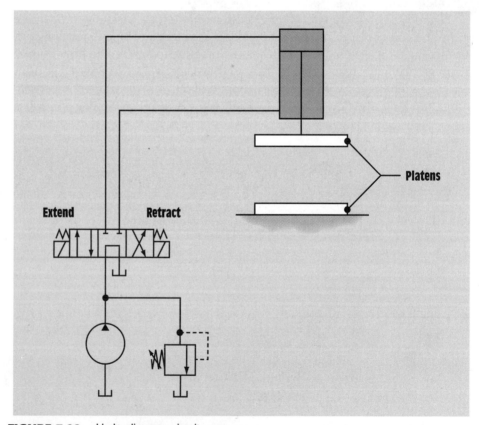

FIGURE 7-23   Hydraulic press circuit.

Whenever the cylinder is bottomed out, flow is going over the relief valve. The DCV cannot be shifted into neutral, even though in neutral the ports to the cylinder are blocked, because leakage across the DCV spool and piston seals will cause the cylinder lines to lose pressure. During the period when the cylinder is fully extended under pressure the pump is supplying full flow to the system, even though only a small amount of flow is required to make up for the leakage, which makes this circuit very wasteful.

Figure 7-25 shows a hydraulic press circuit with a pressure-compensated pump. The operation of this circuit is the same as that of the previous circuit. When the cylinder is fully extended, however, flow will not be sent over the pressure relief valve. Instead, pressure will rise until it reaches the setting of the compensator. The compensator will then reduce the output of the pump to prevent a further rise in system pressure. Any leakage in the system will cause the pressure to drop and the pump displacement will increase again to make up the leakage. A pressure-compensated pump greatly improves the efficiency of this system by supplying only enough flow to keep the pressure up to a

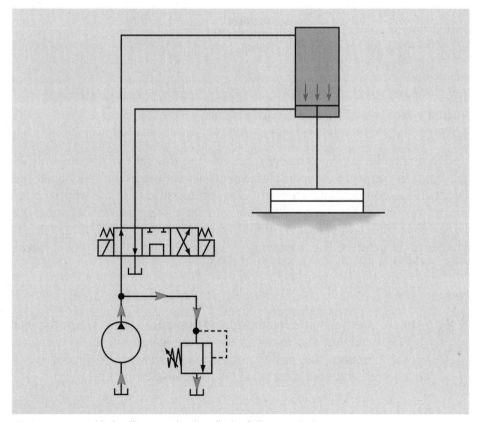

FIGURE 7-24   Hydraulic press circuit, cylinder fully extended.

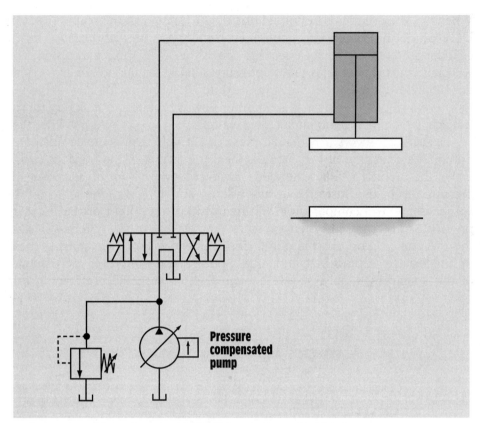

FIGURE 7-25 Hydraulic press circuit with a pressure-compensated pump.

predetermined level. The pressure relief valve in this system is acting only as a backup to the pressure compensator. It would therefore be set to a higher pressure than the compensator so that it does not open during the normal operation of the circuit. One rule of thumb is to set the pressure relief valve 10% higher than the compensator setting, or a minimum of 200 psi higher, whichever is greater. The following example illustrates the power savings obtained with a pressure-compensated pump.

EXAMPLE 7-2.

A hydraulic press circuit operates at 2500 psi and 8 gpm. When the cylinder is fully extended and under pressure, the leakage in the system totals 0.3 gpm. Compare the power consumption when the cylinder is fully extended with and without a pressure-compensated pump.

SOLUTION:

1. Without: 7.7 gpm is forced over the relief valve and 0.3 gpm is lost due to leakage, both at 2500 psi. The total wasted energy is therefore:

$$HP_H = \frac{p \cdot Q}{1714} = \frac{2500 \text{ psi} \cdot (8 \text{ gpm})}{1714} = 11.67 \text{ hp}$$

2. With: When the cylinder bottoms out, the pump's displacement will be reduced to 0.3 gpm. The total wasted energy in this case is:

$$HP_H = \frac{p \cdot Q}{1714} = \frac{2500 \text{ psi} \cdot (0.3 \text{ gpm})}{1714} = 0.44 \text{ hp}$$

This example shows that using a pressure-compensated pump results in far less energy consumption in circuits in which pressure, but not flow, is required for a significant portion of the machine cycle. Systems such as these are much less expensive to operate and far less likely to overheat with a pressure-compensated pump.

Pressure-compensated pumps are also commonly found in systems that use DCVs with closed neutrals, such as the parallel circuit shown in Figure 7-26. This circuit cannot use tandem center neutral DCVs. Closed neutrals are required for pressure to build when less than all of the cylinders are being actuated. A tandem center valve sends the pump flow back to tank under low pressure when in neutral, and no flow will ever go to the cylinders if at least one DCV is in neutral because flow follows the path of least resistance. Without a pressure-compensated pump, flow will be forced over the relief valve whenever all three DCVs are shifted to neutral. This is very wasteful because the pump is producing full flow and pressure even though they are not needed by the system. With the pressure-compensated pump, the displacement is reduced to zero and the pump remains under pressure but does not produce flow. The pump is said to be in *high-pressure standby mode.* Very little power is consumed because the pump is not producing any flow. Using a pressure-compensated pump in a circuit such as this will greatly improve the efficiency of the system.

A pressure-compensated pump may also be equipped with *load sensing,* which provides flow control as well as pressure control. These pumps are used in applications in which a constant actuator speed is required, even if the load fluctuates. Flow control is achieved through the arrangement shown in Figure 7-27, which

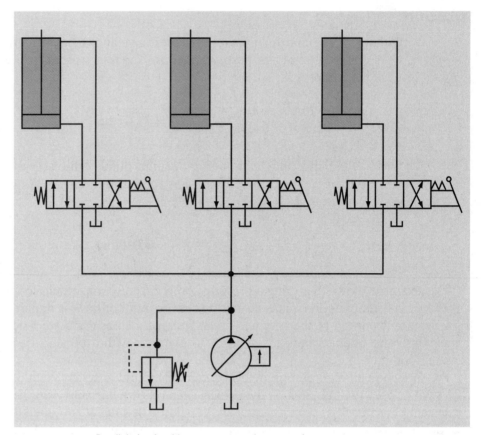

FIGURE 7-26   Parallel circuit with a pressure-compensated pump.

shows a pressure-compensated pump with load sensing driving a bi-directional motor. Instead of sensing the pressure only at the pump outlet, the compensator with load sensing determines the pressure at the pump outlet and after the directional control valve. This allows it to control the pressure drop across the directional control valve. The flow rate through a valve is directly proportional to the pressure drop across it. A pump with load sensing can therefore maintain a constant flow by maintaining a constant pressure drop across the directional control valve. The shuttle valve allows the pressure to be sensed by the compensator at whichever line is under pressure, which changes as the motor direction changes.

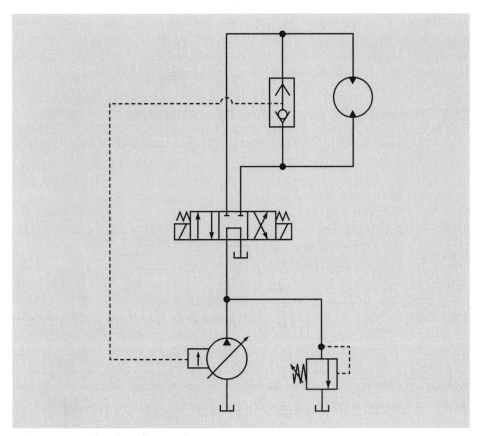

FIGURE 7-27   Load-sensing pump.

7.9   Pressure Control Valve Mounting

Like directional control valves, pressure control valves are available in either *inline* or *subplate* mounting styles. Recall from the previous chapter that inline-mounted valves have fittings that are screwed directly into the valve. Subplate-mounted valves, on the other hand, have unthreaded connections on the bottom of the valve. The valve is then attached to a subplate that has matching connections. The subplate has the threaded connections to which the fittings are attached. Multiple subplate valves can be mounted on a manifold to create a compact, integrated hydraulic circuit. The advantages of subplate mounting and manifolding were discussed in Chapter 6.

Pressure control valves, like directional control valves, are available as cartridge valves. Recall that these small valves that screw directly into a cavity in a manifold. They therefore do not require separate valve port and mounting holes. The advantages of cartridge valve systems were discussed in Chapter 6.

7.10   Pressure Control Valve Specifications

As is the case with all hydraulic components, pressure control valve manufacturers provide specifications for:

1. Maximum pressure,
2. Maximum flow,
3. Filtration level,
4. Fluid type and viscosity range, and
5. Physical size, mounting, and porting.

The pressure setting range is also an important specification for all pressure control valves. A variety of ranges are generally available for any particular model. For pressure relief valves specifically, the pressure override characteristics are important. The pressure override is the difference between the cracking and full-flow pressure at a given relief valve setting. Figure 7-6 compares the pressure override characteristics of direct-acting and pilot-operated pressure relief valves. Figure 7-28 shows a typical pressure override graph that a manufacturer would provide for a particular model. Curves are provided for multiple pressure settings. For example, the top curve in Figure 7-28 shows that if the full-flow pressure is set to 3000 psi, the cracking pressure will be about 2650 psi. The pressure override is therefore 3000 psi -2650 psi $= 350$ psi at this setting. The bottom curve shows that if the full-flow pressure is set to 500 psi, the cracking pressure will be about 125 psi. The pressure override is 500 psi -125 psi $= 375$ psi at this setting. These curves will be provided for a particular fluid viscosity.

7.11   Review Questions and Problems

1. What is the purpose of a pressure relief valve?
2. Describe the operation of a direct-acting relief valve.
3. Describe the operation of a pilot-operated relief valve.
4. List three advantages of a pilot-operated PRV over a direct-acting PRV.
5. Define the terms *cracking pressure, full-flow pressure,* and *pressure override.*
6. Why are the pressure override characteristics of a pressure relief valve important?
7. What is the purpose of a cross-over PRV? Describe their operation.

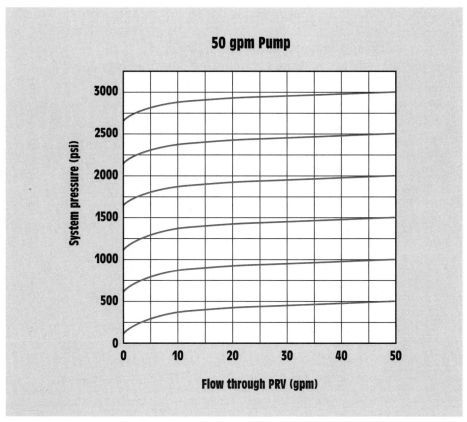

FIGURE 7-28   Pressure override graph for a pilot-operated pressure relief valve.

8. How is an unloading valve different from a pressure relief valve?

9. What is the purpose of a pressure-reducing valve? Describe its operation.

10. What is the purpose of a sequence valve? Describe its operation.

11. What is the purpose of the check in a sequence valve?

12. Make two copies of Figure 7-16 and trace the flow path when: (1) the clamp cylinder is extending, and (2) the clamp cylinder has bottomed out and the bending cylinder is extending.

13. Give an example of an application for a sequence valve.

14. What is the purpose of a counterbalance valve? Describe its operation.

15. What is the purpose of a brake valve? Describe its operation.

16. Define the term *overrunning load.*

17. Describe the operation of a pressure-compensated pump. What is the advantage of a system with a pressure-compensated pump over a system with a relief valve alone?

18. What are the two mounting options for pressure control valves?
19. Using Figure 7-28, determine the cracking pressure of the relief valve if the full-flow pressure is set to 2000 psi.
20. Identify the pressure control valves shown in Figure 7-29.

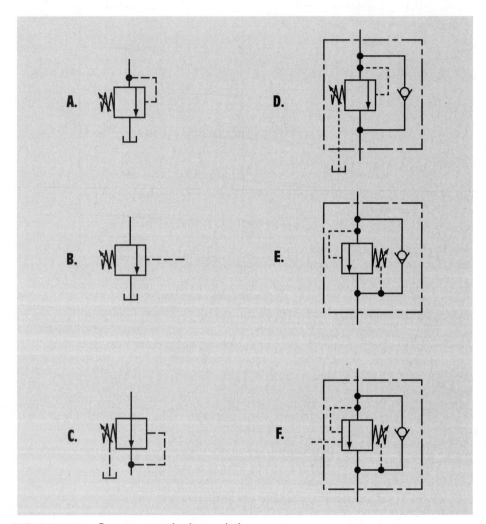

FIGURE 7-29   Pressure control valve symbols.

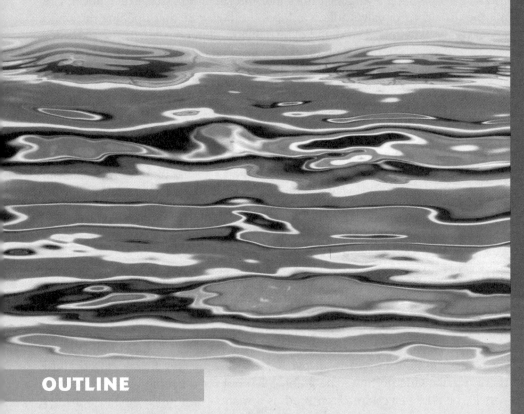

CHAPTER **8**

Hydraulic Flow Control

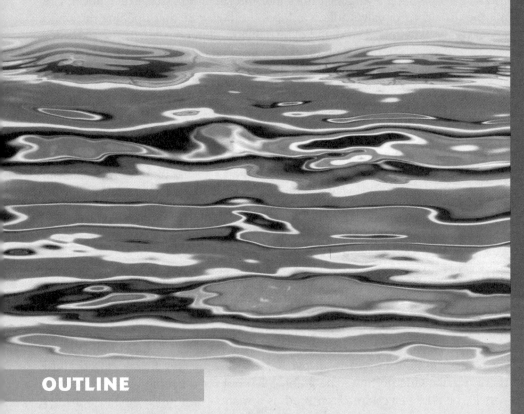

OUTLINE

8.1 Introduction

8.2 Flow Control Valve Types

8.3 Flow Coefficient

8.4 Circuits

8.5 Cushioned Cylinders

8.6 Flow Dividers

8.7 Flow Control Valve Specifications

8.8 Equations

8.9 Review Questions and Problems

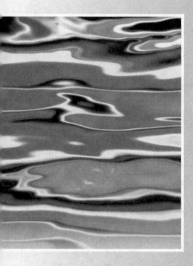

8.1   Introduction

Flow control valves control the flow rate of fluid in a circuit. They accomplish this by incorporating a variable orifice into the circuit that acts like a faucet; closing the flow control valve orifice reduces the flow rate and opening the orifice increases the flow rate. The speed of an actuator depends directly upon the flow rate in the system. Controlling the flow rate therefore allows us to control the speed of actuators.

Using flow control valves is not the only way to control the flow rate in a hydraulic circuit, however. As stated in Chapter 3, pumps can be variable displacement. A variable displacement pump's flow output can be varied, even while it is being driven at a constant speed. This will also control the actuator's speed. In spite of this, flow control valves are commonly used because they are much less expensive and easier to control than variable pumps. The disadvantage of using flow control valves is that they cause increased inefficiency, which is discussed later in this chapter.

8.2   Flow Control Valve Types

The simplest type of flow control is a needle valve (Figure 8-1). This valve is basically just an adjustable orifice that can be closed to reduce the flow rate in a circuit. The orifice size is adjusted by turning the adjustment knob, which raises or lowers the valve stem and needle. Figure 8-1A shows the valve fully open, allowing nearly unrestricted flow. The valve is partially closed and is restricting the flow in part B. In part C, the valve is completely closed and is therefore allowing no flow. The graphic symbol for a needle valve is shown in part D. Needle valves are often used as a manual shut-off in applications that require good *metering* characteristics. Metering simply means to control the flow rate. Needle valves usually have a preferred direction of flow, which is from *A* to *B* in Figure 8-1. This keeps the pressure from being applied to the valve stem when the valve is completely closed, as in part C.

In most fluid power applications, a needle valve with an integral check valve is used to control the flow rate (Figure 8-2). Part A shows the flow going through the valve from *A* to *B*. In this direction, it cannot go through the check and must therefore go through the restriction. In part B, the flow is coming from the opposite direction (*B* to *A*) and can pass through the check valve. The flow is virtually unrestricted in this direction. This flow control valve therefore only controls the flow rate from *A* to *B*. From *B* to *A*, the flow is uncontrolled because the restriction is bypassed through the check. Flow controls of this type may also have ports labeled *IN* and *OUT*, rather than *A* and *B*. In this case, the controlled direction is from *IN* to *OUT*, while flow going from *OUT* to *IN* is uncontrolled. Another labeling method simply puts an arrow on the valve that indicates the

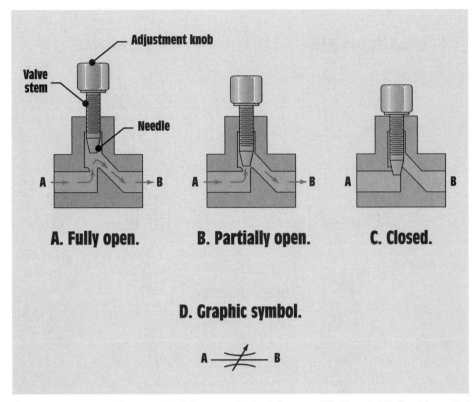

FIGURE 8-1   Needle valve. (A) Fully open. (B) Partially open. (C) Closed. (D) Graphic symbol.

direction of controlled flow. Needle valves without the check are often referred to as *metering valves,* while the term *flow control valve* is usually reserved for needle valves with the integral check. Figure 8-2C shows the graphic symbol for a flow control valve. Part D shows a photo of an actual flow control valve of this type. The arrow indicates the direction of controlled flow.

The flow control valve designs just discussed function by creating a pressure drop across the valve that is proportional to the size of the orifice. The smaller the orifice, the larger the pressure drop across the valve. The flow rate through the valve is then proportional to the pressure drop across the valve. One drawback to the previous designs is that a change in system pressure will affect the pressure drop across the valve, and consequently change the flow rate through the valve. Changes in pressure can therefore cause the speed of the actuator to change when using flow controls, even though the valve setting has not changed. This can be troublesome because the load and, consequently, the pressure, may change frequently in a hydraulic circuit. A *pressure-compensated* flow control valve virtually eliminates this problem. This type of flow control valve automatically adjusts the size of the orifice in response to changes in system pressure. It accomplishes this through the use of

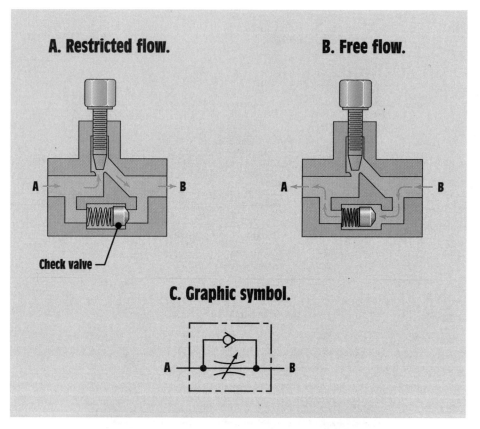

FIGURE 8-2   Flow control valve (needle valve with integral check valve). (A) Restricted flow. (B) Free flow. (C) Graphic symbol. (D) Easy Read® flow control valve, from Deltrol Fluid Products, Bellwood, IL, USA.

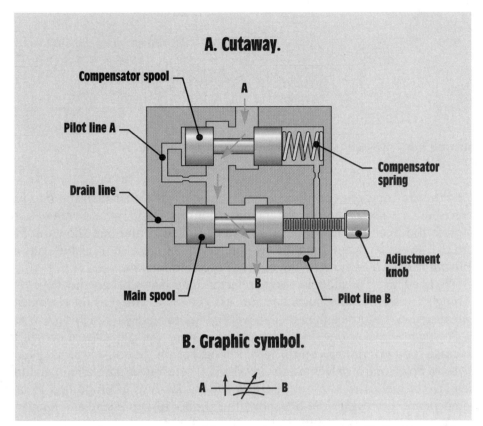

FIGURE 8-3   Pressure-compensated flow control valve. (A) Cutaway. (B) Graphic symbol.

a spring-loaded compensator spool that reduces the size of the orifice when the upstream pressure increases relative to the downstream pressure. Once the valve is set, the pressure compensator will act to keep the pressure drop across the valve nearly constant. This in turn keeps the flow rate through the valve nearly constant.

Figure 8-3A shows a cutaway of a pressure-compensated flow control valve. This valve consists primarily of a main spool and a compensator spool. The adjustment knob controls the main spool's position, which controls the orifice size at the outlet. The pressure upstream of (before) the main spool is ported to the left side of the compensator spool through pilot line A. Pressure downstream of (after) the main spool is ported to the right side of the compensator spool through pilot line B. The compensator spring biases the compensator spool to the fully open (left) position. If the pressure upstream of the main spool increases too much relative to the downstream pressure (i.e., the pressure drop becomes too high), the compensator spool will move to the right against the force of the spring. This acts to keep the pressure drop across the main spool and, consequently, the flow rate, nearly constant.

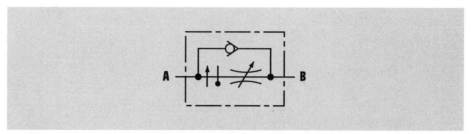

FIGURE 8-4   Pressure- and temperature-compensated FCV symbol.

Pressure-compensated flow control valve designs other than the one just described are also common, but the basic concept is the same. Some designs replace the main spool with a needle valve similar to the one shown in Figure 8-1. Pressure-compensated flow control valves are also available with an integral check valve for free flow in the reverse direction.

Temperature variation is another factor that may change the flow rate through a valve because fluids become less viscous (thinner) as their temperature increases. Because a thinner fluid will flow more readily through a given size orifice than a thicker one, increases in temperature will cause the flow rate to increase for a given flow control valve setting. For this reason, *temperature-compensated* flow control valves are also available. These valves have a temperature-sensitive element that causes the orifice size to decrease in proportion to any temperature increase. Figure 8-4 shows the symbol for a pressure-and temperature-compensated flow control valve with return check.

Another way to control the flow in a hydraulic circuit is to *throttle* (partially close) a directional control valve. This can be done with either tandem or closed neutral directional control valves, which close the outlets to the actuator when in neutral. Shifting the valve so that it is between one of the open positions and the neutral position allows the flow to be restricted, just as it is with a flow control valve. Directional control valves that are intended to be used in this fashion usually have small grooves, called *metering notches,* cut into the spool, which improve the metering characteristics of the valve. Throttling a directional control valve is a commonly used method to control the flow rate on mobile equipment. The advantage of throttling in these applications is to allow the operator to control the direction and the speed of a particular actuator with one lever. Valve throttling is also often used to control the flow in electronically controlled circuits (see Chapter 13).

8.3   Flow Coefficient

The flow rate through a valve is proportional to the pressure drop across it. The *flow coefficient* describes the exact relationship between pressure drop and flow rate for a given valve mathematically. It is used in the following equation:

$$Q = C_V \cdot \sqrt{\frac{\Delta p}{S_g}}$$ **(8-1)**

where: C_v = flow coefficient $\left(\frac{\text{gpm}}{\sqrt{\text{psi}}}, \frac{\text{lpm}}{\sqrt{\text{kPa}}}\right)$

Δp = pressure drop across the valve (psi, kPa)
Q = flow rate through the valve (gpm, lpm)
S_g = specific gravity of the fluid

The C_V value is determined experimentally by the valve manufacturer. Specific gravity is the density of a liquid divided by the density of water. Because it is density divided by density, it has no units. The specific gravity is necessary in the equation because the test to determine the C_V is performed using water. It is basically a correction factor so that the equation is accurate for other fluids. The following examples will illustrate how the flow coefficient is used.

EXAMPLE 8-1.

A valve with a C_V of 1.9 gpm/$\sqrt{\text{psi}}$ has a pressure drop of 30 psi. What must the flow rate be through the valve? The system uses standard hydraulic oil ($S_g = 0.9$).

SOLUTION:

$$Q = C_V \cdot \sqrt{\frac{\Delta p}{S_g}} = 1.9 \frac{\text{gpm}}{\sqrt{\text{psi}}} \cdot \sqrt{\frac{30 \text{ psi}}{0.9}} = 10.97 \text{ gpm}$$

EXAMPLE 8-1M.

A valve with a C_V of 2.5 lpm/$\sqrt{\text{kPa}}$ has a pressure drop of 200 kPa. What must the flow rate be through the valve? The system uses standard hydraulic oil ($S_g = 0.9$).

SOLUTION:

$$Q = C_V \cdot \sqrt{\frac{\Delta p}{S_g}} = 2.5 \frac{\text{lpm}}{\sqrt{\text{kPa}}} \cdot \sqrt{\frac{200 \text{ kPa}}{0.9}} = 37.27 \text{ lpm}$$

EXAMPLE 8-2.

A valve with a C_V of 1.5 gpm/$\sqrt{\text{psi}}$ is being considered for use in a system that has a flow rate of 15 gpm. What will the pressure drop be across this valve? The system uses standard hydraulic oil ($S_g = 0.9$).

SOLUTION:

$$\Delta p = \frac{Q^2}{C_V^{\,2}} \cdot S_g = \frac{(15\ \text{gpm})^2}{\left(1.5\ \dfrac{\text{gpm}}{\sqrt{\text{psi}}}\right)^2} \cdot 0.9 = 90\ \text{psi}$$

EXAMPLE 8-2M.

A valve with a C_V of 3.0 lpm/$\sqrt{\text{kPa}}$ is being considered for use in a system that has a flow rate of 60 lpm. What will the pressure drop be across this valve? The system uses standard hydraulic oil ($S_g = 0.9$).

SOLUTION:

$$\Delta p = \frac{Q^2}{C_V^{\,2}} \cdot S_g = \frac{(60\ \text{lpm})^2}{\left(3.0\ \dfrac{\text{lpm}}{\sqrt{\text{kPa}}}\right)^2} \cdot 0.9 = 360\ \text{kPa}$$

EXAMPLE 8-3.

What would the pressure drop be for the valve in Example 8-2 if the flow rate were doubled?

SOLUTION:

$$\Delta p = \frac{Q^2}{C_V^{\,2}} \cdot S_g = \frac{(30\ \text{gpm})^2}{\left(1.5\ \dfrac{\text{gpm}}{\sqrt{\text{psi}}}\right)^2} \cdot 0.9 = 360\ \text{psi}$$

This example illustrates an important point. If the flow rate through a valve is doubled, the pressure drop across it is quadrupled! This is because the pressure drop is proportional to the *square* of the flow rate. This is an extremely important point to remember when selecting valves of any type.

EXAMPLE 8-4.

A valve tested in the lab is found to have a pressure drop of 20 psi at a flow rate of 25 gpm. The valve is tested using water. What is the C_V value for this valve?

SOLUTION:

$$C_V = Q \cdot \sqrt{\frac{S_g}{\Delta p}} = 25\ \text{gpm} \cdot \sqrt{\frac{1}{20\ \text{psi}}} = 5.59\ \frac{\text{gpm}}{\sqrt{\text{psi}}}$$

A larger C_v represents a smaller pressure drop at a given flow rate. Larger valves have larger C_v values because a larger valve will have less resistance to flow, and therefore a smaller pressure drop. While the flow coefficient is most often used in hydraulics with flow control valves, it can in fact be used to relate pressure drop and flow for any orifice.

8.4   Circuits

A simple cylinder circuit illustrates the use of flow controls in hydraulic circuits. Figure 8-5 shows such a circuit when the cylinder is extending. Fluid is flowing from the pump to the blind end of the cylinder. Fluid is also flowing from the rod end of the cylinder to the tank. This provides two opportunities to control the speed of the cylinder. We can control the flow rate going into the cap end of the cylinder (*meter-in*), or we can control the flow rate coming out of the rod end of the cylinder (*meter-out*). A cylinder with meter-in flow control of the extend stroke is shown in Figure 8-6. When the cylinder is extending,

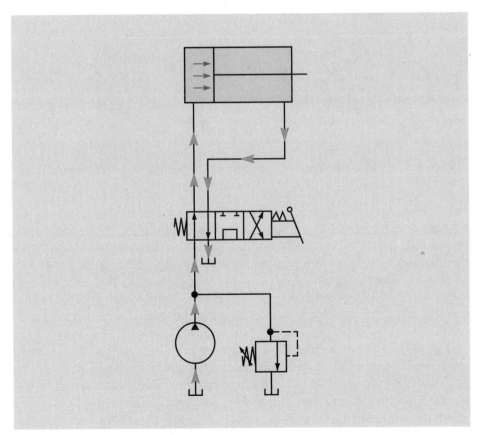

FIGURE 8-5   Cylinder extending.

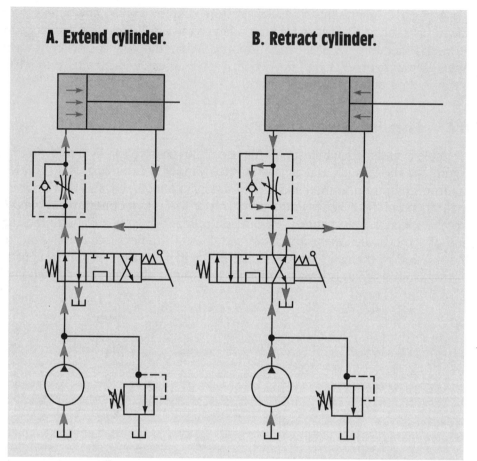

A. Extend cylinder.   **B. Retract cylinder.**

FIGURE 8-6   Meter-in flow control, extend stroke. (A) Extend cylinder. (B) Retract cylinder.

the flow coming from the pump cannot pass through the check valve and is forced to go through the metering orifice (part A). When the cylinder is retracting, the needle valve is being bypassed through the check (part B). The net result is that the flow control valve is controlling the extend speed, while the retract speed of the cylinder is uncontrolled. It is common to control only the working stroke of a cylinder, while allowing the return stroke move at full speed.

Figure 8-7 shows a cylinder with meter-out flow control of the extend stroke. The flow control valve in this circuit is placed in the rod end line. When the cylinder is extending, the flow coming from the cylinder cannot pass through the check and is forced to go through the metering orifice (part A). When the cylinder is retracting, the metering orifice is being bypassed through the check

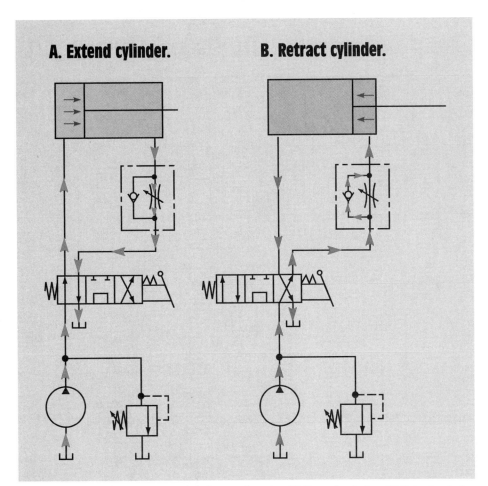

FIGURE 8-7   Meter-out flow control, extend stroke. (A) Extend cylinder. (B) Retract cylinder.

(part B). The net result is the same as with the previous circuit. The extend speed is controlled, while the retract speed is uncontrolled. However, in this circuit we control the flow rate out of the cylinder, while in the previous circuit we controlled the flow rate into the cylinder.

We may also use meter-in or meter-out flow control to regulate the retract speed of the cylinder. A circuit that has meter-in control of both strokes is shown in Figure 8-8. When the cylinder is extending, flow control valve 1 is metering the flow into the blind end of the cylinder and flow control valve 2 is bypassed (part A). When the cylinder is retracting, flow control valve 2 is metering the flow into the rod end and flow control valve 1 is bypassed (part B). The net result is that flow control valve 1 controls the extend speed, while flow control valve 2 controls the retract speed.

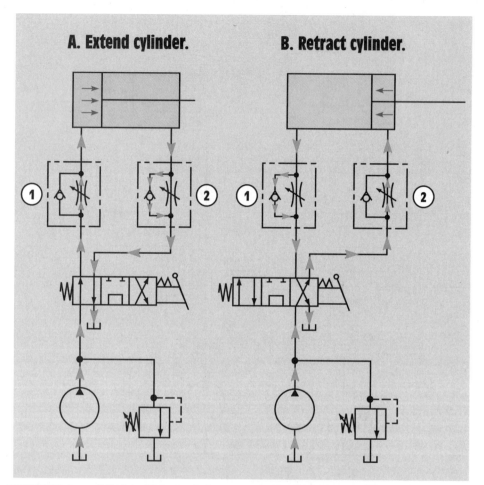

FIGURE 8-8   Cylinder with meter-in flow control of both strokes. (A) Extend cylinder. (B) Retract cylinder.

Figure 8-9 shows a circuit that has meter-out control of both strokes. When the cylinder is extending, flow control valve 1 is bypassed and flow control valve 2 is metering the flow out of the rod end of the cylinder (part A). When the cylinder is retracting, flow control valve 2 is bypassed and flow control valve 1 is metering the flow out of the blind end (part B). The net result is that flow control valve 2 controls the extend speed, while flow control valve 1 controls the retract stroke. This is the opposite of the previous circuit. The difference between the two previous circuits is the orientation of the check valves on the flow control valves.

Actual flow control valves frequently have an arrow pointing in the direction of controlled flow displayed prominently on the valve body. This makes it easy

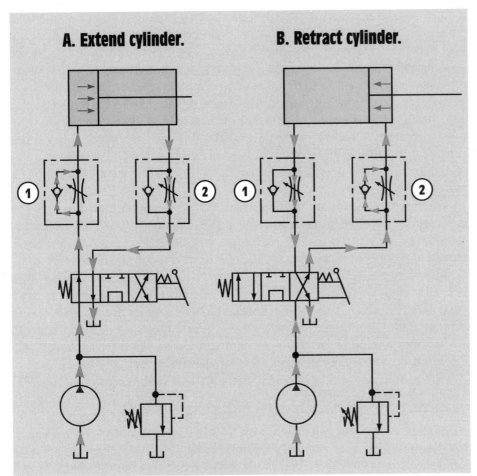

A. Extend cylinder.   ## B. Retract cylinder.

FIGURE 8-9   Cylinder with meter-out flow control of both strokes. (A) Extend cylinder.
(B) Retract cylinder.

to hook up flow controls. If meter-in is desired, point the arrow toward the cylinder port. If meter-out is desired, point the arrow away from the cylinder port.

To explain how flow control valves limit the flow rate in a hydraulic system, we must answer the following question: If the pump is producing a steady flow, how can the flow control valve slow down the flow in the system? The fluid in a hydraulic system is virtually incompressible, so whatever flow comes out of the pump must be flowing somewhere in the system. It cannot simply disappear. For example, if a pump is producing a flow of 10 gpm and a flow control valve is partially closed so that a flow of only 7 gpm is passing to the actuator, where is the other 3 gpm going? To answer this question, we must understand that by partially closing the flow control valve we are introducing a restriction in the system, which causes increased pressure upstream of the valve due to the increased resistance. If the pressure is high enough, it will cause the pressure relief valve to crack and some of the flow will be relieved to the tank, thereby reducing the amount of flow that goes to the actuator. This is how a flow control valve limits flow in a hydraulic system: Partially closing the valve causes a pressure increase, which in turn causes some of the flow to be relieved to the tank through the pressure relief valve. This is why a flow control valve will cause increased inefficiency in a system. The flow that is forced over the pressure relief valve is lost energy that is converted to heat. Another source of wasted energy is the pressure drop across the flow control valve, which is also converted to heat.

When flow controls are used with pressure-compensated pumps, the function is slightly different. Instead of the increased pressure causing the excess flow to be forced over the pressure relief valve, it causes the pressure-compensated pump to reduce its flow output. Recall from the previous chapter that a pressure-compensated pump reduces its flow when the pressure reaches a preset level, called the firing pressure. It will reduce its flow to zero if necessary to prevent the pressure from rising above the maximum setting. The pressure-compensated pump therefore makes the system more efficient by reducing the flow without forcing oil over the pressure relief valve. There is, however, still a pressure drop across the flow control valve that is converted to heat.

How does one decide between meter-in and meter-out flow control? To explain this we must understand two important concepts in hydraulics: *pressure intensification* and *tractive* versus *resistive* loads. Pressure intensification can occur in cylinders with meter-out flow controls. It occurs if the flow control valve is closed completely, as in the circuit shown in Figure 8-10. In this situation, the cylinder will not move because no flow is allowed to exit the rod end. The pressure in the pump line will immediately build to the pressure relief valve setting because the flow in the system has nowhere else to go. What will the pressure be in the rod end? The pressure is *not* the same as the pressure in the blind end because the piston is sealed. In fact, the pressure will be considerably higher in the rod end. This is because the pressure in the blind end is applying a force on the piston, and this force is then applied to the smaller area on the rod side. The pressure will therefore be higher on the rod side and is said to be *intensified*. The following example illustrates this concept.

EXAMPLE 8-5.

The circuit shown in Figure 8-10 has a cylinder with a 2 in diameter bore and a 1 in diameter rod. The pressure relief valve is set to 2500 psi. If the meter-out flow control valve is closed completely, what will the pressure be in the rod end? The piston diameter is assumed to be the same as the bore diameter.

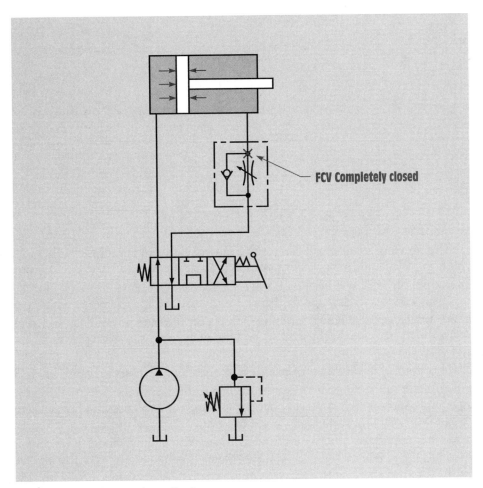

FIGURE 8-10 Pressure intensification.

SOLUTION:

1. Calculate the piston and rod areas:

$$A_P = \frac{\pi \cdot D_P^{\,2}}{4} = \frac{3.142 \cdot (2 \text{ in})^2}{4} = 3.142 \text{ in}^2$$

$$A_R = \frac{\pi \cdot D_R^{\,2}}{4} = \frac{3.142 \cdot (1 \text{ in})^2}{4} = 0.7854 \text{ in}^2$$

As stated earlier, the pressure in the blind end will be the pressure relief setting.

2. Calculate the force on the piston:

$$F = p_{CAP} \cdot A_p = 2500 \frac{\text{lbs}}{\text{in}^2} \cdot (3.142 \text{ in}^2) = 7855 \text{ lbs}$$

This force is then applied over the smaller area on the rod end $(A_p - A_R)$.

3. Calculate the pressure in the rod end:

$$p = \frac{F}{A_p - A_R} = \frac{7855 \text{ lbs}}{3.142 \text{ in}^2 - 0.7854 \text{ in}^2} = 3333 \frac{\text{lbs}}{\text{in}^2}$$

The previous example illustrates the drawback to using meter-out flow control: If the flow control valve on the rod end was inadvertently closed, the pressure in the rod end will be significantly higher than the pressure relief valve setting. This can cause damage to the rod end seals on the cylinder, which may create a dangerous situation. Chapter 9 discusses pressure intensifiers, which are devices that use this principle to intentionally intensify the pressure in a hydraulic system.

In spite of the possibility of pressure intensification, there are applications in which meter-out flow control is a must. Figure 8-11 shows a schematic of one such application. In this circuit, a cylinder is being used to raise and lower a large weight. We will almost certainly need to control the rate of descent when the cylinder is retracting, otherwise the weight may cause the cylinder to accelerate uncontrollably. Will meter-in do the job? Almost certainly not. Controlling how fast fluid enters the cylinder may do nothing at all if the weight begins to drive the cylinder downward. The only way to have complete control of the rate of descent in this circuit is to control how fast the fluid exits the blind end. In this situation, we must meter-out. Figure 8-12 shows this circuit with meter-out flow control of the rate of descent. Meter-out is preferred in situations such as this in which the load is acting in the direction of movement. Loads that act in the direction of movement are known as *tractive* loads.

When the cylinder in the previous circuit is extending, however, the load is opposing the motion. Loads that oppose the direction of motion are known as *resistive* loads. With resistive loads, meter-out has no particular advantage over meter-in. It does, however, have the disadvantage of the possibility of pressure intensification. We would therefore use meter-in flow control when controlling a resistive load. Figure 8-13 shows the previous circuit with meter-in flow control on the extend stroke and meter-out flow control on the retract stroke. When

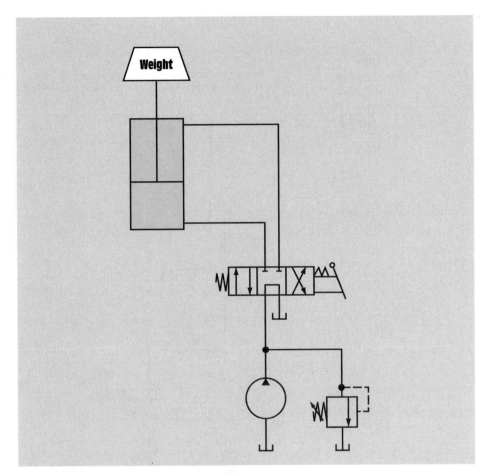

FIGURE 8-11   Cylinder lifting a large weight.

the cylinder is extending, the speed is controlled by flow control valve 1 and flow control valve 2 is bypassed (part A). When the cylinder is retracting, the speed is controlled by flow control valve 2 and flow control valve 1 is bypassed (part B).

When controlling the speed of hydraulic motors with flow control valves, the same rule applies: Use meter-in flow control for resistive loads and meter-out flow control for tractive loads.

In addition to meter-in and meter-out flow control, there is a less commonly used flow control configuration known as *bleed-off* (Figure 8-14). In this type of flow control, an additional line is run through a flow control back to the

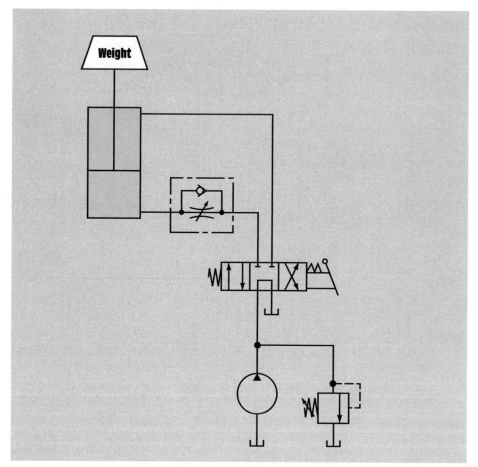

FIGURE 8-12   Cylinder with meter-out control of the retract stroke.

tank. To slow down the actuator, some of the flow is bled off through this line, thereby reducing the flow to the actuator. Figure 8-14A shows bleed-off control of the extend stroke; part B shows control of the retract stroke. Note that the operation of a bleed-off flow control valve is opposite to a meter-in or meter-out flow control valve. Opening a bleed-off flow control valve slows down an actuator, while opening a meter-in or meter-out flow control valve increases actuator speed.

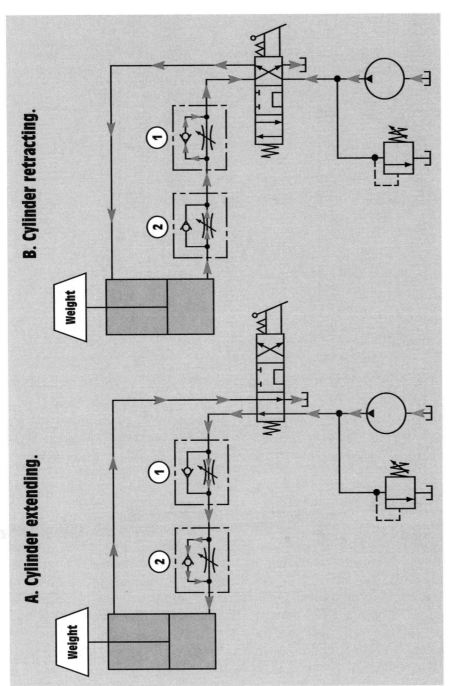

FIGURE 8-13 Cylinder with meter-in control of the extend stroke and meter-out control of the retract stroke. (A) Cylinder extending. (B) Cylinder retracting.

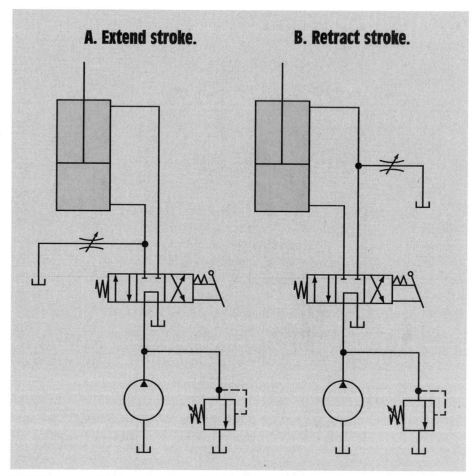

FIGURE 8-14   Cylinders with bleed-off flow control. (A) Extend stroke. (B) Retract stroke.

8.5   Cushioned Cylinders

In some applications, it is desirable to slow a cylinder down just before it reaches the end of its stroke. This may be necessary in applications in which the cylinder is moving a delicate load that would be damaged by a rapid deceleration at the end of its stroke. It may be necessary to prevent a clamp cylinder from causing damage when it contacts the workpiece. It may also be necessary to prevent a fast-moving cylinder from causing damage to itself when it bottoms out. Cushioned cylinders have a built-in flow control valve that acts only near the end of its stroke. A simplified cutaway of a cushioned cylinder is shown in Figure 8-15A. As this cylinder reaches the end of its extend stroke, the cushion spear enters the cushion recess, which blocks the direct path to the rod end port. The fluid must now flow through the

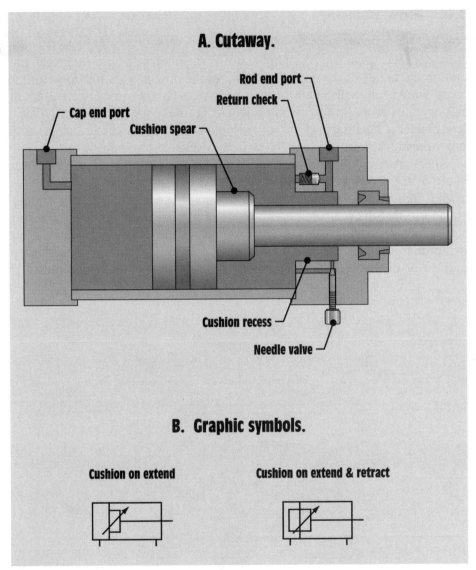

A. Cutaway.

Rod end port
Return check
Cap end port
Cushion spear
Cushion recess
Needle valve

B. Graphic symbols.

Cushion on extend

Cushion on extend & retract

FIGURE 8-15   Hydraulic cylinder with cushioning. (A) Cutaway. (B) Graphic symbols.

passageway controlled by the needle valve to get to the rod end port. The needle valve allows the speed at the end of the stroke to be adjusted. The return check allows the cushioning to be bypassed when the cylinder is retracted. Cylinders may have cushions on one stroke, such as the one shown, or they may have cushions on both strokes. Figure 8-15B shows the symbols for cushioned cylinders.

8.6   Flow Dividers

Flow dividers divide the flow from a pump into two or more streams of equal flow rates. They maintain equal flow rates in the branch circuits even if the pressures in the branches are not equal. Without the flow divider, the flow from the pump would follow the path of least resistance (lowest pressure). The branch that operates at the lowest pressure would normally receive all of the flow. A flow divider is necessary whenever multiple branch circuits that operate at different pressures must divide the flow from a single pump equally.

There are two commonly used flow divider designs: *balanced spool* and *rotary*. Figure 8-16A shows a simplified cutaway of a balanced spool flow divider. The spool is free to slide back and forth in the housing and will naturally assume a position so that the pressure on either side of the spool will be equal. The spool is therefore pressure balanced. For example, if the pressure at outlet 1 was greater than the pressure at outlet 2, the spool would slide to the right to partially cover outlet 2. By partially restricting the more lightly loaded outlet, the

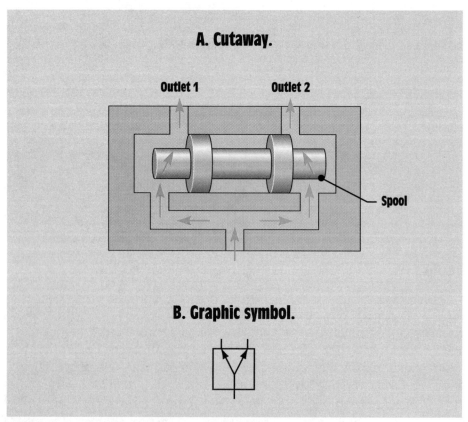

FIGURE 8-16   Balanced spool flow divider. (A) Cutaway. (B) Graphic symbol.

flow divider adds more resistance to this path. This acts to equalize the resistance of each path, thereby ensuring that equal flow will go to each path. Figure 8-16B shows the graphic symbol for a flow divider.

A rotary flow divider is basically two gear pumps in one housing whose inlets are joined together. Their shafts are also coupled together so that they must turn at the same speed. Because they are forced to turn at the same speed, they will supply equal flow to their outlets when placed in a pump line (Figure 8-17).

One common application for a flow divider is the synchronization of two cylinders (Figure 8-18). This type of circuit is used whenever two cylinders must move simultaneously *and* at the same speed. This circuit is just two cylinders connected in parallel, with a flow divider placed between the pump and the two cylinder branches. A flow divider is required in a circuit such as this even if the cylinders are thought to have equal loads. This is because slight variations in the load and cylinder friction will most likely cause the cylinders to *not* move simultaneously and at the same speed. With this configuration, the cylinders may be operated together or one at a time. Each branch will receive half the pump flow in all situations. Thus, whether one or both of the cylinders is operated, the speed is the same.

Notice that each branch circuit shown in Figure 8-18 is equipped with its own pressure relief valve. This is necessary because the two branches will not

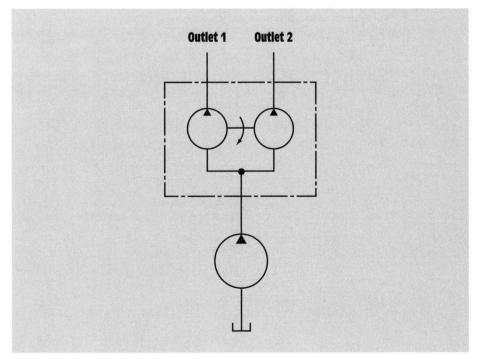

FIGURE 8-17   Rotary flow divider (shown with pump).

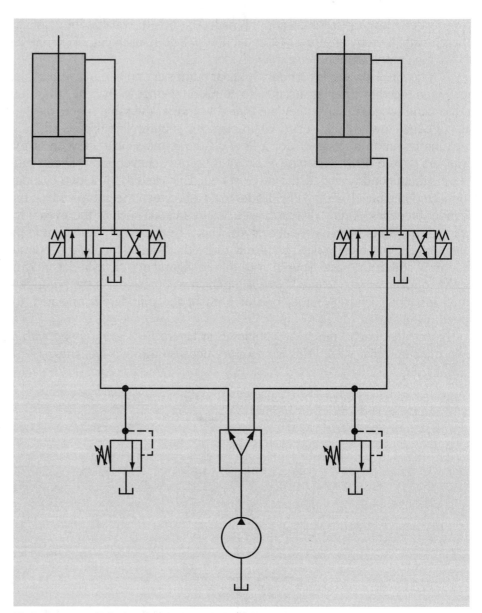

FIGURE 8-18   Cylinder synchronization with a flow divider.

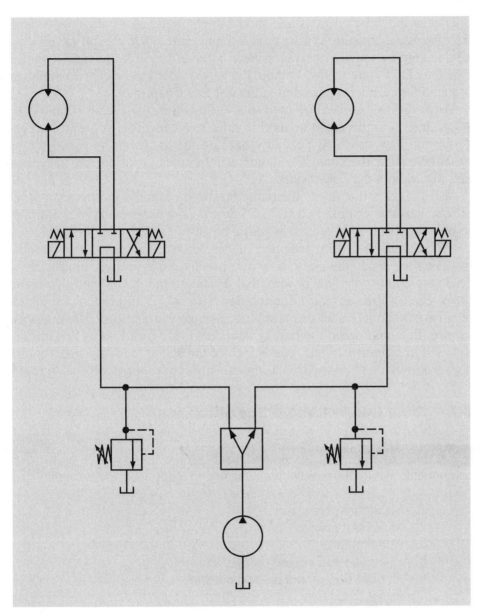

FIGURE 8-19   Motor synchronization with a flow divider.

operate at the same pressure, and should therefore be protected independently from excessive pressure. In this circuit, we may use tandem neutral valves, which save energy by sending the pump flow back to the tank under low pressure when the DCV is in neutral. Without the flow divider, closed center neutrals are required for cylinders connected in parallel (see Chapter 6).

Flow dividers can also be used to split flow in motor circuits (Figure 8-19). A circuit such as this could be used to drive two conveyors at the same speed. Again, the flow divider is necessary because the loads on the conveyors will never be exactly the same. The motors will most likely not operate in synchronization without the flow divider.

Flow dividers that divide the pump flow into proportions other than 50/50 are also available. For example, a 75/25 flow divider will provide 75% of its flow to one outlet and 25% of its flow to the other.

The flow dividers discussed previously are known as *proportional* flow dividers because they divide the pump flow into two fixed proportions, usually 50/50. There are also *priority* flow dividers that divide the pump flow into two branch flows. These valves do not divide the flow into two fixed proportions; instead, they have a priority outlet that always receives flow and a secondary outlet that receives flow only when the priority flow demand is met. Priority flow dividers are used in systems in which one branch of the system contains essential functions that must be performed, while the other branch contains less vital functions.

8.7   Flow Control Valve Specifications

Like directional and pressure control valves, flow control valves are available in inline, subplate, and cartridge mounting types. As is the case with all hydraulic components, flow control valve manufacturers will provide specifications for:

1. Maximum pressure,
2. Maximum flow,
3. Filtration level,
4. Fluid type and viscosity range, and
5. Physical size, mounting, and porting.

The manufacturer will also supply flow (Q) versus pressure drop (Δp) information for each model. This information is provided for at least two situations:

1. When the flow control valve is fully open and flow is going through the needle valve (the *controlled flow* direction), and
2. When the flow control valve is fully closed and fluid is flowing in the reverse direction through the check valve (the *free-flow* direction).

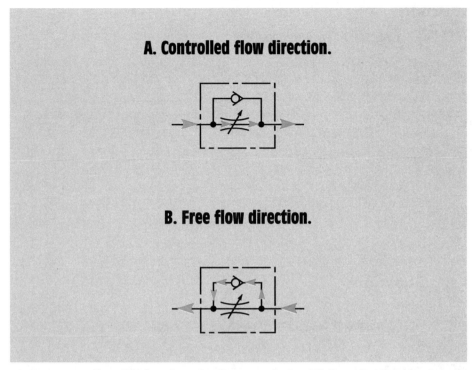

A. Controlled flow direction.

B. Free flow direction.

FIGURE 8-20   Flow directions through a flow control valve. (A) Controlled flow direction.
(B) Free flow direction.

Flow directions are illustrated in Figure 8-20. One way to supply pressure drop information is with the C_v value which was discussed earlier in this chapter. This is the most compact way to supply pressure drop information because only two values must be provided, one for each of the two situations just listed. The pressure drop can then be calculated for any particular flow rate. Pressure drop versus flow information may also be provided with a table or graph.

EXAMPLE 8-6.

A flow control valve has a controlled flow C_v of 0.74 gpm/$\sqrt{\text{psi}}$ and a free flow C_v of 2.25. gpm/$\sqrt{\text{psi}}$. Determine the pressure drop across the valve in both the controlled flow and free-flow directions. The system has a flow rate of 5 gpm and uses standard hydraulic oil ($S_g = 0.9$).

SOLUTION:

1. Controlled flow direction:

$$\Delta p = \frac{Q^2}{C_v{}^2} \cdot S_g = \frac{(5 \text{ gpm})^3}{\left(0.74 \dfrac{\text{gpm}}{\sqrt{\text{psi}}}\right)^2} \cdot 0.9 = 41.1 \text{ psi}$$

2. Free-flow direction:

$$\Delta p = \frac{Q^2}{C_v{}^2} \cdot S_g = \frac{(5 \text{ gpm})^2}{\left(2.25 \dfrac{\text{gpm}}{\sqrt{\text{psi}}}\right)^2} \cdot 0.9 = 4.44 \text{ psi}$$

8.8 Equations

EQUATION NUMBER	EQUATION	REQUIRED UNITS
8-1	$Q = C_v \cdot \sqrt{\dfrac{\Delta p}{S_g}}$	Any consistent units

8.9   Review Questions and Problems

1. What effect does a flow control valve have on an actuator?
2. What is another way to control the flow rate to an actuator?
3. What is the purpose of the check valve in the flow control valve?
4. What is the purpose of pressure compensation in flow control valves?
5. What is the purpose of temperature compensation in flow control valves?
6. What does the flow coefficient (C_V) reveal about a valve?
7. Describe the difference between meter-in and meter-out flow control.
8. Describe the operation of bleed-off flow control.
9. Draw a hydraulic motor with meter-in flow control of both directions of rotation.
10. Draw a hydraulic cylinder with meter-in flow control of both strokes.
11. Draw a hydraulic motor with meter-out flow control of both directions of rotation.
12. Draw a hydraulic cylinder with meter-out flow control of both strokes.
13. Describe the difference between a resistive and a tractive load. What type of flow control is preferred for each?
14. Define *pressure intensification.*
15. A flow control is used in a system with a pump that is *not* pressure compensated. What happens to the excess flow that does not pass through the partially closed flow control valve?
16. Describe the operation of a cushioned cylinder.
17. What is the purpose of a flow divider?
18. What is the difference between a proportional flow divider and a priority flow divider?
19. A valve with a C_V of 2.5 gpm/$\sqrt{\text{psi}}$ has a pressure drop of 50 psi. What is the flow rate through the valve? The system uses standard hydraulic oil ($S_g = 0.9$).
20. A valve with a C_V of 2.75 lpm/$\sqrt{\text{kPa}}$ has a pressure drop of 320 kPa. What is the flow rate through the valve? The system uses standard hydraulic oil ($S_g = 0.9$).
21. A valve with a C_V of 1.6 gpm/$\sqrt{\text{psi}}$ is being considered for use in a system that has a flow rate of 10 gpm. What will the pressure drop be across this valve? The system uses standard hydraulic oil ($S_g = 0.9$).

22. A valve with a C_v of 1.75 lpm/$\sqrt{kPa}$ is being considered for use in a system that has a flow rate of 35 lpm. What will the pressure drop be across this valve? The system uses standard hydraulic oil ($S_g = 0.9$).

23. A valve being tested in the lab was found to have a pressure drop of 25 psi at a flow rate of 10 gpm. The valve was tested using water. What is the C_v value for this valve?

24. A valve being tested in the lab was found to have a pressure drop of 175 kPa at a flow rate of 30 lpm. The valve was tested using water. What is the C_v value for this valve?

Ancillary Hydraulic Components

OUTLINE

9.1 Introduction

9.2 Accumulators

9.3 Intensifiers

9.4 Reservoirs

9.5 Heat Exchangers

9.6 Filters

9.7 Instrumentation and Measurement

9.8 Conduits and Fittings

9.9 Seals and Bearings

9.10 Hydraulic Fluids

9.11 Equations

9.12 Review Questions and Problems

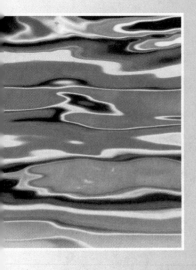

9.1   Introduction

Chapter 9 covers the supporting devices used in hydraulic circuits. The fact that they are termed "ancillary" or "supporting" should not diminish their importance in a hydraulic system. Most of the devices discussed in this chapter are critical components in any circuit in which they are used. Failure of many of these components will result in failure of the circuit just as surely as failure of a pump or actuator. Proper understanding of the basic operation of these devices is critical to obtain a complete understanding of hydraulic circuits. The reader is again encouraged to seek out catalogs for each of these components to get a better understanding of their selection and use. Many companies have their catalogs available on the Internet for viewing and/or downloading. The National Fluid Power Association's web site (www.nfpa.com) is a good place to start a search.

9.2   Accumulators

Accumulators are devices that store hydraulic fluid under pressure. Storing hydraulic fluid under pressure is a way of storing energy for later use. Perhaps the most common application for an accumulator is supplementing the pump flow in a hydraulic system in which a high flow rate is required for a brief period of time. Accumulators also have other important applications that will be discussed shortly. Most accumulators operate by storing energy in the form of a compressed gas. These are known as *gas-charged* accumulators.

The basic construction of a piston-type, gas-charged accumulator is shown in Figure 9-1A. Its operation begins when the gas chamber is filled with a gas to some predetermined pressure, called the *precharge,* which causes the free-sliding piston to move down. Once the accumulator is precharged, hydraulic fluid can be pumped into the hydraulic fluid port. As the hydraulic fluid enters the accumulator, it causes the piston to slide up, thereby compressing the gas. Compressing the gas increases its pressure, and this pressure is then applied to the hydraulic fluid through the piston. Because the piston is free sliding, the pressure on the gas and the hydraulic fluid is always equal. A photograph of an actual piston accumulator is shown in Figure 9-1B.

When the accumulator is filled with hydraulic fluid, usually to the point where the gas volume has been reduced to half its original volume, the accumulator is *fully charged.* We now have a reservoir of pressurized hydraulic fluid from which the system can draw. Whenever the pressure in the system drops below the pressure in the accumulator, fluid will flow out of the accumulator and into the system. As the hydraulic fluid flows out of the accumulator, the gas decompresses and loses pressure, which in turn causes the pressure on the hydraulic fluid to be reduced. Therefore as we draw fluid out, the pressure is constantly decreasing. This concept must be understood in order to understand the operation of an

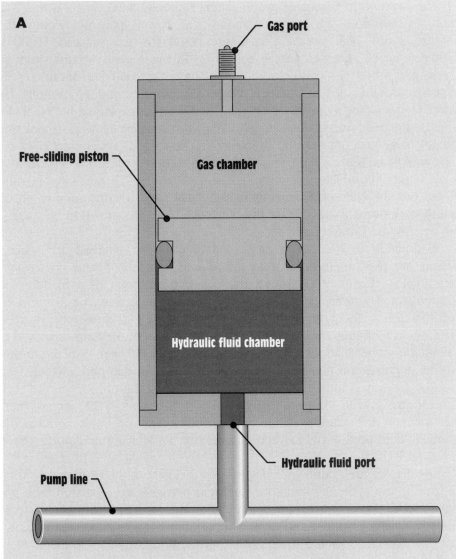

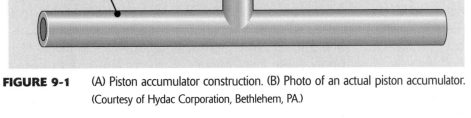

FIGURE 9-1   (A) Piston accumulator construction. (B) Photo of an actual piston accumulator. (Courtesy of Hydac Corporation, Bethlehem, PA.)

accumulator. The system will continue to draw fluid from the accumulator until the pressure in the accumulator equals the pressure in the system. When the pressure in the system rises above the pressure in the accumulator, fluid will flow from the system into the accumulator, thereby recharging it. Accumulators frequently operate on this type of cycle—filling with fluid during part of the cycle to store energy, then discharging fluid to the system during the other part of the cycle.

The gas used to precharge accumulators is usually nitrogen because it is an inert gas and does not support combustion. This is particularly important because at the high precharge pressures used for accumulators (1000 to 1500 psi), air becomes extremely dangerous in this regard. Accumulators are precharged with a nitrogen bottle before being put into service. Because of the high pressures involved, this process can be dangerous if not done carefully. The manufacturer's procedure for precharging an accumulator should be strictly followed. After the accumulator is initially charged, it must be allowed to cool, then the precharge pressure must be checked because a significant amount of heat is generated when a gas is compressed. As the gas cools, its pressure will decrease (see Chapter 10, which covers the behavior of gases). The precharge pressure should then be checked frequently in the initial stages of operation to ensure that there are no gas leaks. After the precharge has been shown to be stable, it can be checked less frequently.

In addition to the piston design, three other gas-charged accumulator designs are used in hydraulic circuits: *bladder, diaphragm,* and *nonseparated* accumulators. Figure 9-2A shows the basic construction of a bladder-type accumulator. These accumulators function in the same way as a piston accumulator, storing energy in the form of a compressed gas. However, instead of the gas and hydraulic fluid being separated by a piston, they are separated by a synthetic rubber bladder. The bladder is filled with nitrogen until the desired precharge pressure is achieved. Hydraulic fluid is then pumped into the accumulator, thereby compressing the gas and increasing the pressure in the accumulator, just as with the piston type. The port cover is a small piece of metal that protects the bladder from damage as it expands and contacts the hydraulic fluid port. A photograph of an actual bladder accumulator is shown in Figure 9-2B.

The diaphragm accumulator (Figure 9-3) is similar to the bladder-type accumulator. In this type, the hydraulic fluid and nitrogen are separated by a synthetic rubber diaphragm. Its operation is the same as that of the bladder type. The advantage of bladder and diaphragm accumulators over the piston type is that they have no sliding surface that requires lubrication and can therefore be used with fluids with poor lubricating qualities. They are also less sensitive to contamination due to the lack of any close-fitting sliding components. The nonseparated type of gas-charged accumulator, as the name implies, has no separator between the gas and the hydraulic fluid. The gas applies pressure directly to the hydraulic fluid. The disadvantage to this type is that the gas can become absorbed into the hydraulic fluid and cause problems downstream in the system.

How do we correctly size a gas-charged accumulator for a given application? Because the pressure in the accumulator drops as hydraulic fluid is drawn out, it will not provide flow to the system at a constant pressure. We must therefore determine minimum and maximum pressures at which the system can operate. We must then choose an accumulator that will provide the volume of fluid that

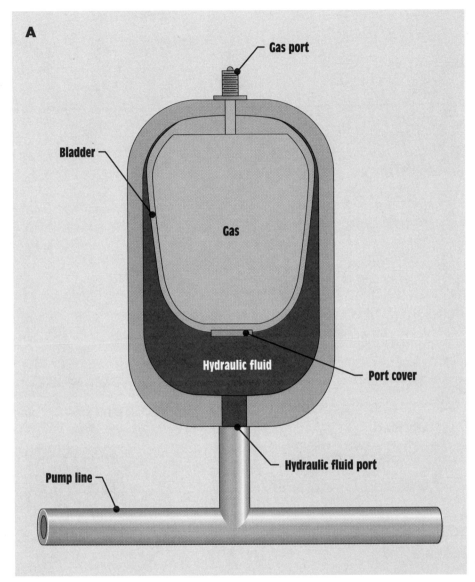

FIGURE 9-2   (A) Bladder accumulator. (B) Photo of an actual bladder accumulator.
(Courtesy of Hydac Corporation, Bethlehem, PA.)

is required between these pressures. Manufacturers provide methods for sizing their accumulators that usually involve graphs and/or equations. To completely understand the sizing process, a basic understanding of the behavior of gases, is necessary (see Chapter 10).

While gas-charged accumulators are by far the most commonly used type, there are also *spring-loaded* and *weight-loaded* accumulators. A spring-loaded

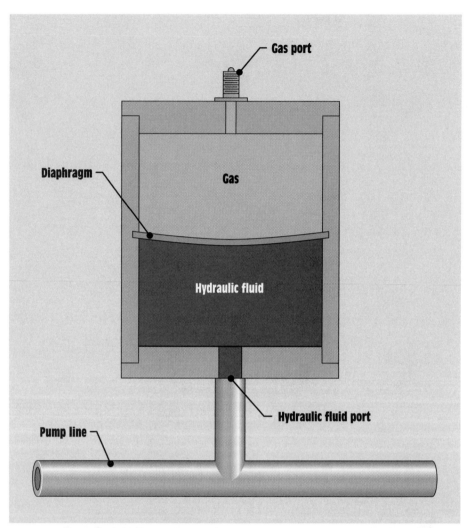

FIGURE 9-3   Diaphragm accumulator.

accumulator (Figure 9-4) stores energy in the form of a compressed spring. Hydraulic fluid is pumped into the accumulator, causing the piston to move up and compress the spring. The spring then applies a force on the piston that exerts a pressure on the hydraulic fluid. Just as with the gas-charged type, the pressure is constantly decreasing as hydraulic fluid is drawn out because the spring decompresses and therefore exerts less force on the piston. Spring-loaded accumulators are not commonly used in hydraulic circuits because a large spring must be used to generate enough pressure. This makes them much larger than a gas-charged type of an equivalent capacity.

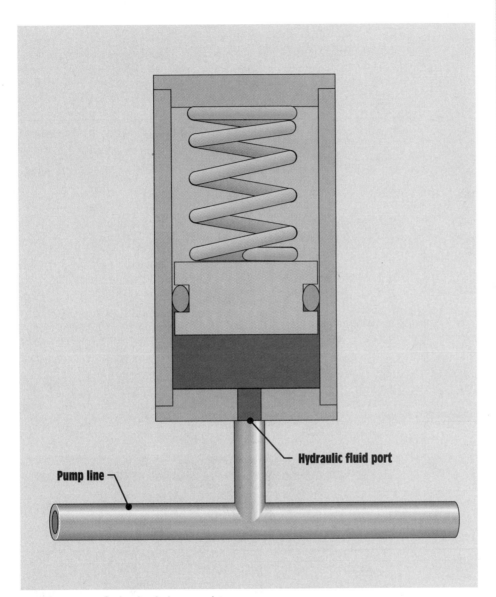

FIGURE 9-4   Spring-loaded accumulator.

A weight-loaded accumulator is basically a vertically mounted cylinder with a large weight (Figure 9-5). When hydraulic fluid is pumped into this accumulator, the weight is raised. The weight then applies a force to the piston, which generates a pressure on the fluid side of the piston. The advantage of this type of accumulator over all of the others discussed is that it applies a constant pressure on the fluid throughout its entire range of motion. The disadvantage is that a very large weight must be used to generate enough pressure. Because of this,

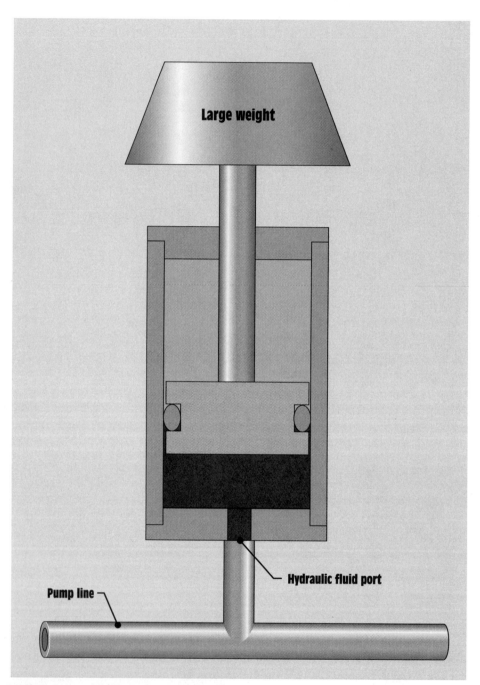

FIGURE 9-5   Weight-loaded accumulator.

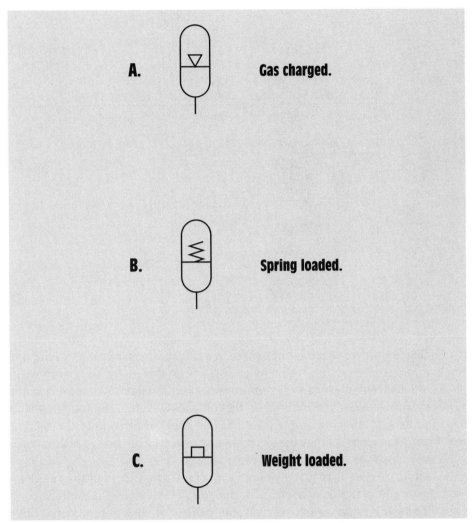

A.   **Gas charged.**

B.   **Spring loaded.**

C.   **Weight loaded.**

FIGURE 9-6   Accumulator graphic symbols. (A) Gas charged. (B) Spring loaded. (C) Weight loaded.

weight-loaded accumulators are seldom used. Figure 9-6 shows the graphic symbols for gas-charged, spring-loaded, and weight-loaded accumulators.

One common application for an accumulator is to supplement the pump flow during intermittent periods of high flow demand. Many applications require large amounts of flow to generate fast actuator speeds, but only for a portion of the operating cycle. To achieve these speeds without an accumulator, we would need to purchase a large pump, even though we are only using its capacity for a brief time period. In a circuit such as this, an accumulator can be used to supplement the pump flow during the high demand period, thereby allowing a smaller pump to be used. The supply side of such a circuit is shown in Figure 9-7. Notice the use

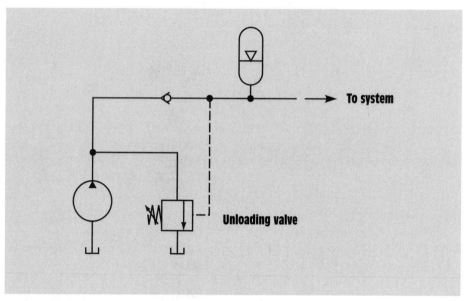

FIGURE 9-7   Accumulator supplementing pump flow.

of an unloading valve, which is different than a relief valve in that it reads the pressure in an external line, rather than in its own line. In this circuit, the pump will fill the accumulator whenever the system does not require flow. When the accumulator pressure reaches the setting of the unloading valve, the pump will be unloaded back to the tank at low pressure. The check valve isolates the accumulator from the pump so that the accumulator side can remain at high pressure while the pump is unloaded at low pressure.

In some circuits, it may be desirable to shut down the pump completely once the accumulator has been fully charged. This can be accomplished by incorporating a *pressure switch* that will shut down the electric motor that drives the pump when the accumulator is fully charged (Figure 9-8). When the accumulator pressure in this circuit reaches the setting of the pressure switch, an electrical signal is sent that will cause the pump to be shut down. The relief valve in this circuit is a safety backup to the pressure switch and would be set to a slightly higher pressure than the pressure switch so that it does not open during the normal operation of the circuit.

Using a pressure switch can be particularly useful in the press circuit shown in Figure 9-9. In circuits such as this, the cylinder may remain fully extended under pressure for a significant portion of the machine's cycle. When the cylinder reaches the end of its stroke, the accumulator will receive all of the pump flow. When the accumulator is fully charged, the pressure switch will trip, causing the pump to be shut down. The accumulator then supplies only enough flow to compensate for leakage. This keeps active pressure on the press without requiring the pump to be

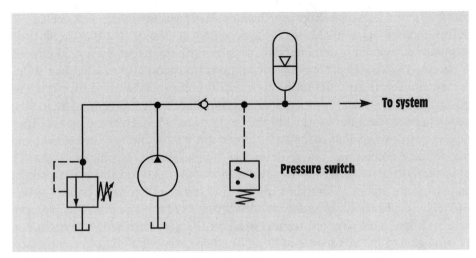

FIGURE 9-8   Accumulator with a pressure switch.

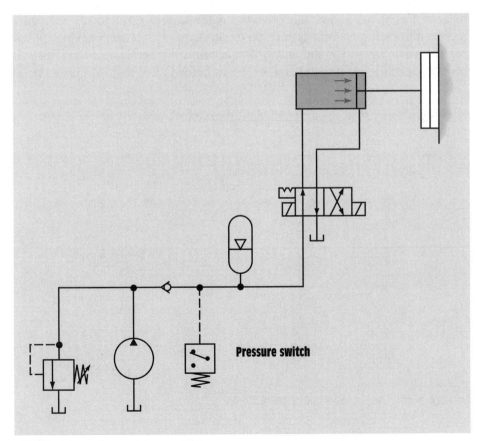

FIGURE 9-9   Accumulator used with a press.

running. If the pressure drops significantly, the pressure switch will activate the pump until the accumulator is fully charged again. Notice that this circuit uses a four-way, two-position, detented directional control valve that is solenoid actuated. A detented valve remains indefinitely in the last position into which it was shifted.

Accumulators are utilized in other applications in addition to supplementing pump flow and leakage compensation. Accumulators can also be used as an emergency pressure and flow source in the event of a loss of electrical power. This is necessary in circuits that may be dangerous if pressure and flow are cut off without notice. Another application for an accumulator is *surge suppression*. In this type of application, an accumulator is basically used as a shock absorber to dampen out large pressure spikes. A pressure spike will occur if a directional control valve is suddenly shifted to the closed position when the system is operating at high pressure and flow. This type of pressure spike can be very damaging to system components due to the vibration it induces. The compressibility of the gas in the accumulator provides a shock-absorbing affect by allowing some of the hydraulic fluid to push into the accumulator and compress the gas when a pressure spike occurs.

A *discharge valve* is an important element of any system that uses an accumulator. This valve allows the pressure in the accumulator to be relieved by opening the hydraulic fluid port to the tank. A two-way, two-position, normally closed discharge valve is shown with an accumulator in Figure 9-10. This valve must be opened to depressurize the system before the circuit can be serviced.

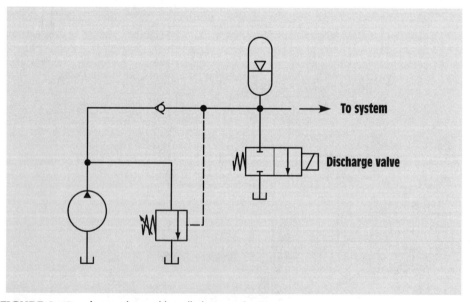

FIGURE 9-10   Accumulator with a discharge valve.

9.3   Intensifiers

Pressure intensifiers, or *boosters,* are devices used to generate pressures greater than those achievable with standard hydraulic pumps alone. They take the inlet flow from the pump and intensify the pressure. As discussed previously, any time we gain something in a physical system, we must sacrifice something in return. Pressure intensifiers increase the pressure and decrease the flow rate by the same factor. A simplified cutaway of an intensifier is shown in Figure 9-11A. The intensifier is shown on the forward stroke. In this situation, the pump flow (Q_{PUMP}) is fed into port A of the intensifier, which applies a pressure (p_{PUMP}) to the piston, causing it to move right. This in turn generates a force that is applied to the rod. The force on the rod then creates pressure (p_{INT}) and flow (Q_{INT}) at the outlet to the system. When the four-way directional control valve is shifted to the opposite position, the pump flow is sent to port B of the intensifier, causing the piston to move left. This causes fluid to be drawn into the rod chamber, which completes one cycle. The graphic symbol for an intensifier is shown in Figure 9-11B.

To correctly size an intensifier, we must calculate the pressure and flow on the outlet of the intensifier, given the inlet flow and pressure from the pump. To determine the outlet pressure, we must first realize that the force on the piston (F_{PISTON}) and the force on the rod (F_{ROD}) must be equal because they are physically connected. Stated mathematically:

$$F_{PISTON} = F_{ROD}$$

Plugging in $F = p \cdot A$, we obtain:

$$p_{PUMP} \cdot A_{PISTON} = P_{INT} \cdot A_{ROD}$$

Solving for p_{INT}, we obtain an equation that allows us to calculate the outlet pressure of an intensifier:

$$p_{INT} = p_{PUMP} \cdot \frac{A_{PISTON}}{A_{ROD}} \tag{9-1}$$

What about the flow out of the intensifier? We can determine this if we realize that the piston and rod must move with the same velocity, again because they are connected. Stated mathematically:

$$v_{PISTON} = v_{ROD}$$

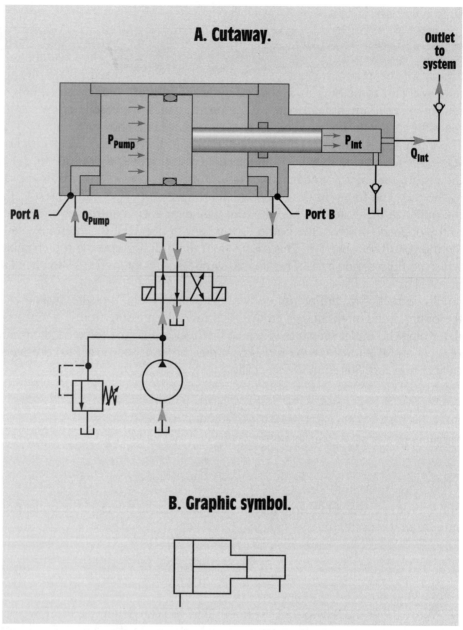

FIGURE 9-11   Pressure intensifier. (A) Cutaway. (B) Graphic symbol.

Plugging in $v = Q/A$, we obtain:

$$\frac{Q_{PUMP}}{A_{PUMP}} = \frac{Q_{INT}}{A_{ROD}}$$

Solving for Q_{INT}, we obtain an equation that allows us to calculate the outlet flow of an intensifier:

$$Q_{INT} = Q_{PUMP} \cdot \frac{A_{ROD}}{A_{PISTON}} \tag{9-2}$$

Equation 9-1 tells us that the pressure is *increased* by a factor of A_{PISTON}/A_{ROD}, which is called the *intensification ratio (IR)*. Equation 9-2 tells us that the flow is *reduced* by the same factor (multiplied by the inverse). This makes sense because the power, which equals pressure times flow, must remain constant. In other words, we cannot get more power out of a device than we put into it. Therefore, to increase the pressure, we must sacrifice flow. The *intensification ratio (IR)* would normally be provided by the manufacturer, rather than the areas. In this case, the following simple equations can be used:

$$p_{INT} = p_{PUMP} \cdot IR \tag{9-3}$$

$$Q_{INT} = \frac{Q_{PUMP}}{IR} \tag{9-4}$$

EXAMPLE 9-1.

The intensifier shown in Figure 9-11 has a piston area of 10 in² and a rod area of 1 in². What is the maximum outlet pressure if the maximum inlet pressure is 1500 psi? What is the flow rate out of the intensifier if the pump has a flow rate of 5 gpm?

SOLUTION:

$$p_{INT} = p_{PUMP} \cdot \frac{A_{PISTON}}{A_{ROD}} = 1500 \text{ psi} \cdot \frac{10 \text{ in}^2}{1 \text{ in}^2} = 15,000 \text{ psi}$$

$$Q_{INT} = Q_{PUMP} \cdot \frac{A_{ROD}}{A_{PISTON}} = 5 \text{ gpm} \cdot \frac{1 \text{ in}^2}{10 \text{ in}^2} = 0.5 \text{ gpm}$$

EXAMPLE 9-1M.

The intensifier shown in Figure 9-11 has a piston area of 0.008 m² and a rod area of 0.001 m². What is the maximum outlet pressure if the maximum inlet pressure is 7000 kPa? What is the flow rate out of the intensifier if the pump has a flow rate of 15 lpm?

SOLUTION:

$$p_{INT} = p_{PUMP} \cdot \frac{A_{PISTON}}{A_{ROD}} = 7000 \text{ kPa} \cdot \frac{0.008 \text{ m}^2}{0.001 \text{ m}^2} = 56,000 \text{ kPa}$$

$$Q_{INT} = Q_{PUMP} \cdot \frac{A_{ROD}}{A_{PISTON}} = 15 \text{ lpm} \cdot \frac{0.008 \text{ m}^2}{0.001 \text{ m}^2} = 120 \text{ lpm}$$

EXAMPLE 9-2.

An intensifier with an intensification ratio of 4:1 is being used in a system with a maximum pump pressure of 2500 psi and a flow rate of 10 gpm. What is the flow rate and the maximum pressure out of the intensifier?

SOLUTION:

$$p_{INT} = p_{PUMP} \cdot IR = 2500 \text{ psi} \cdot 4 = 10,000 \text{ psi}$$

$$Q_{INT} = \frac{Q_{PUMP}}{4} = \frac{10 \text{ gpm}}{4} = 2.5 \text{ gpm}$$

One common application for an intensifier is to use compressed air on the low-pressure side of an intensifier and hydraulic fluid on the high-pressure side. These are called *hydropneumatic,* or *air-over-oil,* intensifiers. Hydropneumatic intensifiers are commonly used to power hydraulic stamping presses, clamps, and other such machinery. These systems are convenient and relatively inexpensive because most shops have compressed air power readily available.

EXAMPLE 9-3.

A hydropneumatic intensifier with an intensification ratio of 25:1 is being used to power the punch press that uses a 3 in diameter cylinder. The shop has compressed air at 100 psi available. What is maximum force output of the press?

SOLUTION:

1. Calculate the intensified pressure:

$$p_{INT} = p_{COMP} \cdot IR = 100 \text{ psi} \cdot 25 = 2500 \text{ psi}$$

2. Calculate the area of the press:

$$A_{PRESS} = \frac{\pi}{4} \cdot D_{PRESS}^2 = \frac{3.142}{4} \cdot (3 \text{ in})^2 = 7.070 \text{ in}^2$$

3. Calculate the force output:

$$F = p_{INT} \cdot A_{PRESS} = 2500 \frac{\text{lbs}}{\text{in}^2} \cdot (7.070 \text{ in}^2) = 17,680 \text{ lbs}$$

It may be desirable to cycle an intensifier automatically to create a continuous flow of fluid. In this case, a double-acting intensifier (Figure 9-12), would most likely be used. This type produces outlet flow on both strokes. The four-way directional control valve is cycled continuously to create continuous outlet flow.

9.4 Reservoirs

A *reservoir*, or *tank*, is used for storing the fluid in a hydraulic system. In addition to storing the oil, the reservoir has several other important functions that must be kept in mind when putting together a hydraulic circuit. The four most important functions of a reservoir are to:

1. Store the hydraulic fluid,
2. Provide heat exchange (cooling of the oil),
3. Allow contaminants to settle, and
4. Allow air to escape.

Heat is generated in a hydraulic circuit due to pressure drops across components and when fluid is forced over the relief valve at high pressure. Some of this heat is dissipated through the components to the atmosphere, while the rest raises the temperature of the fluid. In a properly designed system, the excess heat is radiated through the reservoir walls and carried away by airflow around the reservoir, thereby cooling the fluid. Heat exchange through fluid flow around the surfaces of an object is known as *convection*. Because of the heat exchange function, the surfaces of the reservoir should be kept clear and away from

326

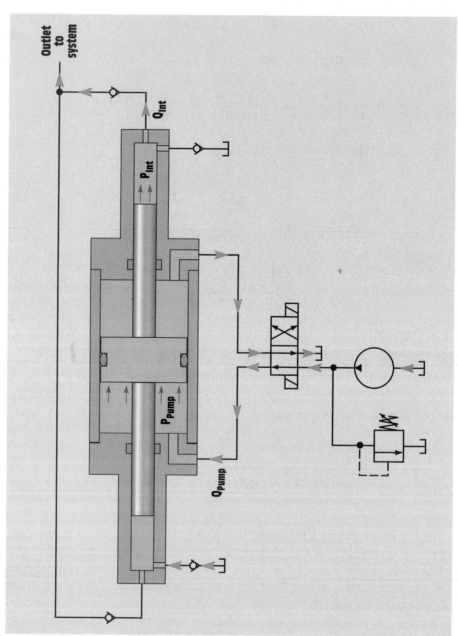

FIGURE 9-12 Double-acting pressure intensifier.

sources of heat. Placing a reservoir too close to another piece of machinery that generates heat may cause the system to run hot.

Air exists in a hydraulic system in three forms: *dissolved air, entrained air,* and *free air.* Dissolved air is in solution with the oil and has no adverse effects on system performance as long as it remains dissolved. Recall from our discussion on cavitation in Chapter 3 that excessive vacuum on the inlet can cause dissolved air to come out of solution and form bubbles, which are harmful (see Chapter 3). Entrained air, or air that is in the form of small bubbles, can be from dissolved air that has come out of solution or from external air leaking into the system. Free air is an air pocket that has formed within the system. The section on cavitation and aeration in Chapter 3 discusses the adverse effects of air and vapor bubbles within a hydraulic system.

The typical features of a hydraulic reservoir are illustrated in Figure 9-13. A reservoir typically has three lines: the *pump inlet* (suction line), the *return line,* and the *drain line.* The pump inlet line has a strainer to filter out contaminants. The return and drain lines are for flow returning from the system back to the reservoir. The return line is the main return path, while the drain line is only for external drain lines from pumps, valves, and motors. The return line typically has a filter, while the drain line does not because drain lines must have as little resistance as possible to keep the drain cavities at low pressure.

The baffle plate prevents the fluid from traveling directly from the return line to the pump inlet. It forces the fluid to *dwell* in the reservoir, which serves several important functions. Most importantly, it allows time for heat to be dissipated through the reservoir walls. Forcing the fluid to travel a circuitous path also promotes heat exchange by bringing it into contact with more surfaces. The dwell time also allows time for entrained air to escape (*deaeration*) and for contaminants to settle to the bottom of the tank.

The access plates, usually located on both sides of the reservoir, provide access to the reservoir for cleaning and component replacement. The filler/breather cap allows the reservoir to be refilled with oil and also has an air vent (breather), which allows the entrained air to escape. The breather also prevents pressure or vacuum from building in the reservoir as the oil level goes up and down. The breather has a filter element that prevents atmospheric contamination from entering the reservoir.

The reservoir is mounted off the ground, allowing air to flow underneath the reservoir. This provides another surface for heat exchange with the atmosphere. In addition to the features shown in Figure 9-13A, a reservoir typically has a fluid level indicator and a thermometer. Reservoirs may also have magnetic plugs at the bottom of the tank to attract and hold iron and steel contamination particles.

Figure 9-13B shows the graphic symbol for a reservoir (tank). Several tank symbols are often used on a schematic to represent one actual reservoir. Five tank symbols were used in Figure 9-12, although there is probably only one reservoir. This avoids having to draw multiple lines back to a single tank symbol,

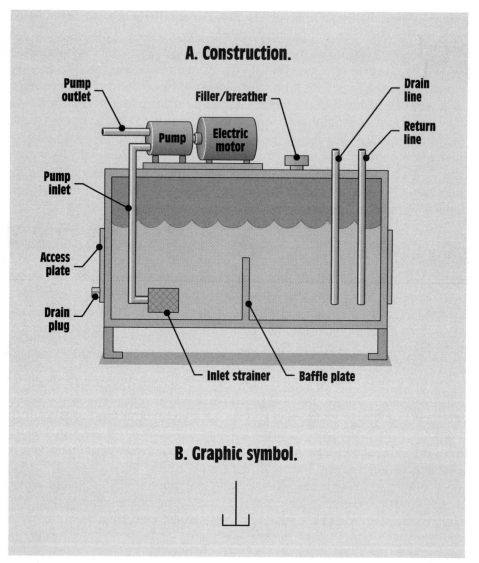

FIGURE 9-13   Hydraulic reservoir. (A) Construction. (B) Graphic symbol.

which would inevitably cause lines to cross and add confusion to the schematic. Schematics represent the logic of a hydraulic system, but do not necessarily represent the actual physical construction of a system.

To ensure that the fluid in a hydraulic system is given time to cool and deaerate, the reservoir must be properly sized. An often-used rule of thumb for industrial systems is to select a reservoir that is three times the flow rate of the system; for example, a system with a 10 gpm flow rate would need a 30 gal reservoir. In a system with a reservoir sized according to this standard, the fluid would make a complete circuit every three minutes. This rule of thumb is only a rough starting point, however, as many factors determine how hot a system will run. The reservoirs in mobile systems are much smaller because of weight and size limitations inherent to any mobile system.

The conventional configuration shown in Figure 9-13, where the pump is mounted atop the reservoir, is the most commonly found in industrial applications. Other configurations are also available, including the overhead reservoir, in which the pump is mounted below the reservoir. The overhead type provides a positive pressure to the pump inlet because the fluid level is above the inlet. In the conventional configuration, the pump must create a vacuum at its inlet to draw the fluid in against the force of gravity. Providing a positive pressure to the pump inlet can be advantageous because it eliminates the possibility of cavitation.

9.5   Heat Exchangers

In some systems, the amount of heat generated within the circuit cannot be dissipated through the walls of the components and reservoir. In these systems a *heat exchanger* is required to prevent the system from running above normal operating temperatures (140° F is the typical maximum). Two types of heat exchangers are used in hydraulic systems: *air-cooled* and *water-cooled*. Air-cooled heat exchangers are similar to an automobile radiator; a fan is used to blow air over a bundle of tubes through which the hydraulic fluid flows (Figure 9-14). A bundle of tubes is used rather than one large tube because the surface area is much greater. The surface area exposed between the hot fluid (hydraulic oil) and the cooling fluid (air) is the most important factor in heat exchange. For this reason, *fins* or other similar devices are often added to increase the surface area (Figure 9-15). Heat travels from the tube to the fin by *conduction* and is carried away by *convection*. Conduction is internal heat flow through a material; convection is heat transfer through fluid flow around the surfaces of an object.

Figure 9-16 shows a water-cooled heat exchanger. This particular design, a *shell and tube* heat exchanger, is commonly used in fluid power as well as in other types of machinery. Cool tap water is sent through the tubes via the water inlet. Hot hydraulic fluid flows into the shell via the hydraulic fluid inlet and flows around the tubes, which cools the fluid. Again, a bundle of tubes is used rather than one large tube because the surface area is much greater. The tubes are often made of bronze because of its high heat conductivity. Notice that the fluids flow in opposite directions, known as *counterflow*. Counterflow is most often used because it maximizes the amount of heat transferred between the fluids. The design shown in Figure 9-16 is known as a single-pass because the hydraulic fluid flows past the tubes once, then exits. In a double-pass unit, the hydraulic fluid flows past the tubes twice before exiting. Four-pass models are also available.

Water-cooled heat exchangers provide better heat transfer characteristics and are preferred when tap water is readily available. Tap water maintains a consistently

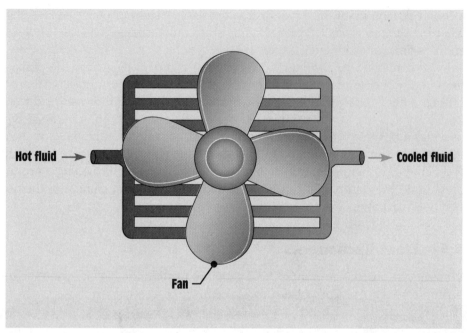

FIGURE 9-14   Air-cooled heat exchanger.

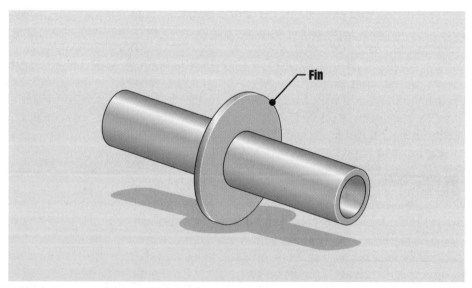

FIGURE 9-15   Fins are used to increase surface area.

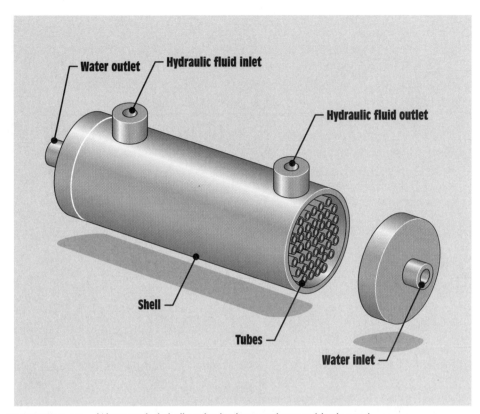

Water outlet   Hydraulic fluid inlet

Hydraulic fluid outlet

Shell

Tubes

Water inlet

FIGURE 9-16   Water-cooled shell and tube heat exchanger (single pass).

cool temperature, while the ambient air temperature is usually warmer and can vary considerably throughout the day and year. In many cases, such as mobile applications, a source of cool water is not readily available and an air-cooled heat exchanger is the more practical choice.

The previously discussed heat exchangers are also called *coolers* because they cool the hydraulic fluid. In some instances, the fluid may need to be heated. In a system that has been inactive for a long period of time in a cold environment, the hydraulic fluid will be cold and consequently very viscous. A fluid that is too viscous will cause sluggish operation and increased heating because of increased flow resistance. To eliminate the damage that may be caused by a cold start, it may be desirable to preheat the fluid. A shell and tube heat exchanger can also be used to heat a fluid by passing a hot fluid (e.g., steam) through the heat exchanger, transferring heat to the hydraulic fluid, rather than taking heat away. It is often simpler and more effective to heat the fluid in a hydraulic system by using an electric element-type heater, however. In these *immersion heaters,* the electric heating element is immersed in the fluid at the reservoir. The graphic symbols used for heat exchangers are shown in Figure 9-17.

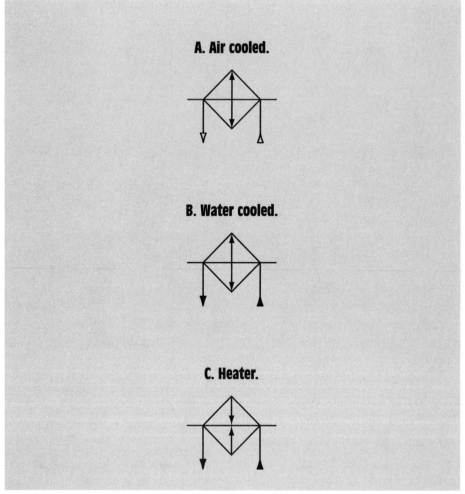

FIGURE 9-17   Heat exchanger symbols. (A) Air-cooled. (B) Water-cooled. (C) Heater.

Hydraulic systems that require heat exchangers do not usually require heating or cooling during their entire operating time. They are frequently coupled with a thermostat and electronic controls so that they can be turned on and off as needed. Figure 9-18 shows a partial circuit with a water-cooled heat exchanger. The heat exchanger is on a return line, which is always at low pressure. Heat exchangers are rarely designed to withstand high pressure. In the position shown, the three-way diverter valve is sending the fluid directly back to the tank without passing it through the heat exchanger. When the system raises above a preset temperature, the three-way diverter valve is shifted by the solenoid, thereby passing the return flow through the heat exchanger as it flows back to the reservoir.

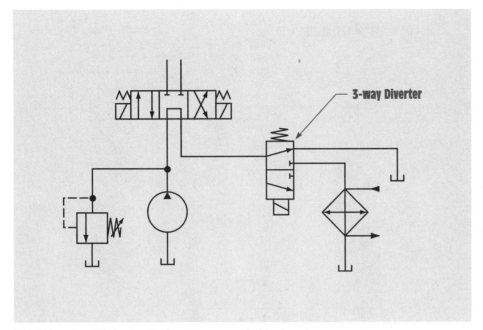

FIGURE 9-18   Hydraulic circuit with water-cooled heat exchanger.

9.6   Filters

The importance of oil cleanliness for the proper operation and longevity of a hydraulic circuit cannot be overstated. Hydraulic components such as pumps and motors are very sensitive to contamination. Operating these components in systems with fluid that is not filtered according to the manufacturer's specifications will cause increased wear, which will increase the leakage and decrease the efficiency of the system. The cause of the majority of hydraulic system failures can be traced back to contamination.

To understand filtration, we must first understand the sources of contamination in a hydraulic circuit. Contamination can be generated either externally or internally. Externally generated contaminants, such as dirt and grime, come from outside the system. Internally generated contaminants come from the wear of components within the system. They range from metal particles from pump and actuator wear, to plastic and synthetic rubber particles from seal and bearing wear. These two types of contamination are collectively known as *particulate contamination* because they consist of solid particles of various materials. Other types of contamination include water, air, and unwanted chemicals. Filters that remove water and other contamination are available. This section will focus on filters designed to remove particulate contamination.

Hydraulic filters are designed to remove particles down to the micrometer (μm) range (1 micrometer is one-millionth of a meter [1 μm = 0.000001 m], which is approximately 0.0000394 inches). This is frequently referred to as a

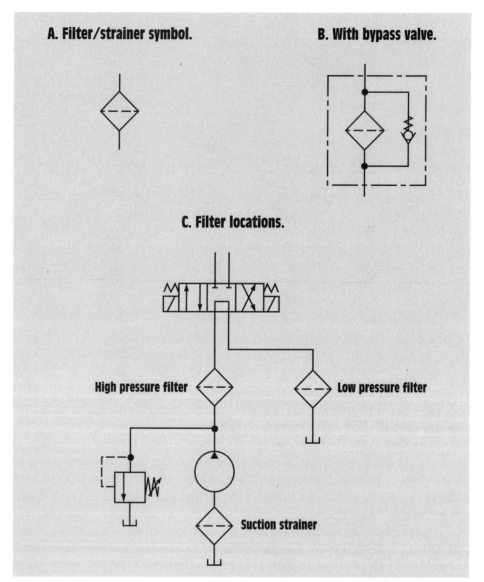

FIGURE 9-19   Filter symbol and possible locations. (A) Filter/strainer symbol. (B) Filter locations.

micron in filtration literature. Hydraulic filtration devices can be divided into two categories: *strainers* and *filters*. A strainer is basically a coarse filter that is made of a wire mesh that is cleaned and reused. Strainers typically remove particles that are 150 μm and larger, although models with finer levels are available. Filters have a much finer element that can filter out particles down to 5 μm and smaller. They typically use a cellulose (paper) type filter element that is disposable.

Figure 9-19 shows the graphic symbols for filters (or strainers), along with the possible locations for filters within a hydraulic circuit. Part A shows the basic

FIGURE 9-20   Spin-on, canister-type, low-pressure filter.
(Courtesy of Hydac Corporation, Bethlehem, PA.)

FIGURE 9-21 High-pressure filter.
(Courtesy of Hydac Corporation, Bethlehem, PA.)

filter symbol. Part B shows the symbol for a filter with a bypass check valve, which allows the filter to be bypassed if it were to become clogged.

Figure 9-19C shows three locations where filtration devices are most commonly placed within a hydraulic circuit. The suction line (pump inlet) usually contains a suction strainer that protects the pump from larger contaminants that may enter the tank. A fine filter is often *not* placed on the suction line because the restriction may cause excessive vacuum (and cavitation) at the pump inlet. Fine filtration is most commonly put on the return line because an inexpensive, low-pressure filter may be used. Figure 9-19 also shows a high-pressure filter, which is put in the pump outlet line. High-pressure filters are more expensive because of their more robust construction. They are necessary in some contamination-sensitive systems to catch the particles generated by pump wear. In many hydraulic systems, however, a suction strainer and return line filter are sufficient.

Low-pressure filters are most commonly the spin-on canister type, such as those shown in Figure 9-20. Low-pressure filters are easy to replace because the hydraulic circuit can remain connected. The entire canister is disconnected, discarded, and replaced while the filter head remains connected. A high-pressure filter is shown in Figure 9-21. High-pressure filters use replaceable elements, such as those shown in Figure 9-22. They must be taken out of the circuit in order for the element to be replaced, which obviously requires more time.

Some systems may also use *offline filtration,* in which the fluid is run through a filtering circuit that is separate from the main system. A low-pressure, low-volume pump moves the fluid through a filter and then immediately back to the

FIGURE 9-22   Filter elements.
(Courtesy of Hydac Corporation, Bethlehem, PA.)

reservoir. This type of filtration can be used for periodic maintenance. Portable units mounted on a cart are available for this purpose (Figure 9-23). These portable filtration units are also commonly used to transfer fluids from one location to another, such as from a container to a hydraulic system. This is a good practice because it ensures that the new fluid does not contaminate the system.

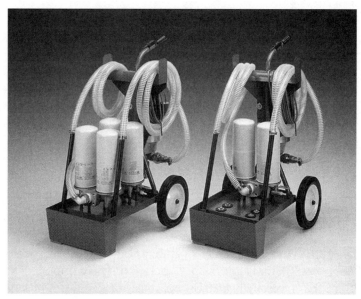

FIGURE 9-23   Portable filter carts.
(Courtesy of Hydac Corporation, Bethlehem, PA.)

To determine what level of filtration is required for a particular circuit, the component that is most sensitive to contamination must first be identified. This will set the level of filtration for the entire circuit. The ISO cleanliness code, the industry standard for specifying filtration levels, consists of two scale numbers that represent the maximum permissible number of particles per milliliter of size 5 μm or larger and 15 μm or larger (see Chapter 3). The ISO cleanliness code specifies how many particles of a given size per milliliter that a component can tolerate. This determines the maximum number of these particles that the filter can allow to pass. The *beta ratio* is the standard that measures the performance of a filter to ensure that the required level of filtration is achieved. The beta ratio is the ratio of the particles per milliliter of a given size or larger before the filter to the particles of that size that are present after passing through the filter. Stated mathematically:

$$\beta_X = \frac{N_U}{N_D} \qquad \textbf{(9-5)}$$

where: β_X = the beta ratio for particle size X
N_U = the number of particles of size X or larger per milliliter upstream of (before) the filter
N_D = the number of particles of size X or larger per milliliter downstream of (after) the filter

The following examples illustrate this concept.

EXAMPLE 9-4.

A fluid is known to have 8000 particles of size 10 μm or larger per milliliter. After being passed through a filter, the fluid is found to have 100 particles of size 10 μm or larger per milliliter. What is the β_{10} ratio for this filter?

SOLUTION:

$$\beta_{10} = \frac{N_U}{N_D} = \frac{8000}{100} = 80$$

EXAMPLE 9-5.

The manufacturer of a particular hydraulic component has specified an ISO cleanliness code of 16/13 for proper operation. The component will be used in a system that typically has 10,000 particles of size 5 μm or larger per milliliter and 2500 particles of size 15 μm or larger per milliliter. What should the β_5 and β_{15} ratios be for a filter that is to be used in this system?

SOLUTION:

1. Using Figure 3-24, determine the number of particles that is acceptable:
   a. Greater than 5 μm: up to 640 particles
   b. Greater than 15 μm: up to 80 particles
2. Calculate the beta ratios:

$$\beta_5 = \frac{N_U}{N_D} = \frac{10,000}{640} = 15.63$$

$$\beta_{15} = \frac{N_U}{N_D} = \frac{2500}{80} = 31.25$$

We must select a filter that has β_5 and β_{15} ratios of *at least* these values because a larger beta ratio means the filter is catching more particles. This can be best understood if we relate the beta ratio to *removal efficiency*. The efficiency of a filter is related to the beta ratio through the following equation:

$$\eta_X = 1 - \frac{1}{\beta_X} \qquad \textbf{(9-6)}$$

The removal efficiency represents the proportion of particles of size X or larger that are caught by the filter. This relationship makes sense mathematically because the inverse of the beta ratio is the proportion of particles that passed through the filter. One minus this value is then the proportion that is retained by the filter.

EXAMPLE 9-6.

■ What are the efficiencies (η_5 and η_{15}) for the filter in Example 9-5?

SOLUTION:

$$\eta_5 = 1 - \frac{1}{\beta_5} = 1 - \frac{1}{15.63} = 0.936 \ (93.6\%)$$

$$\eta_{15} = 1 - \frac{1}{\beta_{15}} = 1 - \frac{1}{31.25} = 0.968 \ (96.8\%)$$

The previous example shows that a beta ratio of 15.63 means that the filter removes 93.6% of the particles, while a beta ratio of 31.25 means that the filter removes 96.8% of the particles. Again, a higher beta ratio means a greater removal efficiency. Table 9-1 relates the beta ratio to removal efficiency for some specific values.

The number of particles present per milliliter is determined by fluid sampling. This can be done in house if the proper sampling equipment is purchased, or it can be outsourced. Filter manufacturers usually sell sampling equipment and offer fluid analysis services.

Flow capacity and pressure drop characteristics are important factors to consider when selecting a filter. Manufacturers specify the maximum permissible

BETA RATIO	REMOVAL EFFICIENCY
2	50%
5	80%
10	90%
20	95%
50	98%
100	99%
1000	99.9%

TABLE 9-1   Beta ratio versus removal efficiency

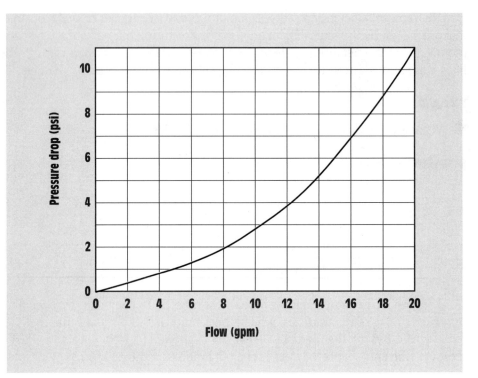

FIGURE 9-24   Flow versus pressure drop graph.

flow rate and pressure drop-versus-flow information, usually in the form of a graph (Figure 9-24). For this filter, a flow of 15 gpm will result in a pressure drop of approximately 6 psi across the filter.

The graph shown in Figure 9-24 is for a new (clean) filter. As dirt accumulates in the filter, the resistance to flow and, consequently, the pressure drop, will increase. This property can be used to determine when a filter element has reached its dirt-holding capacity. The pressure drop at which the filter is full is known as the *terminal pressure drop*. Filters usually have some sort of indicator to warn the operator or maintenance personnel that the filter element must be replaced. These indicators may be visual, such as a gauge, or electronic sensors that send a signal when the terminal pressure drop is reached. Figure 9-25 shows some typical pressure drop indicators.

As mentioned earlier, filters also may have an integral bypass valve. These valves allow the filter to be bypassed when the pressure drop across them becomes too large. They are used because bypassing the filter when it becomes clogged is preferable to forcing flow through it, which may cause the element to collapse and release all of the contamination it contains into the system. Bypass valves are basically a nonadjustable pressure relief valve (see Figure 9-19B).

FIGURE 9-25   Filter pressure drop indicators.
(Courtesy of Hydac Corporation, Bethlehem, PA)

System startup is perhaps the most critical time to consider fluid cleanliness. One should never assume that the new system components, including the hydraulic fluid, are clean. New components often have particles left over from the manufacturing processes. Likewise, new hydraulic fluid is often filled with contamination particles that entered during processing and transport. Hydraulic fluid transported to a new system should be run through fine filtration. Portable filtration units are often used in this process. All components should be filled with filtered fluid before the system is started for the first time. This is critical because the hydraulic fluid is also the lubricant that protects the components from wear. When the entire system is filled with filtered fluid, the system should be run under no load for a trial run and observed closely. After it has undergone this break-in stage, the filters should be replaced before beginning normal operation. The system can then be run with gradually increasing pressure. Close supervision of the machine is critical during these initial stages of operation. This startup procedure should also be followed whenever a new component is put into service in an existing circuit.

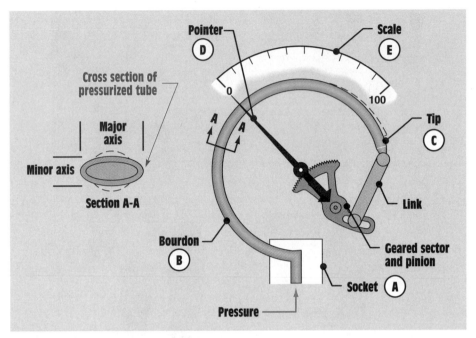

FIGURE 9-26   Bourdon tube pressure gauge.
(Courtesy of *Hydraulics and Pneumatics Magazine*. From Norvelle, *Fluid Power Technology:* West, 1993, p. 484.)

To ensure that the system continues to function properly for as long as possible, the filter pressure drop should be monitored so that filters can be replaced when necessary. Periodic sampling and analysis of the fluid is also important. This will not only provide information on the levels of contamination, but also identify the types of particles that are present. This will help identify any sources of contamination that can be reduced or eliminated.

9.7   Instrumentation and Measurement

9.7.1 PRESSURE GAUGES

It is often desirable to monitor the pressure at various points in the hydraulic circuit. Knowing the pressure at different points in the circuit is indispensable information when troubleshooting a problem. Two categories of devices read pressure: mechanical pressure gauges and electronic pressure sensors. The most commonly used mechanical gauge is a *bourdon tube* gauge (Figure 9-26). This device uses a hollow tube that is bent into a circular shape. The interior of the tube is subjected to pressure from one end and is sealed at the other. As the pressure in the tube increases, it begins to straighten out because the area

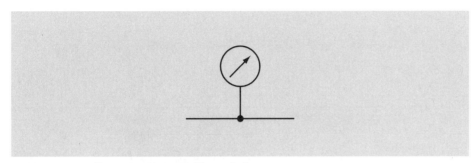

FIGURE 9-27 Pressure gauge graphic symbol.

subjected to pressure on the outside diameter is larger than the area on the inside diameter. This can be used to detect the pressure because the amount of deflection of the tube is proportional to the pressure. The motion of the tube is transferred to the pointer through a mechanical linkage.

Other, less commonly used mechanical pressure gauges use the same principle: Fluid pressure is used to cause the deflection of some mechanical device and the motion is then transmitted to a pointer. The face of mechanical gauges are often filled with a clear liquid, typically glycerine, which acts to smooth out the motion and prevent vibration of the pointer. The symbol for a pressure gauge is shown in Figure 9-27.

The bourdon tube design can also be coupled with electronics to provide a digital readout or to send the information to a computer. Devices that convert mechanical motion to an electrical signal are known as *electromechanical*. Purely electronic pressure sensors, which produce an electrical signal (voltage) that is proportional to the pressure, are also available. These *pressure transducers* are discussed in Chapter 13, which covers electrical control of fluid power.

9.7.2 FLOWMETERS

Pressure and flow are the two most important operating parameters in a fluid power system. Therefore, it may also be desirable to be able to read the flow rate at different points in a hydraulic circuit. This is accomplished with a flowmeter. Three types of flowmeters are commonly used to measure flow rates in fluid systems: *orifice flowmeters, rotameters,* and *turbine flowmeters.*

An orifice flowmeter, also called a *differential pressure* flowmeter, is shown in Figure 9-28. It operates on the principle that the pressure drop across an orifice is proportional to the flow through it. The pressure is read before and after the orifice (*p1* and *p2,* respectively). The flow rate is then determined from the pressure drop ($\Delta p = p1 - p2$). The relationship between the flow rate and the pressure drop through an orifice is described by the flow coefficient (C_v) (see

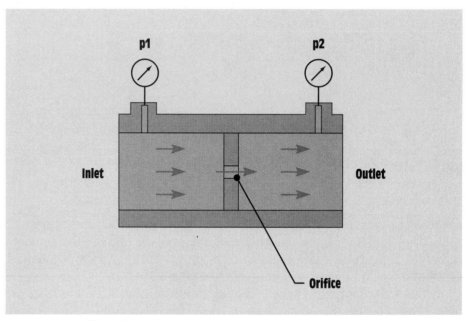

FIGURE 9-28   Orifice flow meter.

Chapter 8). If we know the flow coefficient for the orifice and the pressure drop, we can calculate the flow rate. This calculation is frequently accomplished with onboard electronics. The practice of inferring one quantity by reading another is used often in instrumentation and measurement. An orifice flowmeter is *not* commonly used in fluid power applications. It is, however, widely used in fluid transfer applications.

A rotameter, also called a *variable area flowmeter,* is shown in Figure 9-29. In this device, a float is allowed to move freely inside a vertically mounted, tapered tube. The flowing fluid enters from bottom and tends to lift the float because of the *flow force,* which is created by fluid drag as the fluid flows around the float. The force of gravity opposes this motion. As the float moves up, the tube diameter increases, causing the flow force on the float to decrease as it moves up because of the larger flow area. The float will stop at a position where the flow force just balances the force of gravity. The faster the flow rate, the greater the flow force on the float and the higher it will rise. The height of the float (and the flow rate) can be read directly through a transparent sight glass (as shown in Figure 9-29), or a magnetic float can be used to raise an indicator that is located outside the tube. A float type rotameter must be vertically mounted because it depends on gravity for its operation. The flowmeter shown in Figure 9-29 is reading a flow rate of 15 gpm.

Another type of rotameter uses a spring, rather than gravity, to oppose the force created by fluid flow (Figure 9-30). The float in this design is replaced by a piston. The operation is otherwise very similar; the greater the flow rate, the

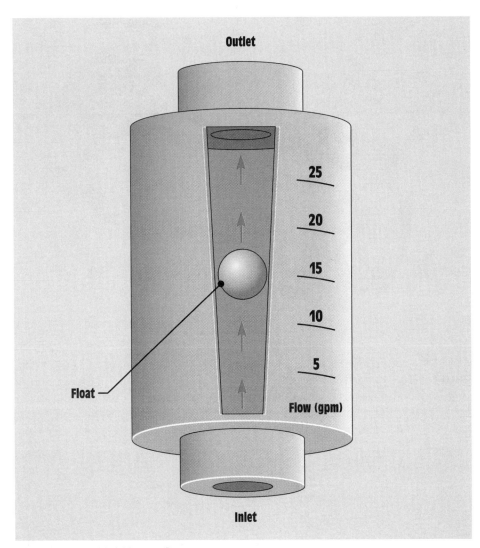

FIGURE 9-29 Variable area flow meter.

farther the piston will move against the force of the spring. The position of the piston is detected through a magnetized indicator on the outside of the tube that follows a magnet on the piston. The advantage of this design is that it need not be vertically mounted, as was the case with the float design.

A turbine flowmeter measures the velocity of the fluid stream and determines the flow rate from this measurement. It measures the fluid velocity by passing the flow through a turbine. The rotational speed of the turbine is proportional to the flow rate. The speed of the turbine is converted to an electrical signal, which can provide a digital readout or send the information to a computer.

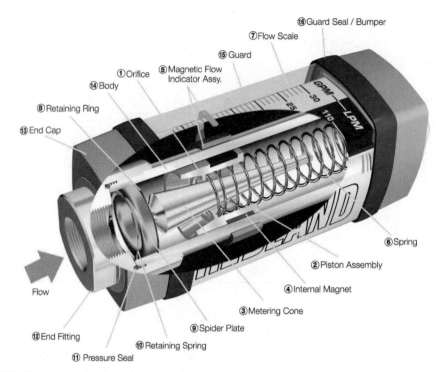

FIGURE 9-30   Hedland flow meter.
(Courtesy of Hedland® Flow Meters.)

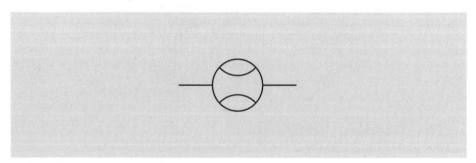

FIGURE 9-31   Flow meter graphic symbol.

The graphic symbol that is used to represent any of the previously discussed flowmeters is shown in Figure 9-31. The rotameter and turbine flowmeters are the most commonly used types in fluid power systems. Orifice flowmeters are more common in fluid transfer applications.

9.7.3 TEMPERATURE GAUGES

The temperature of a hydraulic system is critical to proper operation due to the effect of temperature variations on fluid viscosity. High temperature also promotes other adverse effects, which we will discuss in the fluids section of this chapter. Hydraulic systems are usually designed to operate at temperatures between 120° F and 140° F. Most reservoirs come with a temperature gauge for monitoring system temperature. It may be necessary to monitor the temperature in other points in the system as well.

9.8 Conduits and Fittings

Three types of conduits are used to transfer fluid from one component to another in a hydraulic circuit: *pipe, tubing,* and *hose.* Flow velocity is one of the most important considerations in selecting conduit. If the conduit chosen for a system is too small for the flow rate, the flow velocity will be excessive. Excessive flow velocities cause large pressure drops because of increased fluid friction, resulting in power loss and heating of the fluid and, consequently, decreased efficiency. The flow velocity (v) is related to the flow rate (Q) according to the following equation, which was introduced in Chapter 2:

$$Q = v \cdot A$$

where: A equals the flow area of the conduit. This area is determined from the inside diameter (*ID*). The previous equation, as well as common sense, tells us that a larger conduit will have a smaller flow velocity for a given flow rate. Lower flow velocities are desirable because they transfer the fluid more efficiently. However, larger components are heavier and more expensive. We must find a balance between smaller conduits that are lighter and less expensive, and larger conduits that are more efficient.

How do we decide what flow velocity will be acceptable? It depends on the system. For instance, on mobile systems it is critical to keep the size and weight of the components down, so larger flow velocities and higher pressure drops will be tolerated. In stationary industrial systems, size and weight are not as important, so these systems will generally have larger components and smaller flow velocities. In addition to being large enough to give an acceptable flow velocity, the conduit must also be able to withstand the operating pressures of the system.

EXAMPLE 9-7.

A conduit size needs to be determined for a system in which the flow rate will be 20 gpm. Determine the conduit *ID* if the flow velocity is not to exceed 20 ft/s.

SOLUTION:

1. Convert gpm to in³/min:

$$Q = 20 \, \frac{\text{gal}}{\text{min}} \cdot \left(\frac{231 \, \text{in}^3}{1 \, \text{gal}} \right) = 4620 \, \frac{\text{in}^3}{\text{min}}$$

2. Convert ft/s to in/min:

$$v = 20 \, \frac{\text{ft}}{\text{s}} \cdot \left(\frac{12 \, \text{in}}{1 \, \text{ft}} \right) \cdot \left(\frac{60 \, \text{s}}{1 \, \text{min}} \right) = 14,400 \, \frac{\text{in}}{\text{min}}$$

3. Calculate the conduit area:

$$A = \frac{Q}{v} = \frac{4620 \, \dfrac{\text{in}^3}{\text{min}}}{14,400 \, \dfrac{\text{in}}{\text{min}}} = 0.3208 \, \text{in}^2$$

4. Calculate the conduit *ID*:

$$D = \sqrt{\frac{4 \cdot A}{\pi}} = \sqrt{\frac{4 \cdot 0.3208 \, \text{in}^2}{3.142}} = 0.6391 \, \text{in}$$

We must choose a conduit that has an *ID* larger than 0.6391 in to ensure that the flow velocity will not exceed 20 ft/s.

EXAMPLE 9-7M.

A conduit size needs to be determined for a system in which the flow rate will be 75 lpm. Determine the conduit *ID* if the flow velocity is not to exceed 6 m/s.

SOLUTION:

1. Convert lpm to m³/min:

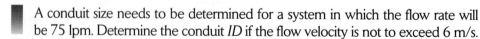

$$Q = 75 \, \frac{1}{\text{min}} \cdot \left(\frac{1 \, \text{m}^3}{1000 \, 1} \right) = 0.075 \, \frac{\text{m}^3}{\text{min}}$$

2. Convert m/s to m/min:

$$v = 6\frac{m}{s} \cdot \left(\frac{60\ s}{1\ min}\right) = 360\frac{m}{min}$$

3. Calculate the conduit area:

$$A = \frac{Q}{v} = \frac{0.075\dfrac{m^3}{min}}{360\dfrac{m}{min}} = 0.0002083\ m^2$$

4. Calculate the conduit *ID*:

$$D = \sqrt{\frac{4 \cdot A}{\pi}} = \sqrt{\frac{4 \cdot 0.0002083\ m^2}{3.142}} = 0.01628\ m\ \ (16.28\ mm)$$

9.8.1 PIPE

Pipe is a thick-walled steel conduit that can withstand high pressures because of its strength. It is also very abuse-resistant and is used for permanent, stationary plumbing in industrial applications. Pipe is specified according to a *schedule* system, which refers to its wall thickness. Schedule 40 pipe is considered the standard size. Its dimensions for a few selected sizes are shown in Table 9-2. Notice that for a nominal size of ½ in, neither the inside diameter (*ID*) nor the outside diameter (*OD*) is ½ inch! For any calculations, we must always use the *ID* of a conduit because this is where the fluid flows. Table 9-3 shows data for schedule 80 pipe. For each nominal size, the *OD* is the same it is with the schedule 40 pipe, but the *ID*s of the schedule 80 pipe are smaller. Schedule 80 pipe has a greater wall thickness and can withstand higher pressures. Other schedules are

NOMINAL SIZE	OD (in)	ID (in)	WALL THICKNESS
½	0.840	0.622	0.109
¾	1.050	0.824	0.113
1	1.315	1.049	0.133
1 ¼	1.660	1.380	0.140
1 ½	1.900	1.610	0.145
2	2.375	2.067	0.154

TABLE 9-2   Schedule 40 Pipe Specifications

NOMINAL SIZE	OD (IN)	ID (IN)	WALL THICKNESS
½	0.840	0.546	0.147
¾	1.050	0.742	0.154
1	1.315	0.957	0.179
1 ¼	1.660	1.278	0.191
1 ½	1.900	1.500	0.200
2	2.375	1.939	0.218

TABLE 9-3   Schedule 80 Pipe Specifications

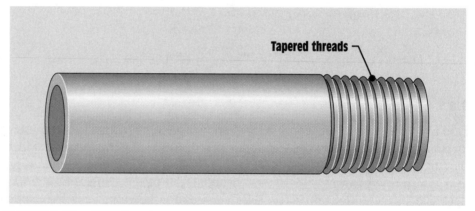

FIGURE 9-32   NPTF threads.

available. Schedule 160 pipe has the same *OD* as schedules 40 and 80 for each nominal size, but has a smaller *ID* than does schedule 80.

Two types of threaded connections are used with pipe: *NPTF* threads and *straight* threads. NPTF (national pipe taper fuel) threads, also called dryseal threads, are tapered slightly (Figure 9-32). When assembled together, the threads deform to provide a seal. The problem with this method is that the threads must be properly deformed to provide a seal, which requires that the assembly torque be within a specific range. They also tend to leak when subjected to vibration and temperature variations, which occur in hydraulic systems. Disassembly and reassembly exacerbate the situation as the threads become over-deformed. There are also NPT threads, which require a sealant, such as Teflon tape or paste, to be added to provide a seal. NPT threads are used in water distribution systems and for other nonindustrial uses. They should never be used with fluid power systems.

Straight threads do not rely on deformation of the threads to provide a seal. Instead, they use a flange with an o-ring or weld to seal the connection (Figure 9-33). O-rings are used for sizes up to 1¼ in (Figure 9-33A). Welded

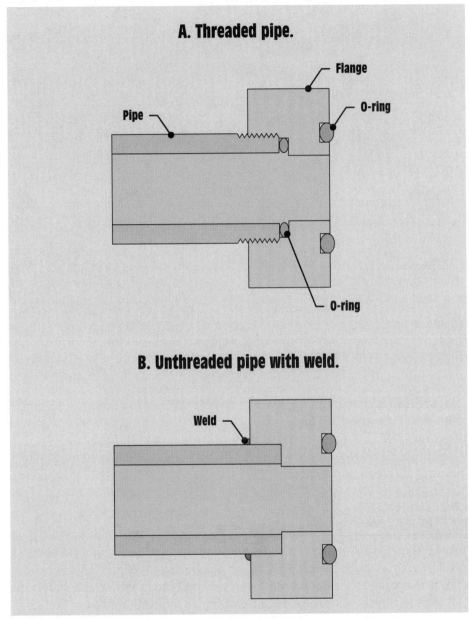

FIGURE 9-33   Straight thread, flange connections. (A) Threaded pipe. (B) Unthreaded pipe with weld.

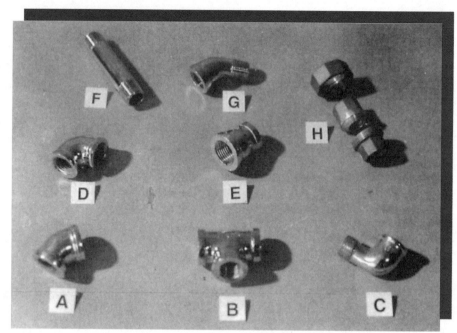

FIGURE 9-34   Typical pipe fittings: (A) 45° elbow, (B) tee, (C) 90° street elbow, (D) 90° elbow, (E) reducing coupling, (F) nipple, (G) 45° street elbow, (H) union (disassembled).

(From Reeves, *Technology of Fluid Power,* Albany, NY: Delmar, 1996, p. 292.)

connections are used for larger sizes (Figure 9-33B). Straight threads are preferred over NPTF threads because they provide a more reliable seal.

Figure 9-34 shows a variety of typical pipe fittings available with both NPTF and straight thread connections.

9.8.2 TUBING

Tubing, like pipe, is made of steel, but has thinner walls. This makes it less abuse-resistant than pipe, but allows it to be bent into shape. Pipe cannot be bent. The fact that tubing can be bent is a big advantage because it eliminates the need for a fitting every time the conduit must change directions. This makes a system that uses tube less prone to leakage because of fewer connections. Fewer fittings also reduces the weight of a hydraulic system, which makes it well-suited to mobile and aerospace applications.

The *OD* is the nominal size of tubing. For each nominal size, a variety of wall thicknesses is available. The ID must be used in flow calculations. The *ID*

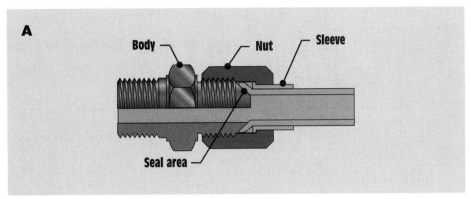

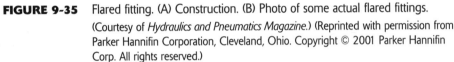

FIGURE 9-35   Flared fitting. (A) Construction. (B) Photo of some actual flared fittings. (Courtesy of *Hydraulics and Pneumatics Magazine*.) (Reprinted with permission from Parker Hannifin Corporation, Cleveland, Ohio. Copyright © 2001 Parker Hannifin Corp. All rights reserved.)

can be determined from the *OD* and the wall thickness (*t*) using the following simple equation:

$$ID = OD - 2 \cdot t \qquad\qquad \textbf{(9-7)}$$

Unlike pipe, tubing is not threaded. It relies solely on fittings to make connections. Three types of connections are used with tube fittings: *flared, flareless,* and *o-ring*. The construction of a flared fitting is shown in Figure 9-35A. When using this type of fitting, the tube must be flared at the end using a flaring tool. A sleeve and nut then slide over the tube. As the nut is screwed onto the fitting, the flare on the tube is pressed firmly against the fitting, providing a seal. Figure 9-35B shows some actual flared fittings.

The construction of a flareless fitting is shown in Figure 9-36A. This type does not require the tube to be flared, which simplifies the assembly of the circuit. In this design, a nut and ferrule are slipped over the tube. When the nut is screwed onto the fitting, the ferrule penetrates the *OD* of the tube, providing a seal. Because the *OD* of the tube is penetrated, flareless fittings cannot be used with thin-walled tubing. Figure 9-36B shows some actual flareless fittings.

Figure 9-37A illustrates the construction of a flat-face, o-ring fitting. This design is more expensive than the flared and flareless designs, but is less prone to leakage. The metal-to-metal seal of flared and flareless fittings will not seal properly if they are over- or under-tightened. Figure 9-37B shows some actual flat-face, o-ring fittings.

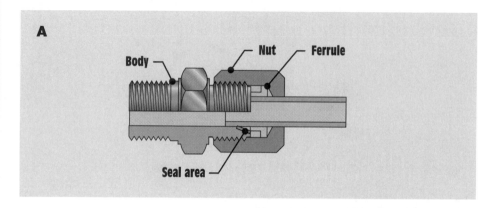

FIGURE 9-36   Flareless fitting. (A) Construction. (B) Photo of some actual flareless fittings. (Courtesy of *Hydraulics and Pneumatics Magazine*.) (Reprinted with permission from Parker Hannifin Corporation, Cleveland, Ohio. Copyright © 2001 Parker Hannifin Corp. All rights reserved.)

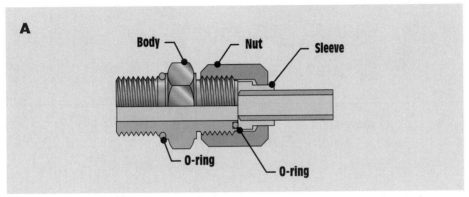

FIGURE 9-37   Flat-face o-ring fitting. (A) Construction. (B) Photo of some actual flat-face o-ring fittings.

(Courtesy of *Hydraulics and Pneumatics Magazine*.) (Reprinted with permission from Parker Hannifin Corporation, Cleveland, Ohio. Copyright © 2001 Parker Hannifin Corp. All rights reserved.)

9.8.3 HOSE

Hose, unlike pipe and tube, is very flexible and is most commonly used in applications where there is relative movement between components. The application shown in Figure 9-38, a cylinder rotating a lever arm, is one example. Because the cylinder must be allowed to pivot as it extends and retracts, flexible hose must be used to connect the directional control valve and the cylinder.

Both tubing and hose are used on the same system in many instances. For example, mobile equipment such as excavators have tubing for the stationary connections and flexible hose to make connections between components that move relative to one another. Hose is also used frequently for stationary connections where the path between the components is particularly circuitous and would be difficult to plumb with tubing. Hose has the disadvantage of being less abuse resistant and has a shorter life than tubing or pipe.

Hose is constructed of three or more layers. The inside and outside layers are typically made of an elastomer (synthetic rubber). In between are one or more layers of reinforcement constructed of a fiber or wire braid. The nominal size of hose is its inside diameter. The outside diameter is determined by the pressure rating of the hose. Hoses with higher pressure ratings have more reinforcement layers, and therefore a larger outside diameter, than those with a lower pressure rating. Because the inside surface of the hose is made of a synthetic rubber, care should be taken to make sure that the material is compatible with the fluid to be used.

Fittings are most commonly attached to hose using *crimping* (Figure 9-39). In crimping, also called *swaging*, parts of the fitting are deformed to grip the hose and provide a seal. In applications where the hoses are to be disconnected frequently, *quick disconnects* are used. Figure 9-40 shows a quick disconnect *socket* and *plug*.

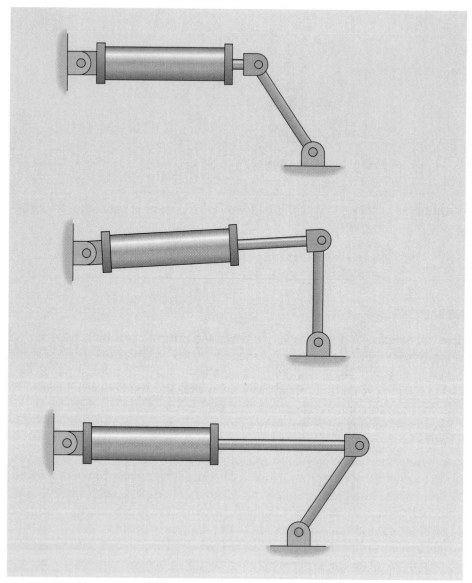

FIGURE 9-38   Cylinder rotating a lever arm.

There are some important things to consider to avoid premature failure when installing hose. First, hose should be installed so that it is not subjected to heat or abrasion. Other concerns relate to the fact that the hose will flex as it is subjected to pressure changes. Some important assembly guidelines are shown in Figures 9-41 through 9-47. Following these guidelines will help keep the hose from being subjected to unnecessary stresses.

357

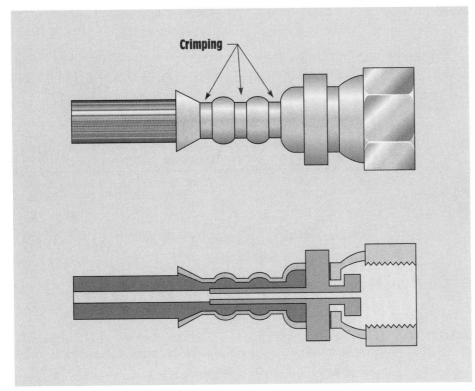

FIGURE 9-39   Crimping is used for permanent hose connections.
(From Norvelle, *Fluid Power Technology:* West, 1993, p. 423.)

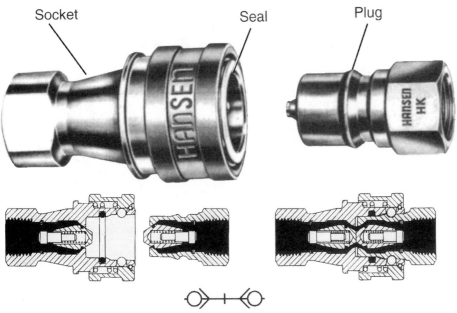

FIGURE 9-40   Quick disconnect socket and plug.
(Courtesy of Tuthill Corp., Hansen Coupling Division. From Norvelle, *Fluid Power Technology:* West, 1993, p. 425.)

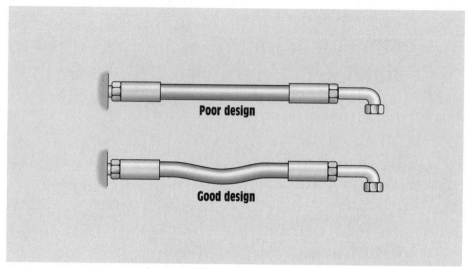

FIGURE 9-41   To prevent excessive strain at hose-to-coupling interfaces, hose assemblies should be made long enough to allow for contraction and expansion.
(Courtesy of *Hydraulics and Pneumatics Magazine*.)

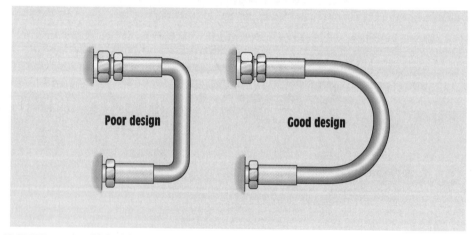

FIGURE 9-42   Make hose assemblies long enough and routed in a manner that prevents exceeding the minimum bend radius recommendations.
(Courtesy of *Hydraulics and Pneumatics Magazine*.)

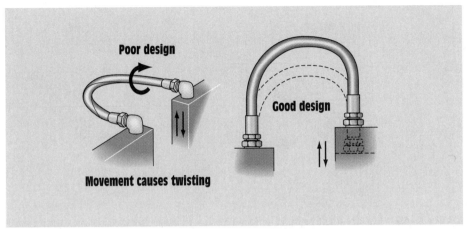

FIGURE 9-43 Left-hand drawing shows how hose twists because it is bent in one plane while oscillating motion bends in a second plane. Rerouting the hose eliminates multi-plane bending.

(Courtesy of *Hydraulics and Pneumatics Magazine*.)

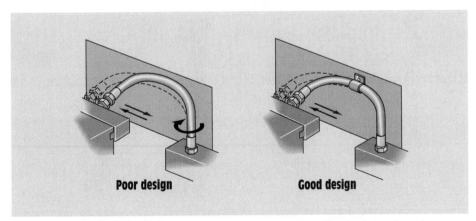

FIGURE 9-44 When multi-plane bending cannot be avoided, install a hose clamp between bends and provide enough hose length on both sides of the clamp to allow torsion to be relaxed and to compensate for hose length contraction.

(Courtesy of *Hydraulics and Pneumatics Magazine*)

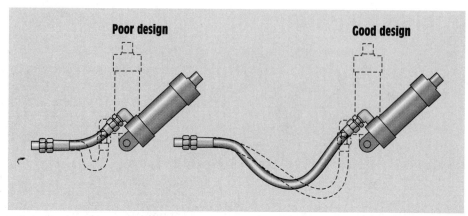

FIGURE 9-45 Design at left provides ample hose length when cylinder is pivoted, but bends hose in too small a radius when cylinder is vertical. Increasing hose length and providing greater clearance produce much greater bend radii.

(Courtesy of *Hydraulics and Pneumatics Magazine*)

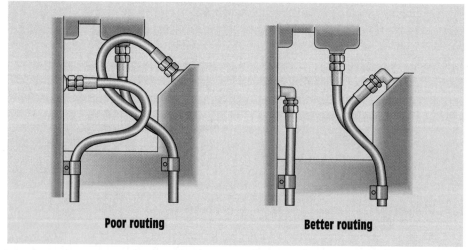

FIGURE 9-46 Lack of planning produces cluttered hose routing, left, that complicates maintenance and can reduce hose life. Well-thought-out routing and choice of end fitting configurations (right) makes assemblies that are more reliable and easier to maintain and troubleshoot.

(Courtesy of *Hydraulics and Pneumatics Magazine*)

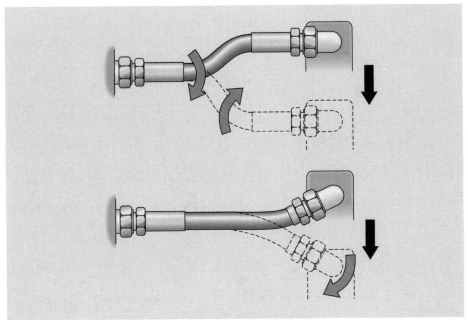

FIGURE 9-47   Swivel joints can extend hose life by reducing the amount of bending caused by relative motion between machine elements. They also aid maintenance by simplifying hose installation and replacement.
(Courtesy of *Hydraulics and Pneumatics Magazine*)

9.9   Seals and Bearings

Modern sealing technology has made possible the use of high pressures, thus making hydraulics a workhorse of modern industry. Because of the high pressures involved, seals are a highly critical element of every hydraulic system. Their importance is magnified by the fact that they are used in virtually every component, including pumps, valves, and actuators. Seal failure in any one of these components can cause severe performance degradation, if not complete failure of the system. Anyone involved in designing fluid power circuits, selecting components, or performing maintenance must have a basic understanding of seal materials and geometries in order to avoid costly mistakes.

Seals can be divided into two broad categories: static seals and dynamic seals. Static seals provide a seal between surfaces that are not moving relative to one another. Dynamic seals provide a seal between moving surfaces. Figure 9-48 shows a cutaway of a hydraulic cylinder that has both static and dynamic seals. The piston and rod both require a dynamic seal because the piston slides back and forth within the barrel. The end cap seal provides a static seal between the cylinder barrel and end caps, which are not moving relative to one another. The

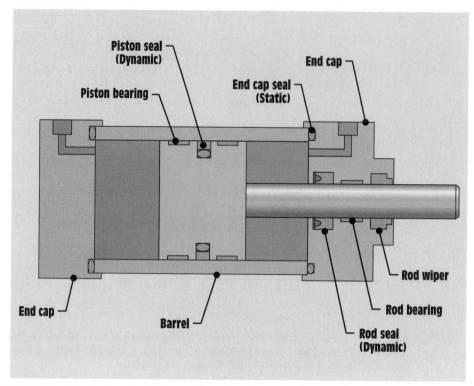

FIGURE 9-48   Hydraulic cylinder seals and bearings.

rod wiper, also a dynamic seal, prevents the ingression of contamination from the environment. Although they serve a different function, bearings are frequently categorized with seals because they are often produced by the same manufacturers. They allow a moderate amount of side loading of the cylinder to be tolerated. Cylinder bearings such as this are also called *guide bushings* or *wear bands*.

The simplest seal geometry is an *o-ring*, a seal with a circular cross section (Figure 9-49). Part A shows an uninstalled o-ring cross section. To install, the o-ring is squeezed into a groove (part B). This applies an initial sealing force, known as *preload*, to the o-ring. The amount of deformation, called the *squeeze*, is critical to proper o-ring function and can vary from 10% to 25% depending on the application. Figure 9-49C shows the o-ring when pressure is applied, in this case from the right. The pressure acts to press the o-ring against the surfaces with more force, thereby providing a tight seal. In the past, o-rings were used extensively as dynamic seals, but this is becoming much less common. The vast majority of o-ring applications are for static sealing, such as the end cap seals shown in Figure 9-48.

The *u-ring* is another commonly used seal geometry (Figure 9-50). This type of seal is used primarily in dynamic applications. Part A shows the u-ring

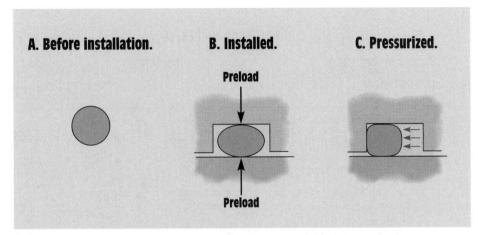

FIGURE 9-49   O-ring seal. (A) Before installation. (B) Installed. (C) Pressurized.

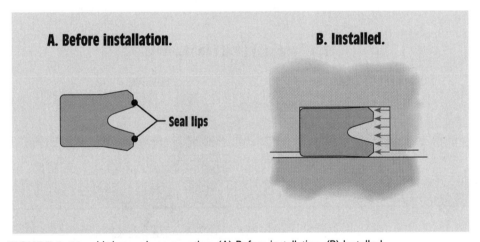

FIGURE 9-50   U-ring seal cross-section. (A) Before installation. (B) Installed.

before installation; part B shows it after installation into its groove. Just as with the o-ring, the amount of squeeze is important for proper functioning. A u-ring is designed to seal pressure from only one direction, which is from the right in this orientation. If pressure is applied from the left, it will push the lip up and allow the fluid to escape. It is therefore critical to install these seals with the "u" facing the pressure, as shown. If a u-ring is to be used as a piston seal, two must be installed back to back, as pressure is usually applied to both sides of the piston. A u-ring is an extremely effective seal design because as pressure is applied, it acts to press the lips more firmly against the surfaces. This type of seal therefore provides a sealing force that increases in proportion to increasing pressure.

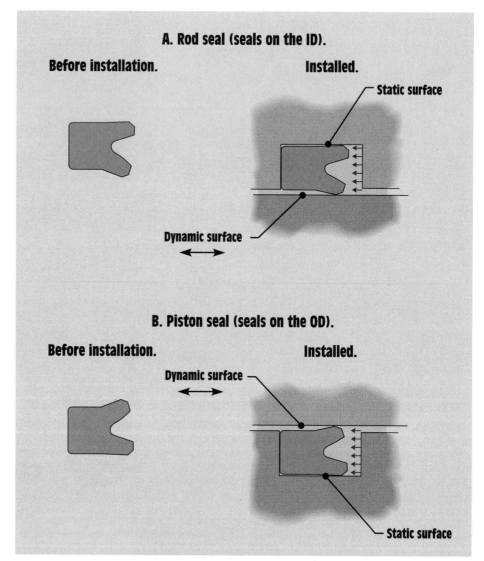

FIGURE 9-51   Asymmetric u-ring seal cross section. (A) Rod seal (seals on the *ID*). (B) Piston seal (seals on the *OD*).

The u-ring shown in Figure 9-50 is called a symmetrical u-ring because the *OD* and *ID* surfaces are identical. The advantage of this design is that the same seal can be used as a rod or piston seal. Although the symmetrical u-ring is fairly common, the asymmetrical u-ring shown in Figure 9-51 is a better design. These seals are designed specifically as a rod seal (seals on the *ID*) or as a piston seal (seals on the *OD*) and cannot be interchanged. Asymmetrical u-ring seals are designed to have more contact area on the static surface, and very little contact area on the dynamic (sliding) surface. This reduces the friction created by the seal.

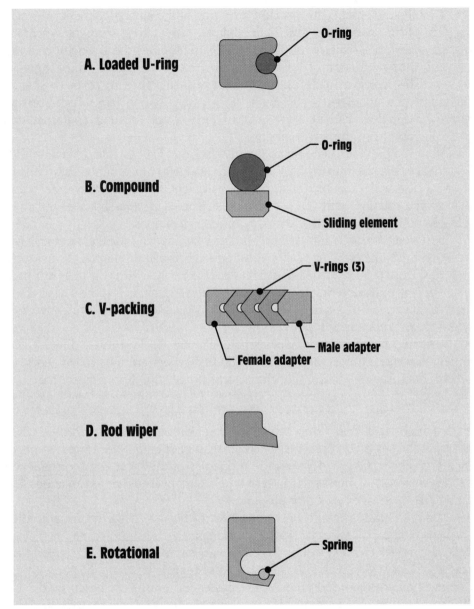

FIGURE 9-52   Other basic seal geometries. (A) Loaded u-ring. (B) Compound. (C) V-packing.
(D) Rod wiper. (E) Rotational.

Seals are also available in a large variety of geometries other than o-rings and u-rings. Most are variations on a few basic designs (Figure 9-52). The *loaded u-ring* (part A) is simply a u-ring with a small o-ring inserted into it. This increases the preload when the seal is squeezed into the groove. Part B shows a *compound* seal. Compound is the general term used to describe a seal made up of multiple elements, in this case an o-ring and a sliding element. The sliding element is made of a low-friction plastic. The sliding element would be placed against the dynamic surface. This seal combines the elasticity of an o-ring and the low friction of the plastic sliding element. Figure 9-52C shows *v-packing*. This seal is a made up of multiple (usually three to seven) v-shaped rings and two adapters. V-packing is frequently used in large, high-pressure hydraulic cylinders. Part D shows a rod wiper, which prevents contamination from entering from the environment.

All of the previous seals are of the translational type, so called because they seal surfaces that are moving back and forth relative to one another. Figure 9-52E shows a *rotational seal,* which seals rotary shafts. The spring helps to apply a preload to the rotating shaft. This type of seal is used on the shafts of hydraulic pumps and motors, as well as in many other applications.

Extrusion, which occurs when the seal begins to squeeze into the gap between the surfaces being sealed, is a frequent problem in high-pressure applications (Figure 9-53A). To eliminate this problem, anti-extrusion rings are used (part B). Anti-extrusion rings fill the gap between the surfaces to be sealed. They are typically made of a hard plastic, such as nylon. Figure 9-53C of shows a u-ring with an anti-extrusion ring.

Seals and bearings are made of materials that fall into three distinct categories: *elastomers, plastics,* and *elastoplastics.* Elastomers are also called synthetic rubber because they have the characteristic elasticity (springiness) that one would expect from a rubber-like material. Natural rubber is never used in hydraulics because it is not compatible with the fluids used. Plastics are materials that are hard at room temperature, but can be molded easily or formed when heated sufficiently. Elastoplastics are softer than plastics, but harder and stronger than elastomers. Table 9-4 shows the properties of the most common seal and bearing materials. Within each material type are many different formulations that may have significantly different properties.

Nitrile, an elastomer also known as *NBR* or *buna-N,* is the most commonly used material for all types of seals. Components such as pumps, motors, cylinders, and valves typically come standard with nitrile seals. One of the primary reasons for its widespread use in hydraulic systems is its compatibility with the two most commonly used fluids: water-based and petroleum-based fluids. It is also compatible with air, and is the most commonly used seal material in pneumatic systems as well.

Fluorocarbon, commonly known as *viton,* is an elastomer that can be used in o-rings and other seals. It is similar to nitrile with respect to its mechanical properties such as elasticity, tensile strength, and hardness. However, it can withstand higher temperatures and has better chemical resistance. In addition to water- and petroleum-based fluids, it is also compatible with *phosphate ester,* a commonly used fire-resistant fluid. Manufacturers of components such as pumps, motors,

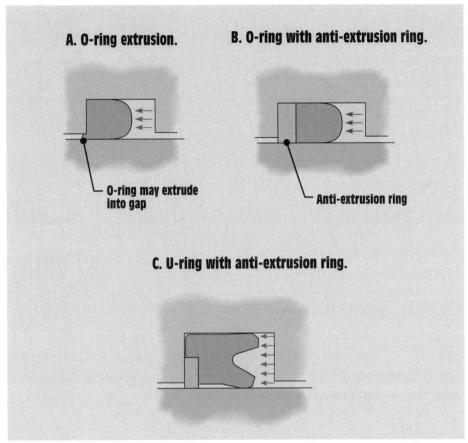

A. O-ring extrusion.   **B. O-ring with anti-extrusion ring.**

O-ring may extrude into gap

Anti-extrusion ring

C. U-ring with anti-extrusion ring.

FIGURE 9-53   Extrusion. (A) O-ring extrusion. (B) O-ring with anti-extrusion ring. (C) U-ring with anti-extrusion ring.

cylinders, and valves usually give the option of viton seals. They are considerably more expensive than nitrile seals, however. Elastomers (both viton and nitrile) are sometimes reinforced with fabric to increase their toughness for high-pressure applications.

Polyurethane is an elastoplastic. Elastoplastics have less elasticity than elastomers but are harder and stronger, making them more resistant to damage caused by extrusion and abrasion. Polyurethane seals can withstand higher pressures and have a longer life than elastomer seals. The disadvantage of standard polyurethane is that it cannot be used in systems that use a water-based fluid. When polyurethane comes into contact with water at temperatures above 140° F, it chemically reacts with the water and breaks apart. This reaction is known as *hydrolysis*. In fact, polyurethane can even be sensitive to the small amount of water found in systems using a nonwater based fluid. Special

MATERIAL	TYPE	TEMPERATURE RANGE	FLUID COMPATIBILITY	APPLICATIONS
Nitrile (NBR, buna-N)	Elastomer	−40° F to 250° F	Air, petroleum-based, water-based	Seals
Fluorocarbon (viton)	Elastomer	−30° F to 450° F	Air, petroleum-based, water-based, phosphate ester	Seals
Polyurethane	Elastoplastic	−50° F to 250° F	Petroleum-based	Seals
PTFE (teflon)	Plastic	−350° F to 500° F	Petroleum-based, water-based, phosphate ester, chemicals	Seals, bearings, anti-extrusion rings
Nylon	Plastic	−50° F to 250° F	Petroleum-based, water-based, phosphate ester	Bearings, anti-extrusion rings

TABLE 9-4   Seal and Bearing Materials

Data courtesy of System Seals Inc., Cleveland, Ohio

polyurethane formulations have been developed that resist hydrolysis up to higher temperatures than the previously stated 140° F.

PTFE (polytetrafluoroethylene), commonly known as Teflon, is a very low-friction plastic. Because it is a plastic, it does not have the elasticity of the elastomers and elastoplastics. When used as a seal, it must be *energized* to provide the necessary sealing force. The compound seal shown in Figure 9-50B is an example. The sliding element is most commonly made of PTFE, which provides very low friction on the dynamic surface. The o-ring *energizes* the sliding element—it provides the elasticity needed to apply an adequate sealing force. PTFE is also commonly used in bearings and anti-extrusion rings. To increase its strength, PTFE is often filled with materials such as bronze, glass, or graphite. In addition to having very low friction, PTFE also has the advantages of a large temperature range and compatibility with a wide variety of hydraulic fluids and chemicals.

Nylon, like PTFE, is a plastic. It does not, however, have the low friction and large temperature range of PTFE. It also does not have the chemical resistance of PTFE. Nylon is widely used in bearings and anti-extrusion rings, however, because its temperature range is adequate for most hydraulic systems and it is compatible with the most commonly used hydraulic fluids. Like PTFE, nylon may be filled with other materials to increase its strength.

Fluid compatibility is of primary importance when selecting seals. Seals that are not compatible with the fluid used will quickly degrade and may cause the entire system to fail. It is also important to consider the temperature and pressure ranges of the system. The seal material must be able to function properly at the expected temperatures. When operated at temperatures below their minimum, seals will normally not have the elasticity to provide a good seal. When operated above their temperature range, the seal material may degrade and fail. The seal material and geometry must be selected to withstand the expected pressures without extrusion (anti-extrusion rings may be necessary).

After selecting the seal, the seal groove and the surface to be sealed must be properly prepared. The manufacturer will specify important parameters such as surface finish and corner radii. Whether the seal is static or dynamic, a smooth surface is critical to achieving a good seal. Surface finish is most critical on a dynamic surface, however, because a surface that is too rough will cause the seal to wear away quickly. The seal must then be properly installed, using the manufacturer's recommended tools and methods. Damage during installation is one of the leading causes of seal failure. Once the right seal is selected and installed properly, the system fluid must be filtered continuously to reduce contamination. Contamination is a common cause of seal failure, so proper filtration is critical.

9.10 Hydraulic Fluids

In addition to its primary purpose of being the media through which the power is transmitted, the fluid in a hydraulic system serves two other important functions: (1) lubricating the components to protect them from wear, and (2) carrying heat away from the components and back to the reservoir where it can be transferred to the environment. Two factors are of primary importance in selecting a fluid: (1) compatibility with seals and bearings, and (2) the physical properties, primarily viscosity. In addition, fire-resistance and/or environmental concerns may be important in particular applications.

Five types of fluids are currently used in hydraulic systems: *petroleum-based, emulsions, water-glycol, synthetics,* and *vegetable-based.* Petroleum-based oil remains the standard fluid for most applications. It is inexpensive, has excellent lubricating properties, and is compatible with most seal materials. It does, however, have two major disadvantages. One is its flammability. It cannot be used in applications with extreme heat or where an open flame may be present. This prevents it from being used in applications such as steel mills, foundaries, and mining. In these applications, a fire-resistant fluid must be used. The other disadvantage is the damage that petroleum oil can cause to the environment. This disadvantage is driving two major trends in the industry: zero-leakage sealing systems and the development and adoption of environmentally safe fluids. Environmentally safe fluids must be biodegradable and nontoxic.

Emulsions are combinations of two fluids that normally will not mix, such as petroleum-based oil and water. An emulsifier is added that forces one of the liquids to form small droplets that remain suspended throughout the other liquid. There are two types of emulsions: *oil-in-water* and *water-in-oil*. Oil-in-water fluids are made up primarily of water with small droplets of oil. Because they are mostly water, they exhibit good fire resistance. Water-in-oil emulsions, also called *invert emulsions,* are primarily oil with small droplets of water dispersed throughout. These fluids have better lubricating characteristics than oil-in water emulsions due to their higher oil content. Fire resistance varies depending on the amount of water. Emulsions are not environmentally safe because of their significant petroleum-based oil content.

Water-glycol is made up of about half water and half glycol. Because of its high water content, it exhibits good fire resistance. It is not environmentally safe, however, because glycol is not biodegradable and is potentially toxic. Another disadvantage to water-glycol is its less than ideal lubricating ability. This causes decreased component life due to increased wear. Systems using water-glycol cannot operate at temperatures as high as those using petroleum-based oil because of the possibility of water evaporation.

Phosphate ester, the most commonly used synthetic fluid, is used because of its excellent fire resistance and good lubricating qualities. It does, however, have two major disadvantages: It is expensive and is not compatible with nitrile, the most commonly used seal material. Systems using phosphate ester must use pumps, motors, and valves with viton seals, which are more expensive. Phosphate ester is also not environmentally safe. Another synthetic, *polyol ester,* claims both fire resistance and environmental safety. Like phosphate ester, it is very expensive relative to the other fluids discussed. In addition to these, there are also less commonly used synthetic formulations used for specific applications.

The increasing trend toward environmentally safe systems has led to the development and use of a new category of hydraulic fluids: *vegetable-based.* These fluids, which are usually rapeseed (canola) or soybean oil, are the most environmentally safe, have good lubricating qualities, can be formulated to exhibit fire resistance, and are much less expensive than synthetics. They have two disadvantages, however: They have a greater tendency to *oxidize* and tend to absorb water. Oxidation, the formation of sludges and harmful chemicals due to reaction with oxygen, is a concern with all hydraulic fluids, but more so with vegetable-based fluids. The tendency to absorb water can cause damage to standard polyurethane seals due to hydrolysis. Special polyurethane formulations have been developed to resist hydrolysis.

All of the preceding fluid formulations have additives designed to improve the performance of the system. These include foaming, oxidation and rust inhibitors and also anti-wear additives that improve the lubrication characteristics of the fluid. Foaming inhibitors cause entrained air to escape quickly from the oil. Oxidation inhibitors slow down the degradation of the fluid through reactions with oxygen. This is particularly important because the by-products of oxidation are sludges that gum up components, and acids that cause corrosion.

The rate of oxidation increases with increasing fluid temperature, so keeping the system as cool as possible also helps to slow down oxidation.

The basic fluid properties that are important in hydraulics are density (specific gravity), viscosity, viscosity index, and bulk modulus (see Chapter 2). Viscosity is of primary importance because the thickness of a fluid largely determines its lubricating ability. In the hydraulics industry, viscosity is most commonly measured in *Saybolt Second Universal* (SSU). Pumps and motors are the components most sensitive to viscosity variations, so their specifications should be followed. Viscosity that is too low or too high affects system performance. If the viscosity of a fluid is too high, it becomes difficult to move through the system, causing increased power loss and heat buildup due to excessive fluid friction. If the viscosity is too low, excessive wear and heat buildup result because the fluid no longer provides good lubrication. Low viscosity also causes increased leakage.

Other important properties of hydraulic fluids include *pour point, flash point,* and *fire point.* The pour point is the lowest temperature at which a fluid will flow. This may be important in systems that will be started at low temperatures. The flash point and fire point are properties that measure fire resistance. The flash point is the temperature at which there is enough fluid vapor at the surface to ignite in the presence of an open flame. The flame will then immediately die down. The fire point is the temperature at which there is enough vapor to support combustion for 5 seconds.

9.11 Equations

EQUATION NUMBER	EQUATION	REQUIRED UNITS
9-1	$P_{INT} = P_{PUMP} \cdot \dfrac{A_{PISTON}}{A_{ROD}}$	Any consistent units
9-2	$Q_{INT} = Q_{PUMP} \cdot \dfrac{A_{ROD}}{A_{PISTON}}$	Any consistent units
9-3	$P_{INT} = P_{PUMP} \cdot IR$	Any consistent units
9-4	$Q_{INT} = \dfrac{Q_{PUMP}}{IR}$	Any consistent units
9-5	$\beta_x = \dfrac{N_U}{N_D}$	Any consistent units
9-6	$\eta_x = 1 - \dfrac{1}{\beta_x}$	Any consistent units
9-7	$ID = OD - 2 \cdot t$	Any consistent units

9.12 Review Questions and Problems

1. What is the purpose of an accumulator?
2. What is the precharge of a gas-charged accumulator?
3. Describe the operation of a gas-charged piston accumulator.
4. Describe the operation of a gas-charged bladder accumulator.
5. What type of gas is used in gas-charged accumulators? Why?
6. Name three applications for an accumulator.
7. What is the purpose of an intensifier? What is gained? What is lost?
8. The intensifier shown in Figure 9-11 has a piston area of 8 in^2 and a rod area of 1 in^2. What is the maximum outlet pressure if the maximum inlet pressure is 750 psi? What is the flow rate out of the intensifier if the pump has a flow rate of 10 gpm?
9. An intensifier with an intensification ratio of 5:1 is being used in a system with a maximum pump pressure of 1000 psi and a flow rate of 10 gpm. What is the flow rate and maximum pressure out of the intensifier?
10. An intensifier with an intensification ratio of 20:1 is being used in a system with a maximum pump pressure of 7000 kPa and a flow rate of 40 lpm. What is the flow rate and maximum pressure out of the intensifier?
11. What are the four functions of a hydraulic reservoir?
12. What is the purpose of a baffle in a reservoir?
13. What size reservoir should be selected for an industrial hydraulic system operating at 20 gpm?
14. What is the purpose of a heat exchanger?
15. What are the two types of heat exchangers used in hydraulic systems?
16. Describe the operation of a shell and tube heat exchanger. Why are many small, bronze tubes used instead of one large one?
17. Why is a hydraulic fluid heater sometimes necessary?
18. Why is filtration important in a hydraulic circuit?
19. What are the sources of particulate contamination in a hydraulic circuit?
20. Describe the difference between a strainer and a filter.
21. What are the three primary locations within a hydraulic circuit where a filter may be placed?
22. Why are strainers, rather than fine filters, often used on the pump inlet line?
23. Why is a filter in the return line generally preferred to a pump outlet line filter?
24. What is the advantage to a filter in the pump outlet line?
25. What is offline filtration?

26. What is the purpose of the ISO cleanliness code?
27. What is the purpose of the beta ratio?
28. A fluid is known to have 10,000 particles of size 15 μm or larger per milliliter. After being run through a filter, the fluid is found to have 200 particles of size 15 μm or larger per milliliter. What is the β_{15} ratio for this filter?
29. What is the removal efficiency (η_{15}) of the filter in problem 28?
30. The manufacturer of a particular hydraulic component has specified an ISO cleanliness code of 18/16 for proper operation. The component will be used in a system that typically has 30,000 particles of size 5μm or larger per milliliter and 10,000 particles of size 15 μm or larger per milliliter. What should the β_5 and β_{15} ratios be for a filter that is to be used in this system?
31. What is the terminal pressure drop of a filter?
32. Describe the operation of a bourdon tube gauge.
33. What are the three types of flowmeters that are commonly used in hydraulic circuits? Describe the operation of each.
34. What are the three types of conduits used in a hydraulic circuit? Name some advantages and disadvantages of each.
35. Why is a slower flow velocity through a conduit desirable?
36. A conduit size needs to be determined for a system in which the flow rate will be 12 gpm. Determine the minimum conduit *ID* if the flow velocity is not to exceed 15 ft/s.
37. A conduit size needs to be determined for a system in which the flow rate will be 60 lpm. Determine the minimum conduit *ID* if the flow velocity is not to exceed 5 m/s.
38. What is the difference between a static and a dynamic seal?
39. Define *squeeze* in relation to o-rings.
40. Why is the u-ring such an effective seal design? Which side faces the pressure?
41. What is the most commonly used seal material?
42. What are the two advantages of fluorocarbon (viton) seals over nitrile seals? What is the disadvantage?
43. What is the advantage of polyurethane seals over viton and nitrile seals?
44. What is the first thing to consider when selecting seals?
45. Other than transmitting the power, what are two other purposes of the fluid in a hydraulic system?
46. Why is the viscosity of a hydraulic fluid so important?

CHAPTER 10

Basic Principles of Pneumatics

OUTLINE

10.1 Introduction

10.2 Absolute Pressure and Temperature

10.3 Gas Laws

10.4 Gas Flow

10.5 Vacuum

10.6 Pneumatic Systems

10.7 Equations

10.8 Review Questions and Problems

10.1   Introduction

Pneumatics is the use of a gas to transmit power from one point to another. Previous chapters focused on hydraulics, but many of the concepts discussed also apply to pneumatics. Pascal's law, which applies to both hydraulics and pneumatics, states that *the pressure exerted on a confined fluid is transmitted undiminished in all directions and perpendicular to the containing surfaces.* Because a confined gas behaves the same as a confined liquid with regard to pressure, we may use the equation:

$$F = p \cdot A$$

to calculate the force generated when a pressure is applied to a given area. In addition to this, many of the components used in both hydraulic and pneumatic systems are similar in construction and function (see Chapters 11 and 12).

The differences between hydraulics and pneumatics all stem from the fact that pneumatics uses a compressible gas, rather than the relatively incompressible liquid that is used in hydraulics. Because hydraulic fluid is relatively incompressible, it is basically like transmitting a force through a solid (Figure 10-1A). The output follows the motion of the input directly, which makes the position and speed of the output easy to control accurately. Because gases are compressible, transmitting a force through a gas is like using a spring (Figure 10-1B). This gives the output motion a jumpy quality because the spring flexes as it moves back and forth. It also makes it more difficult to precisely control the position and speed of the output.

Although a pneumatic system is jumpy and less precise than a hydraulic system, it has the advantage of being quicker acting. This is because when a directional control valve is shifted, the compressed air tends to expand quickly, just as a compressed spring does when released. This causes the air to flow rapidly to the actuator. Another major difference is that while hydraulic systems are self-lubricating, pneumatic systems are not. A lubricant must be added to protect the components from wear. Finally, hydraulic systems operate at high pressures (500 to 5000 psi) and can generate thousands of pounds of force, while pneumatic systems operate at lower pressures (usually around 100 psi) and can generate forces in the hundreds of pounds range. Table 10-1 summarizes the major differences between hydraulics and pneumatics.

The gas used in pneumatics is nearly always air, due to its obvious availability. Air is a mixture of gases and is composed of about 21% oxygen, 78% nitrogen, and 1% other gases such as carbon dioxide and argon. Some systems use pure nitrogen, such as the hydropneumatic accumulators discussed in Chapter 9. Nitrogen gas is used in applications in which the combustibility of air can be dangerous, such as very high-pressure applications. Because the vast majority of pneumatic systems use air, the terms *pneumatic* and *air* will be used interchangeably in this text, as is the case in industry.

A. Transmission through a liquid is like using a solid.

Input → Output

B. Transmission through a gas is like using a spring.

Input → Output

FIGURE 10-1   Transmitting forces through liquids and gases. (A) Transmission through a liquid is like using a solid. (B) Transmission through a gas is like using a spring.

HYDRAULICS	PNEUMATICS
Uses a relatively incompressible liquid	Uses a compressible gas (nearly always air)
Slow, smooth motion	Quick, jumpy motion
Very precise	Not as precise as hydraulics
Self-lubricating	Lubricant must be added
Not as clean (some leakage usually exists)	Generally cleaner
Pressures of 500 to 5000 psi	Pressures of around 100 psi

TABLE 10-1   Differences between Hydraulics and Pneumatics

10.2 Absolute Pressure and Temperature

Before discussing the behavior of gases, we must first discuss *absolute pressure* and *absolute temperature*. When we speak of pressure in a fluid system, we are usually referring to *gauge pressure*, which is the pressure above atmospheric pressure. Atmospheric pressure is the pressure due to the weight of the air in the atmosphere above us. Gauge pressure is what a gauge will always read because the atmospheric pressure is on all sides of a system and therefore cancels out. *Absolute pressure is the total pressure exerted on the system, including atmospheric pressure.* In equation form:

$$p_{ABS} = p_{GAUGE} + p_{ATM} \tag{10-1}$$

where: p_{ABS} = absolute pressure of the gas (psia, kPa abs)
$\quad\quad p_{GAUGE}$ = gauge pressure of the gas (psig, kPa gauge)
$\quad\quad p_{ATM}$ = atmospheric pressure (psia, kPa abs)

We will use *psig* (kPa gauge) to signify gauge pressure and *psia* (kPa abs) to signify absolute pressure. It should be noted that wherever psi or kPa is used, it is assumed to mean the gauge pressure. This is true not only in this text, but in most other fluid power literature as well. The pressure of the atmosphere, p_{ATM}, is equal to 14.7 psia (101 kPa abs) at sea level. The atmospheric pressure is lower at higher elevations because there is less air above to exert pressure.

The temperature scales we use in everyday life, such as °F and °C, are not practical for calculations because their 0° point is arbitrary and does not represent a true zero temperature point. Temperature is actually a measure of molecular movement, with higher temperatures representing faster movement. It would therefore make sense to have a temperature of zero represent no movement, or the point at which molecular movement ceases. This is called *absolute zero,* and it occurs at −460°F (−273°C). Absolute temperature is a scale that uses absolute zero as its zero point. In the U.S. customary system of units, absolute temperature is measured in degrees Rankine (°R). To convert between °F and °R, use the following equation:

$$°R = °F + 460 \tag{10-2}$$

This makes sense because 0°R = absolute zero = −460°F.

In metric units, absolute temperature is measured in kelvin (K). To convert between °C and K, use the following equation:

$$K = °C + 273 \tag{10-2M}$$

The kelvin scale, unlike the other temperature scales discussed, does not use degrees.

10.3 Gas Laws

To understand pneumatic systems, we must first understand the behavior of gases. Their behavior is described by the *perfect gas laws.* These laws describe the relationships among pressure, temperature, and volume for most gases under a wide range of conditions. Air follows these laws very closely under the conditions that are found in pneumatic systems.

Boyle's law, which states that *the absolute pressure of a confined gas is inversely proportional to its volume, provided its temperature remains constant,* is the most useful of the perfect gas laws for pneumatic systems. Boyle's law tells us that changing the volume of a gas by a factor causes its pressure to change by the inverse of that factor. For example, if we compress air (reduce its volume) by two times, its pressure will increase by two times. If we expand air (increase its volume) by two times, its pressure will decrease by two times. Assuming that the temperature remains constant is a simplifying assumption. An *isothermal* process is a process in which the temperature remains constant. The following equation states Boyle's law mathematically:

$$p_1 \cdot V_1 = p_2 \cdot V_2 \qquad\qquad \textbf{(10-3)}$$

where: p_1 = absolute pressure of the gas at state 1
V_1 = volume of the gas at state 1
p_2 = absolute pressure of the gas at state 2
V_2 = volume of the gas at state 2

Absolute pressure *must* be used in the preceding equation. Failure to do so will result in answers that are off considerably.

EXAMPLE 10-1.

Figure 10-2A shows a cylinder that is initially filled with 4 ft³ (V_1) of air at a pressure of 20 psig (p_1). Figure 10-2B shows the air is compressed to a volume of 1 ft³ (V_2). What is the pressure (p_2) in psig after compression?

SOLUTION:

1. Convert to psia:

$$p_1 = p_{GAUGE} + p_{ATM} = 20 \text{ psig} + 14.7 \text{ psia} = 34.7 \text{ psia}$$

2. Calculate the new pressure:

$$p_2 = \frac{p_1 \cdot V_1}{V_2} = \frac{34.7 \text{ psia} \cdot (4 \text{ ft}^3)}{1 \text{ ft}^3} = 138.8 \text{ psia}$$

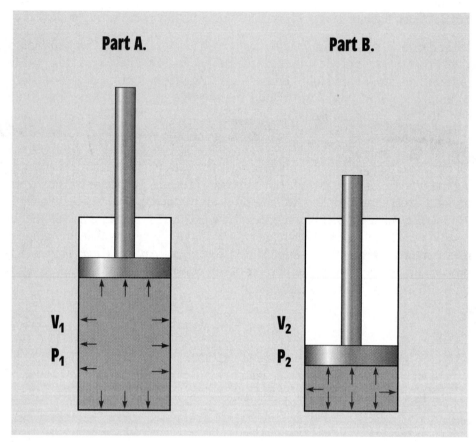

FIGURE 10-2   Example 10-1.

3. Convert back to psig:

$$p_2 = p_{ABS} - p_{ATM} = 138.8 \text{ psia} - 14.7 \text{ psia} = 124.1 \text{ psig}$$

EXAMPLE 10-1M.

Figure 10-2A shows a cylinder that is initially filled with 2 m³ (V_1) of air at a pressure of 140 kPa gauge (p_1). Figure 10-2B shows the air is compressed to a volume of 0.5 m³ (V_2). What is the pressure (p_2) in kPa gauge after compression?

SOLUTION:

1. Convert to kPa abs:

$$p_1 = p_{GAUGE} + p_{ATM} = 140 \text{ kPa gauge} + 101 \text{ kPa abs} = 241 \text{ kPa abs}$$

2. Calculate the new pressure:

$$p_2 = \frac{p_1 \cdot V_1}{V_2} = \frac{241 \text{ kPa abs} \cdot (2 \text{ m}^3)}{0.5 \text{ m}^3} = 964 \text{ kPa abs}$$

3. Convert back to kPa gauge:

$$p_2 = p_{ABS} - p_{ATM} = 964 \text{ kPa abs} - 101 \text{ kPa abs} = 863 \text{ kPa gauge}$$

EXAMPLE 10-2.

Air initially at a volume of 1 ft³ and a pressure of 100 psig is allowed to expand to 2 ft³. What is the gauge pressure after expansion?

SOLUTION:

1. Convert to psia:

$$p_1 = p_{GAUGE} + p_{ATM} = 100 \text{ psig} + 14.7 \text{ psia} = 114.7 \text{ psia}$$

2. Calculate the final pressure:

$$p_2 = \frac{p_1 \cdot V_1}{V_2} = \frac{114.7 \text{ pisa} \cdot (1 \text{ ft}^3)}{2 \text{ ft}^3} = 57.4 \text{ psia}$$

3. Convert back to psig:

$$p_2 = p_{ABS} - p_{ATM} = 57.4 \text{ psia} - 14.7 \text{ psia} = 42.7 \text{ psig}$$

Gay-Lussac's law is another perfect gas law that relates temperature and pressure at constant volume. It states that *the absolute pressure of a confined gas is proportional to its temperature, provided its volume remains constant.* For example, if we raise the temperature of a gas while holding its volume constant, its pressure will increase by the same proportion. This principle applies to the very practical situation in which a gas is held in a rigid container and heated or cooled. The following equation states Gay-Lussac's law mathematically:

$$\frac{p_1}{T_1} = \frac{p_2}{T_2} \tag{10-4}$$

where: p_1 = absolute pressure of the gas at state 1
T_1 = absolute temperature of the gas at state 1

p_2 = absolute pressure of the gas at state 2
T_2 = absolute temperature of the gas at state 2

EXAMPLE 10-3.

Air in a fixed volume container is initially at atmospheric pressure (0 psig) and 70°F. Its temperature is then raised to 150°F. What is the gauge pressure at the final temperature?

SOLUTION:

1. Convert to absolute temperatures (°R = °F + 460):

$$T_1 = 70 + 460 = 530°R$$

$$T_2 = 150 + 460 = 610°R$$

2. Convert to psia:

$$p_1 = p_{GAUGE} + p_{ATM} = 0 \text{ psig} + 14.7 \text{ psia} = 14.7 \text{ psia}$$

3. Calculate the final pressure:

$$p_2 = p_1 \cdot \frac{T_2}{T_1} = 14.7 \text{ psia} \cdot \left(\frac{610°F}{530°F} \right) = 16.9 \text{ psia}$$

4. Convert back to psig:

$$p_2 = p_{ABS} - p_{ATM} = 16.9 \text{ psia} - 14.7 \text{ psia} = 2.2 \text{ psig}$$

EXAMPLE 10-3M.

Air in a fixed volume container is initially at atmospheric pressure (0 kPa gauge) and 20°C. Its temperature is then raised to 80°C. What is the gauge pressure at the final temperature?

SOLUTION:

1. Convert to absolute temperatures (K = °C + 273):

$$T_1 = 20 + 273 = 293 \text{ K}$$

$$T_2 = 80 + 273 = 353 \text{ K}$$

2. Convert to kPa abs:

$$p_1 = p_{GAUGE} + p_{ATM} = 0 \text{ kPa gauge} + 101 \text{ kPa abs} = 101 \text{ kPa abs}$$

3. Calculate the final pressure:

$$p_2 = p_1 \cdot \frac{T_2}{T_1} = 101 \text{ kPa abs} \cdot \left(\frac{353 \text{ K}}{293 \text{ K}}\right) = 122 \text{ kPa abs}$$

4. Convert back to kPa gauge:

$$p_2 = p_{ABS} - p_{ATM} = 122 \text{ kPa abs} - 101 \text{ kPa abs} = 21 \text{ kPa gauge}$$

Charles's law, also a perfect gas law, relates temperature and volume at constant pressure. It states that *the volume of a confined gas is proportional to its temperature, provided its pressure remains constant.* This law describes the tendency of a gas to expand when heated, if allowed to do so. If a gas is not allowed to expand when heated, its pressure will increase according to Gay-Lussac's law. The following equation states Charles's law mathematically:

$$\frac{V_1}{T_1} = \frac{V_2}{T_2} \qquad\qquad \textbf{(10-5)}$$

where: V_1 = volume of the gas at state 1
$\quad\quad\ T_1$ = absolute temperature of the gas at state 1
$\quad\quad\ V_2$ = volume of the gas at state 2
$\quad\quad\ T_2$ = absolute temperature of the gas at state 2

EXAMPLE 10-4.

Air is held in a container with a free sliding piston, such as the one shown in Figure 10-3. It is initially at a temperature of 70°F (T_1) and a volume of 0.5 ft³ (V_1) (part A). Its temperature is then raised to 250°F (T_2) (part B). What is the volume of the air after being heated?

SOLUTION:

1. Convert to absolute temperatures (°R = °F + 460):

$$T_1 = 70 + 460 = 530°R$$

$$T_2 = 250 + 460 = 710°R$$

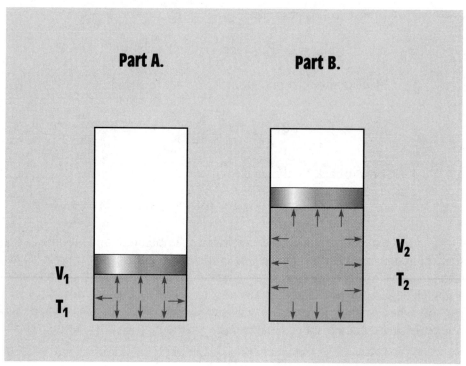

Part A.

Part B.

V_1

T_1

V_2

T_2

FIGURE 10-3   Example 10-4.

2. Calculate the final volume:

$$V_2 = V_1 \cdot \frac{T_2}{T_1} = 0.5 \text{ ft}^3 \cdot \left(\frac{710°F}{530°F}\right) = 0.670 \text{ ft}^3$$

EXAMPLE 10-4M.

Air is held in a container with a free sliding piston, such as the one shown in Figure 10-3. It is initially at a temperature of 20°C (T_1) and a volume of 0.25 m³ (V_1) (part A). Its temperature is then raised to 120°C (T_2) (part B). What is the volume of the air after being heated?

SOLUTION:

1. Convert to absolute temperatures (K = °C + 273):

$$T_1 = 20 + 273 = 293 \text{ K}$$

$$T_2 = 120 + 273 = 393 \text{ K}$$

2. Calculate the final volume:

$$V_2 = V_1 \cdot \frac{T_2}{T_1} = 0.25 \text{ m}^3 \cdot \left(\frac{393 \text{ K}}{293 \text{ K}} \right) = 0.335 \text{ m}^3$$

All of the previously discussed perfect gas laws are actually derived from the *General gas law*, which relates all three parameters (temperature, pressure, and volume), while holding none of them fixed. It is given by the following equation:

$$\frac{p_1 \cdot V_1}{T_1} = \frac{p_2 \cdot V_2}{T_2} \qquad \textbf{(10-6)}$$

Boyle's law is derived from Equation 10-6 by setting $T_1 = T_2$, because the temperature is assumed to be constant. This causes T_1 and T_2 to disappear from the equation, and we are left with Equation 10-3. Similarly, Gay-Lussac's law is derived by setting $V_1 = V_2$ and Charles's law is derived by setting $p_1 = p_2$.

EXAMPLE 10-5.

Air initially at 50°F and 100 psig occupies a volume of 5 ft³. If the air is heated to 300°F while its volume is reduced to 2 ft³, what is the resulting final pressure?

SOLUTION:

1. Convert to absolute temperature:

$$T_1 = 50 + 460 = 510°R$$

$$T_2 = 300 + 460 = 760°R$$

2. Convert to absolute pressure:

$$p_{ABS} = p_{GAUGE} + p_{ATM} = 100 \text{ psig} + 14.7 \text{ psia} = 114.7 \text{ psia}$$

3. Calculate the final pressure:

$$p_2 = \frac{T_2 \cdot p_1 \cdot V_1}{T_2 \cdot V_2} = \frac{760°R \cdot (114.7 \text{ psia}) \cdot 5 \text{ ft}^3}{510°R \cdot (2 \text{ ft}^3)} = 427.3 \text{ psia}$$

4. Convert back to psig:

$$p_2 = p_{ABS} - p_{ATM} = 427.3 \text{ psia} - 14.7 \text{ psia} = 412.6 \text{ psig}$$

EXAMPLE 10-5M.

Air initially at 10°C and 700 kPa occupies a volume of 2 m³. If the air is heated to 150°C while its volume is reduced to 0.2 m³, what is the resulting final pressure?

SOLUTION:

1. Convert to absolute temperature:

$$T_1 = 10 + 273 = 283 \text{ K}$$

$$T_2 = 150 + 273 = 423 \text{ K}$$

2. Convert to absolute pressure:

$$p_{ABS} = p_{GAUGE} + p_{ATM} = 700 \text{ kPa gauge} + 101 \text{ kPa abs} = 801 \text{ kPa abs}$$

3. Calculate the final pressure:

$$p_2 = \frac{T_2 \cdot p_1 \cdot V_1}{T_1 \cdot V_2} = \frac{423 \text{ K} \cdot (801 \text{ kPa abs}) \cdot 2 \text{ m}^3}{283 \text{ K} \cdot (0.2 \text{ m}^3)} = 11{,}973 \text{ kPa abs}$$

4. Convert back to kPa gauge:

$$p_2 = p_{ABS} - p_{ATM} = 11{,}973 \text{ kPa abs} - 101 \text{ kPa gauge} = 11{,}872 \text{ kPa gauge}$$

The General gas law is *not* frequently used in calculations for pneumatic systems because it is usually impractical to measure temperature, pressure, and volume simultaneously. Boyle's law simplifies the situation by assuming that the temperature remains constant when a gas is compressed or expanded. In real systems, however, the temperature of a gas increases when it is compressed and decreases when it is expanded. In spite of this, Boyle's law is often used to predict the behavior of pneumatic systems. It is fairly accurate as long as the temperature does not change too much from one state to the next, as the following example illustrates.

EXAMPLE 10-6.

Air initially at 70°F and atmospheric pressure (14.7 psia) is compressed from 8 ft³ to 1 ft³. The temperature of the air is measured at 90°F after compression. Determine the final pressure in two situations: (a) using the General gas law and (b) using Boyle's law, assuming the temperature remains at 70°F after compression.

SOLUTION:

1. Convert to absolute temperature:

$$T_1 = 70 + 460 = 530°R$$

$$T_2 = 90 + 460 = 550°R$$

2. Calculate the final pressure for situation (a)

$$p_2 = \frac{T_2 \cdot p_1 \cdot V_1}{T_1 \cdot V_2} = \frac{550°R \cdot (14.7\ \text{psia}) \cdot 8\ \text{ft}^3}{530°R \cdot (1\ \text{ft}^3)} = 122.0\ \text{psia}$$

Convert to psig:

$$p_2 = p_{ABS} - p_{ATM} = 122.0\ \text{psia} - 14.7\ \text{psia} = 107.3\ \text{psig}$$

3. Calculate the final pressure for situation (b)

$$p_2 = \frac{p_1 \cdot V_1}{V_2} = \frac{(14.7\ \text{psia}) \cdot 8\ \text{ft}^3}{1\ \text{ft}^3} = 117.6\ \text{psia}$$

Convert to psig:

$$p_2 = p_{ABS} - p_{ATM} = 117.6\ \text{psia} - 14.7\ \text{psia} = 102.9\ \text{psig}$$

We see that in this case assuming the temperature remained fixed resulted in a relatively small difference (about 4%) in the final answer. This is typical because the temperatures before and after compression are often close when compared on the absolute scale.

By assuming the process is isothermal, Boyle's law assumes that all of the heat generated when a gas is compressed is transferred to the atmosphere, which is why the temperature of the gas does not change. An *adiabatic* process is the opposite of an isothermal process. In this process, none of the heat is transferred to the atmosphere, so the temperature of the gas will increase to a maximum when compressed. An adiabatic process is described by the following equations:

$$p_1 \cdot V_1^{k} = p_2 \cdot V_2^{k}$$

$$\frac{p_1}{T_1^{\frac{k}{k-1}}} = \frac{p_2}{T_2^{\frac{k}{k-1}}}$$

where: k = a constant that measures the heat absorption properties of the gas

For air, $k = 1.4$, so these equations become:

$$p_1 \cdot V_1^{1.4} = p_2 \cdot V_2^{1.4}$$
(10-7)

$$\frac{p_1}{T_1^{3.5}} = \frac{p_2}{T_2^{3.5}}$$
(10-8)

As with the previous equations, absolute temperatures and pressures must be used in Equations 10-7 and 10-8. The following example illustrates how these equations are utilized. The same amount of compression and initial temperature used in the previous example will be used so that the results can be compared.

EXAMPLE 10-7.

Air that is initially at 70°F and atmospheric pressure is compressed from 8 ft³ to 1 ft³. Assuming the process is adiabatic, what is the final pressure and temperature of the gas?

SOLUTION:

1. Convert to absolute temperature:

$$T_1 = 70 + 460 = 530°R$$

2. Calculate the final pressure:

$$p_2 = \frac{p_1 \cdot V_1^{1.4}}{V_2^{1.4}} = \frac{14.7 \text{ psia} \cdot (8 \text{ ft}^3)^{1.4}}{(1 \text{ ft}^3)^{1.4}} = 270.2 \text{ psia}$$

$$p_2 = p_{ABS} - p_{ATM} = 270.2 \text{ psia} - 14.7 \text{ psia} = 255.5 \text{ psig}$$

3. Calculate the final temperature:

$$T_2 = \sqrt[3.5]{\frac{p_2}{p_1} \cdot T_1^{3.5}} = \sqrt[3.5]{\frac{270.2 \text{ psia}}{14.7 \text{ psia}} \cdot (530°R)^{3.5}} = 1218°R$$

$$T_2 = 1218°R - 460 = 758°F$$

EXAMPLE 10-7M.

Air that is initially at 20°C and atmospheric pressure is compressed from 3 m³ to 0.5 m³. Assuming the process is adiabatic, what is the final pressure and temperature of the gas?

SOLUTION:

1. Convert to absolute temperature:

$$T_1 = 20 + 273 = 293 \text{ K}$$

2. Calculate the final pressure:

$$p_2 = \frac{p_1 \cdot V_1^{1.4}}{V_2^{1.4}} = \frac{101 \text{ kPa abs} \cdot (3 \text{ m}^3)^{1.4}}{(0.5 \text{ m}^3)^{1.4}} = 1240 \text{ kPa abs}$$

$$p_2 = p_{ABS} - p_{ATM} = 1240 \text{ kPa abs} - 101 \text{ kPa abs} = 1139 \text{ kPa gage}$$

3. Calculate the final temperature:

$$T_2 = {}^{3.5}\sqrt{\frac{p_2}{p_1} \cdot T_1^{3.5}} = {}^{3.5}\sqrt{\frac{1240 \text{ kPa abs}}{101 \text{ kPa abs}} \cdot (293 \text{ K})^{3.5}} = 600 \text{ K}$$

$$T_2 = 600 - 273 = 327°C$$

The values from the previous two examples satisfy the General gas law, as they must. In reality, any time a gas is compressed or expanded, the process will be somewhere between isothermal and adiabatic, depending on the situation. Experience with the particular machinery and conditions is the best guide. The compression process is often much closer to isothermal than adiabatic, however, as the temperatures do not often rise to near the level that Equation 10-8 predicts (758°F in the previous example!). Much of heat of compression is quickly dissipated through the walls of the components to the atmosphere.

10.4 Gas Flow

Due to the compressibility of gases, flow rate measurement in pneumatics is much more complicated than with hydraulics. Flow rate is measured in units of volume per unit time. In hydraulics, the flow rate is most often measured in gal/min (gpm). In pneumatics, flow rate is measured in ft³/min (cfm, cubic feet per minute). The problem with this unit is that because air is compressible, 1 cubic foot of air at a higher pressure is actually more air than 1 cubic foot at a lower pressure. It is therefore necessary to use a unit that is independent of pressure, so that air flow rates can be compared accurately. The unit for doing this is standard cubic feet per minute (scfm). "Standard" refers to standard atmospheric conditions for temperature (68°F), pressure (14.7 psia), and relative humidity (36%). Air flow in scfm is basically an equivalent flow of atmospheric air, rather than the actual flow of pressurized air. Measuring flow rates in scfm assures that we are always referring to the same amount of air, regardless of pressure. This is necessary to accurately determine the flow demand of a pneumatic system.

In the metric system, the unit for flow is standard cubic meters per minute (standard m^3/min). Like scfm, this unit represents the equivalent flow of air under atmospheric conditions.

The terms scfm and standard m^3/min are often replaced with free cfm and free m^3/min, which essentially refer to the same quantity—an equivalent flow of uncompressed air. This term makes no reference to temperature and humidity, however. This makes the term *free* more ambiguous than *standard*.

10.5 Vacuum

Vacuum systems use pressures below atmospheric pressure to create suction. They work by evacuating air from one side of an object, which allows atmospheric air to create a net force on the opposite side. This is best explained by looking at the most common vacuum device, the suction cup (Figure 10-4). Part A shows the suction cup with the object to be lifted, a plate. Under normal conditions, atmospheric pressure presses equally on all sides of any object and therefore has no effect. Part B shows the suction cup pressed against the object while a vacuum pump evacuates the air from inside the suction cup. There is now no air inside the suction cup, so atmospheric pressure does not act on this part of the object, which allows atmospheric pressure to push the plate and the suction cup together. Suction cups, also called vacuum cups, are commonly used in pick and place applications to lift, manipulate, and assemble parts together.

Vacuum is often measured in units of inches of mercury (in Hg), an unusual unit that requires some explanation. This unit stems from a device that is used to measure atmospheric pressure known as a *mercury barometer* (Figure 10-5). The tube is laid flat in a pool of mercury, which expels all of the air from the tube. The tube is then raised into the vertical position, without allowing any air to enter the tube. There is now a perfect vacuum (no air) at the top of the tube and, consequently, no air pressure. Atmospheric pressure on the surface of the pool pushes the mercury up into the tube. The greater the air pressure, the higher the mercury is pushed up into the tube. Mercury is used because it is an extremely heavy liquid, and therefore requires a shorter tube. At sea level, the air pressure will push the mercury to a height of 30 inches. 30 in Hg therefore represents a perfect vacuum. Values of less than 30 in Hg represent less than a perfect vacuum (there is still some air trapped in the tube).

At elevations higher than sea level, a perfect vacuum will be less than 30 inch Hg because the atmospheric pressure is lower. This is an important point because it means that at higher elevations, less force can be generated using vacuum.

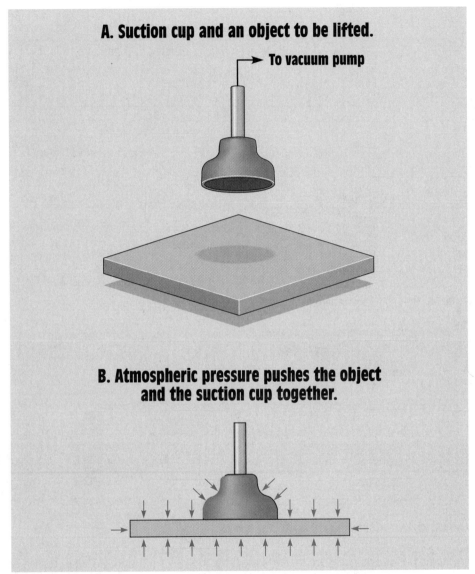

A. Suction cup and an object to be lifted.

→ **To vacuum pump**

**B. Atmospheric pressure pushes the object
and the suction cup together.**

FIGURE 10-4   Suction cup. (A) Suction cup and an object to be lifted. (B) Atmospheric
pressure pushes the object and the suction cup together.

Vacuum pumps will not be able to generate a perfect vacuum because they
can never evacuate all of the air. Industrial vacuum systems typically operate at levels of 20 in Hg or less, although values as high as 27 in Hg are easily achievable.

Table 10-2 compares in Hg to psig and psia. We see that atmospheric pressure = 0 psig = 14.7 psia = 0 in Hg. Notice that increasing in Hg values correspond to decreasing psia values. As air is removed, the amount of vacuum is increasing and the pressure is decreasing. The psig values represent the pressure

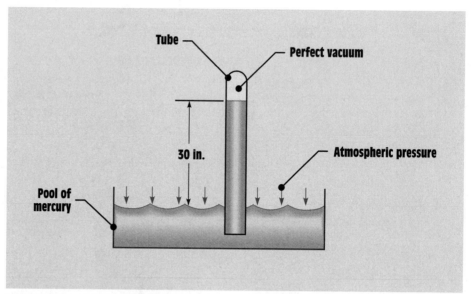

FIGURE 10-5   Mercury barometer.

	IN HG	PSIA	PSIG
Atmospheric pressure	0	14.7	0
	5	12.3	−2.46
	10	9.82	−4.91
	15	7.37	−7.37
	20	4.91	−9.82
	25	2.46	−12.3
Perfect vacuum	30	0	−14.7

TABLE 10-2   Comparison of *in Hg, psia,* and *psig*

below atmospheric pressure, as indicated by the negative sign. As air is removed, the pressure drops farther and farther below atmospheric pressure. When all of the air is removed, a perfect vacuum is achieved (30 in Hg = 0 psia = −14.7 psig). A useful conversion factor is that 1 in Hg = −0.491 psig at sea level.

Negative psig values may also be referred to as positive values, followed by the term *vacuum* to indicate that it is the amount below atmospheric pressure. For example, −10 psig may also be written 10 psi vacuum.

In metric units, vacuum is most commonly measured in *millimeters of mercury* (mm Hg). The same principles apply. Table 10-3 compares mm Hg to kPa gauge and kPa abs. A useful conversion factor is that 1 mm Hg = −0.1333 kPa gauge at sea level.

	MM HG	KPA ABS	KPA GAUGE
Atmospheric pressure	0	101	0
	150	81.0	−20.0
	300	61.0	−40.0
	450	41.0	−60.0
	600	21.0	−80.0
Perfect vacuum	762	0	−101

TABLE 10-3   Comparison of mmHg, kPa abs, and kPa gauge

EXAMPLE 10-8.

A suction cup with an inside diameter of 2 in is used in a vacuum system that pulls 18 in Hg. What is the maximum lifting force?

SOLUTION:

The lifting force is calculated by multiplying the area by the pressure in psig.

1. Convert to psig:

$$18 \text{ in Hg} \cdot \left(\frac{-0.491 \text{ psig}}{1 \text{ in Hg}} \right) = -8.84 \text{ psig}$$

The inside diameter of the suction cup must be used to calculate the area, as this is the area over which the vacuum acts.

2. Calculate the area:

$$A = \frac{\pi \cdot D^2}{4} = \frac{3.142 \cdot (2 \text{ in})^2}{4} = 3.142 \text{ in}^2$$

3. Calculate the force:

$$F_E = p \cdot A_p = -8.84 \frac{\text{lbs}}{\text{in}^2} \cdot (3.142 \text{ in}^2) = -27.8 \text{ lbs}$$

The negative sign signifies that the object is being drawn toward the suction cup.

EXAMPLE 10-8M.

A suction cup with an inside diameter of 50 mm is used in a vacuum system that pulls 500 mm Hg. What is the maximum lifting force?

SOLUTION:

The lifting force is calculated by multiplying the area by the pressure in psig.

1. Convert to kPa gauge:

$$500 \text{ mm Hg} \cdot \left(\frac{-0.1333 \text{ kPa gage}}{1 \text{ mm Hg}} \right) = -66.7 \text{ kPa gage}$$

2. Calculate the area:

$$A = \frac{\pi \cdot D^2}{4} = \frac{3.142 \cdot (0.050 \text{ m})^2}{4} = 0.001964 \text{ m}^2$$

3. Calculate the force:

$$F_E = p \cdot A_p = -66,700 \frac{N}{m^2} \cdot (0.001964 \text{ m}^2) = -131 \text{ N}$$

Suction cups are used widely in industry to lift small to moderate loads. In the automotive industry, for example, two or more large suction cups are used to lift sheet metal in and out of large hydraulic stamping presses that produce the body panels. At the other end of the size spectrum, suction cups are used to pick and place small integrated circuits onto circuit boards.

Vacuum forming is another application area for vacuum (Figure 10-6). In this application, air is evacuated from a mold, allowing atmospheric pressure on the outside to push the plastic sheet into the mold. The plastic sheet is heated to make it pliable so that it will acquire the shape of the mold.

Vacuum pumps and other devices used to generate a vacuum will be discussed in the next chapter, which covers pneumatic power supply.

10.6 Pneumatic Systems

Pneumatic systems are similar to hydraulic systems in that they can be divided into three segments (1) *power supply*, (2) *control valves*, and (3) *output*. The output segment consists of the actuator (cylinder, motor, etc.) and the load. The

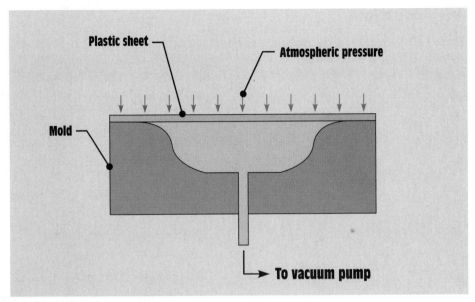

FIGURE 10-6   Vacuum forming.

control valve segment consists of directional, pressure, and flow control valves. Many of the components in the control and output segments of a pneumatic system are very similar in construction and operation to those in a hydraulic system, although they are designed for lower pressures.

Hydraulic and pneumatic systems differ considerably in the way they supply power. The power supply segment in a hydraulic system consists of a prime mover (electric motor or engine) and a pump. The pneumatic power supply segment consists of a prime mover, a *compressor,* and a *receiver tank.* The prime mover in a pneumatic system is most commonly an electric motor, but gas or diesel engines are used in some outdoor systems. The compressor is basically an air pump. Instead of supplying flow directly to the system, as is usually the case with a hydraulic pump, the compressor pumps air into the receiver tank. Because the air is compressible, large volumes of air can be compressed into the receiver tank. By doing this, we are storing energy for later use, just as energy is stored in a compressed spring. The system then draws air from the receiver tank as needed. Another difference between hydraulic and pneumatic power supply is that while each hydraulic machine usually has its own pump, pneumatic systems typically use one compressor and receiver tank to supply multiple machines.

10.7 Equations

EQUATION NUMBER	EQUATION	REQUIRED UNITS
10-1	$p_{ABS} = p_{GAUGE} + p_{ATM}$	Any consistent units
10-2	$°R = °F + 460$	Units as stated
10-2M	$K = °C + 273$	Units as stated
10-3	$p_1 = V_1 + p_2 \cdot V_2$	Any consistent units (p in absolute)
10-4	$\dfrac{p_1}{T_1} = \dfrac{p_2}{T_2}$	Any consistent units (p, T in absolute)
10-5	$\dfrac{V_1}{T_1} = \dfrac{V_2}{T_2}$	Any consistent units (T in absolute)
10-6	$\dfrac{p_1 \cdot V_1}{T_1} = \dfrac{p_2 \cdot V_2}{T_2}$	Any consistent units (p, T in absolute)
10-7	$p_1 = V_1^{1.4} + p_2 \cdot V_2^{1.4}$	Any consistent units (p in absolute)
10-8	$\dfrac{p_1}{T_1^{3.5}} = \dfrac{p_2}{T_2^{3.5}}$	Any consistent units (p, T in absolute)

10.8 Review Questions and Problems

1. List some of the major differences between hydraulics and pneumatics.
2. Describe the basic operation of the power supply segment (compressor and receiver) of a pneumatic system.
3. Differentiate between absolute pressure and gauge pressure.
4. Convert 90 psig to absolute pressure.
5. Convert 800 kPa gauge to absolute pressure.
6. Convert 70°F to absolute temperature (°R).
7. Convert 19°C to absolute temperature (K).
8. State Boyle's law in your own words.
9. 10 ft³ of air that is initially at 50 psig is compressed to a volume of 5 ft³. Assuming the temperature remains constant, what is the final pressure in psig?
10. 2.5 m³ of air that is initially at 400 kPa gauge is compressed to a volume of 1 m³. Assuming the temperature remains constant, what is the final pressure in kPa gauge?

11. 8 ft³ of air that is initially at 0 psig is compressed to a volume of 1 ft³. Assuming the temperature remains constant, what is the final pressure in psig?
12. Air initially at a volume of 2 ft³ and a pressure of 90 psig is allowed to expand to 5 ft³ at constant temperature. What is the gauge pressure after expansion?
13. Air initially at a volume of 0.5 m³ and a pressure of 700 kPa gauge is allowed to expand to 2 m³ at constant temperature. What is the gauge pressure after expansion?
14. State Gay-Lussac's law in your own words.
15. Air in a fixed volume container is initially at atmospheric pressure (0 psig) and 70°F. Its temperature is then raised to 250°F. What is the gauge pressure at the final temperature?
16. Air in a fixed volume container is initially at atmospheric pressure (0 kPa gauge) and 21°C. It's temperature is then raised to 120°C. What is the gauge pressure at the final temperature?
17. State Charles's law in your own words.
18. Air is held in a container with a free sliding piston, such as the one shown in Figure 10-3. It is initially at a temperature of 80°F and a volume of 1 ft³. Its temperature is then raised to 300°F. What is the volume of the air after being heated?
19. Air is held in a container with a free sliding piston, such as the one shown in Figure 10-3. It is initially at a temperature of 25°C and a volume of 0.75 m³. Its temperature is then raised to 100°C. What is the volume of the air after being heated?
20. In a real system, what happens to the temperature of a gas when it is compressed? When it is expanded?
21. Air initially at 70°F and 0 psig initially occupies a volume of 10 ft³. If its volume is reduced to 1 ft³ while its temperature increases to 90°F, what is the resulting final pressure?
22. Air initially at 22°C and 0 kPa abs initially occupies a volume of 3 m³. If its volume is reduced to 0.5 m³ while its temperature increases to 50°C, what is the resulting final pressure?
23. Air initially at 75°F and atmospheric pressure is compressed from 100 ft³ to 10 ft³. The temperature of the air is measured at 95°F after compression. Determine the final pressure in two situations: (a) using the general gas law, and (b) using Boyle's law, assuming the temperature remained at 75°F after compression.
24. Describe the difference between an isothermal process and an adiabatic process.
25. Air initially at 68°F and atmospheric pressure is compressed from 60 ft³ to 10 ft³. Assuming the process is adiabatic, what is the final pressure and temperature of the gas?

26. Air initially at 22°C and atmospheric pressure is compressed from 30 m³ to 5 m³. Assuming the process is adiabatic, what is the final pressure and temperature of the gas?

27. What are the U.S customary and metric units used to measure flow rate in pneumatic systems? Explain why these units are used.

28. Describe how vacuum systems work.

29. Describe the operation of a mercury barometer.

30. A suction cup with an inside diameter of 1 in is used in a vacuum system that pulls 20 in Hg. What is the maximum lifting force?

31. A suction cup with an inside diameter of 20 mm is used in a vacuum system that pulls 600 mm Hg. What is the maximum lifting force?

CHAPTER 11

Pneumatic Power Supply

OUTLINE

11.1 Introduction

11.2 Compressor Types

11.3 Compressor Sizing

11.4 Vacuum Pumps

11.5 Equations

11.6 Review Questions and Problems

11.1   Introduction

Pneumatic power supply consists of a prime mover, a compressor, and a receiver tank. The prime mover is most commonly an electric motor, but a gas or diesel engine may be used in some applications. The compressor is basically an air pump that compresses air into the receiver tank. The receiver tank acts as a pressure source from which the system can draw as dictated by the flow demand. Its function is to store energy and provide a smooth supply of power to the system.

In a typical pneumatic power unit, the compressor fills the receiver tank with air up to some predetermined maximum pressure and then shuts down. The system draws air from the receiver tank as needed. As air is drawn out, the pressure in the receiver tank gradually drops. When the pressure drops to a predetermined minimum, the compressor automatically starts up and charges the receiver back up to the maximum pressure. This process is known as *cycling* a compressor.

The minimum operating pressure of a pneumatic system is determined by the minimum pressure requirements of the machinery being supplied. It is desirable to have the maximum pressure as high as possible. A higher pressure means more air in the receiver and, consequently, a longer time before the receiver must be recharged by the compressor. The maximum pressure is usually determined by the maximum permissible pressure of the power unit or the distribution system, whichever is lower. The minimum pressure is often called the *cut-in* pressure; this is the pressure at which the compressor turns on. The maximum pressure is called the *cut-out* pressure; this is the pressure at which the compressor turns off. Cycling of the compressor is controlled by a *pressure switch,* a device that electronically senses the pressure. A 50% on, 50% off operating cycle is typical for a compressor.

Starting the compressor under load when the cut-in pressure is reached will cause premature wear of the electric motor. To prevent this, an unloader mechanism is usually incorporated into the compressor control, which allows the electric motor to come up to speed before being subjected to the pressure load on the compressor.

In some systems, the frequent starting and stopping of the prime mover may be damaging or impractical. For these systems there are at least two other options that can be used to cycle the system: (1) a clutch can be used to disengauge the electric motor from the compressor when the cut-out pressure is reached, or (2) a valve can be used to discharge the compressor outlet flow to the atmosphere when the cut-out pressure is reached.

In hydraulic systems, each machine typically has its own pump. In pneumatic systems, on the other hand, one centrally located compressor and receiver tank usually supplies multiple machines. The compressor is normally located in its own room in medium-to-large manufacturing facilities. This isolates the noise of the compressor from the rest of the facility. The air is then fed to the individual machines through a distribution system (see Chapter 12).

11.2   Compressor Types

Like hydraulic pumps, compressors may be positive or nonpositive displacement. Positive displacement pumps operate by opening up a cavity to draw in fluid, then closing the cavity to expel the fluid. Positive displacement hydraulic pumps are usually of the gear, vane, or piston type (see Chapter 3). Nonpositive displacement pumps function by imparting a velocity to the fluid. One common type of nonpositive displacement pump is a centrifugal type, which uses an impeller to spin the fluid. Centrifugal force causes the fluid to move toward the outlet. Positive displacement hydraulic pumps are used in fluid power applications because they are capable of producing flow against high pressures. Nonpositive displacement hydraulic pumps are low-pressure, high-flow pumps that are used in fluid transfer systems.

Nonpositive displacement compressors are also called *dynamic* compressors. Like hydraulic nonpositive displacement pumps, they generate high flow at low pressure. The most common type is a centrifugal compressor, which like the hydraulic pump of the same type, uses an impeller to increase the kinetic energy of the air by increasing its velocity. The kinetic energy is then converted to pressure energy in the *diffuser,* which sends the flow into the receiver tank. While a single-stage centrifugal compressor is capable of generating pressures of only 50 to 60 psi, *multistage* designs can generate higher pressures. In a multistage design, the air is passed from one impeller to the next to increase the pressure incrementally. A four-stage design is capable of generating pressures of about 150 psi.

Three designs of positive displacement compressors are prevalent in industry: *reciprocating piston, rotary vane,* and *rotary screw.* Reciprocating piston compressors are the most common design for small- to medium-sized commercial compressors. They consist of one or more pistons arranged radially around a crankshaft. As the crankshaft is rotated, the pistons reciprocate, alternately drawing in atmospheric air and pushing out compressed air. Single-piston designs can generate pressures up to 150 psi (1034 kPa). At higher pressures, the heat generated by compression causes them to become dangerously hot and inefficient. To generate pressures higher than 150 psi, a multistage design is required. Figure 11-1 shows a two-stage piston compressor. In this design, the air is drawn into the first-stage piston, compressed, passed through an intercooler to a second-stage piston where it is compressed further, and then sent to the receiver tank. The intercooler removes some of the heat of compression from the first stage. It is basically a pipe with fins that increase the surface area and promote heat transfer to the

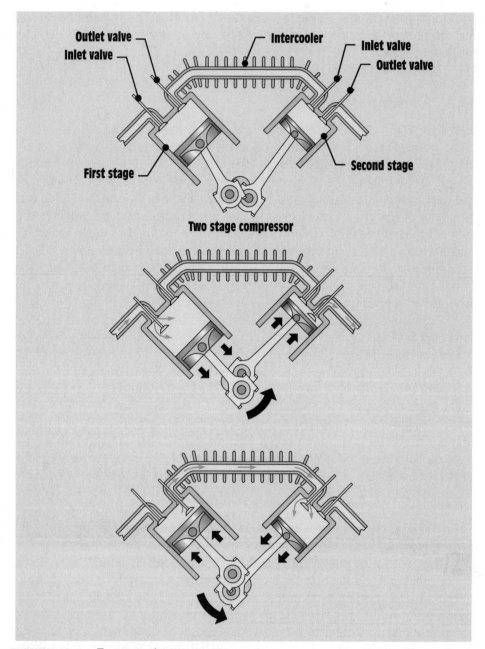

FIGURE 11-1   Two-stage piston compressor.

(Courtesy of Parker Hannifin Corporation. From Norvelle, *Fluid Power Technology:* West, 1993, p. 511.)

FIGURE 11-2   Two-stage piston compressor.
(Photo courtesy of Ingersoll-Rand Company.)

surrounding atmosphere. Most piston compressors also have fins on the outside of the cylinder walls to promote heat transfer at each stage. Figure 11-2 shows a photo of a small, two-stage piston compressor with these features.

A rotary vane compressor is very similar to the unbalanced vane pump that was discussed in Chapter 3. A simplified cutaway is shown Figure 11-3. In this design, a rotor is mounted eccentrically to the housing of the compressor. As the rotor spins, the vanes are kept in contact with the housing by centrifugal force and therefore extend and retract. This causes the chambers between the vanes to increase in volume near the inlet and decrease in volume by the outlet. The increase in volume at the inlet creates a vacuum, allowing atmospheric air to push into the inlet. The decrease in volume at the outlet compresses the air and exhausts it to the receiver tank. Rotary vane compressors typically generate pressures up to 150 psi. They are less common than single- and two-stage piston compressors.

A rotary screw compressor compresses air between two intermeshing screws. Side and top views of this type of compressor are shown in Figure 11-4. In this particular design, oil is used as a coolant to remove some of the heat of compression. Oil-free designs are also available. Rotary screw compressors are

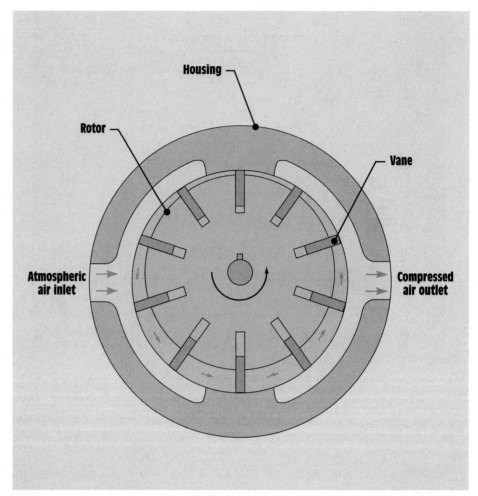

FIGURE 11-3   Rotary vane compressor.

generally capable of generating pressures up to 150 psi. They are available in small models that supply several machines, to very large models that supply entire manufacturing facilities. Rotary screw compressors have become the most commonly used type for industrial applications. Figure 11-5 shows a small, rotary screw model. Figure 11-6 shows a large model that is encased in an enclosure, which isolates the compressor noise and keeps out heat and dirt.

The graphic symbol for a compressor is shown in Figure 11-7. The same symbol is used to represent a hydraulic pump, except the arrow is not shaded for a compressor. This symbol is used to represent all types of compressors.

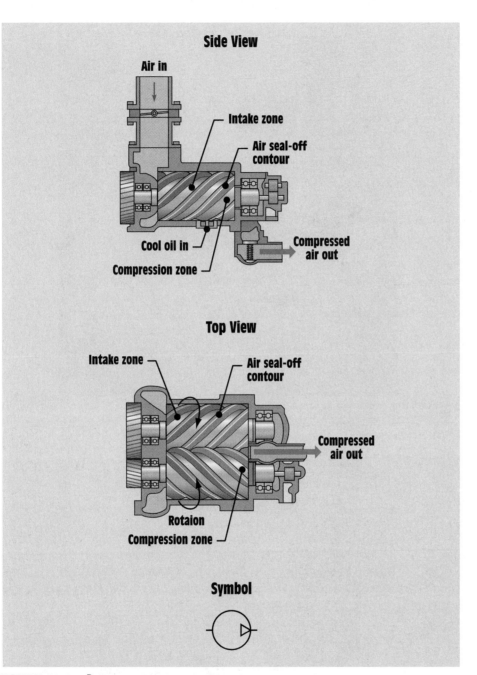

Side View

Air in

Intake zone

Air seal-off
contour

Cool oil in

Compression zone

Compressed
air out

Top View

Intake zone

Air seal-off
contour

Compressed
air out

Rotaion

Compression zone

Symbol

FIGURE 11-4   Rotary screw compressor.
(Courtesy of *Hydraulics and Pneumatics Magazine*. From Norvelle, *Fluid Power Technology:* West, 1993, p. 510.)

FIGURE 11-5   Small rotary screw compressor.
(Photo courtesy of Ingersoll-Rand Company.)

FIGURE 11-6   Large, enclosed rotary screw compressor.
(Photo courtesy of Ingersoll-Rand Company.)

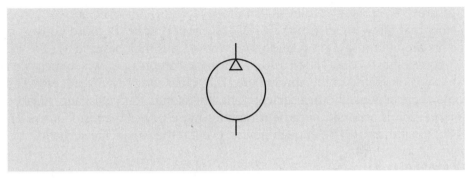

FIGURE 11-7   Compressor graphic symbol.

11.3 Compressor Sizing

Compressor units must be sized to supply sufficient pressure and flow to the system under all expected conditions. The compressor unit includes the electric motor (or other prime mover), the compressor, and the receiver tank. The electric motor must have enough power to drive the compressor at the maximum flow and pressure conditions. The compressor and receiver tank must be sized to supply the required flow and pressure, while allowing the compressor to remain idle for a portion of the cycle.

Compressor input power, like hydraulic pump input power, depends on pressure and flow. Due to the compressibility of air, however, the relationship is more complicated. The following equations can be used to calculate the required input power to a compressor:

$$HP_1 = \frac{p_{IN} \cdot Q}{65.5 \cdot \eta_O} \cdot \left[\left(\frac{p_{OUT}}{p_{IN}} \right)^{0.286} - 1 \right] \qquad \textbf{(11-1)}$$

$$kW_1 = \frac{p_{IN} \cdot Q}{17.14 \cdot \eta_O} \cdot \left[\left(\frac{p_{OUT}}{p_{IN}} \right)^{0.286} - 1 \right] \qquad \textbf{(11-1M)}$$

where: HP_1, kW_1 = drive power (hp, kW)
$\quad\quad\quad Q$ = compressor flow rate (scfm, standard m³/min)
$\quad\quad\quad p_{IN}$ = inlet pressure of the air (psia, kPa abs)
$\quad\quad\quad p_{OUT}$ = outlet pressure of the air (psia, kPa abs)
$\quad\quad\quad \eta_O$ = overall efficiency of the compressor

As stated in Chapter 10, both scfm (standard cubic feet per minute) and standard m³/min are the equivalent flow of atmospheric air, as opposed to the actual flow of pressurized air. The inlet pressure is normally atmospheric pressure (14.7 psia at sea level). The factors 65.5 and 17.14 are conversion factors that allow us to use the previously stated units. Because conversion factors are built into the equation, only these units may be used.

Equations 11-1 and 11-1M assume adiabatic compression, which means that no heat is dissipated to the atmosphere. The actual compression process will be somewhere between isothermal (constant temperature) and adiabatic. Adiabatic compression is assumed because it requires more power than isothermal. This assures that there will be enough power, even in the worst case scenario.

EXAMPLE 11-1.

Determine the input power required to drive a compressor that delivers 200 scfm at 120 psig. The overall efficiency of the compressor is 70%.

SOLUTION:

1. Convert to psia:

$$p_{IN} = p_{GAUGE} + p_{ATM} = 0 \text{ psig} + 14.7 \text{ psia} = 14.7 \text{ psia}$$

$$p_{OUT} = p_{GAUGE} + p_{ATM} = 120 \text{ psig} + 14.7 \text{ psia} = 134.7 \text{ psia}$$

2. Calculate the input power:

$$HP_1 = \frac{p_{IN} \cdot Q}{65.5 \cdot \eta_O} \cdot \left[\left(\frac{p_{OUT}}{p_{IN}} \right)^{0.286} - 1 \right] = \frac{14.7 \cdot (200)}{65.5 \cdot (0.70)} \cdot \left[\left(\frac{134.7}{14.7} \right)^{0.286} - 1 \right]$$
$$= 56.7 \text{ hp}$$

EXAMPLE 11-1M.

Determine the input power required to drive a compressor that delivers 5.7 standard m³/min at 830 kPa. The overall efficiency of the compressor is 75%.

SOLUTION

1. Convert to kPa abs:

$$p_{IN} = p_{GAUGE} + p_{ATM} = 0 \text{ kPa gauge} + 101 \text{ kPa abs} = 101 \text{ kPa abs}$$

$$p_{OUT} = p_{GAUGE} + p_{ATM} = 830 \text{ kPa gauge} + 101 \text{ kPa abs} = 931 \text{ kPa abs}$$

2. Calculate the input power:

$$kW_1 = \frac{p_{IN} \cdot Q}{17.14 \cdot \eta_O} \cdot \left[\left(\frac{p_{OUT}}{p_{IN}} \right)^{0.286} - 1 \right] = \frac{101 \cdot (5.7)}{17.14 \cdot (0.75)} \cdot \left[\left(\frac{931}{101} \right)^{0.286} - 1 \right]$$

$$= 39.7 \text{ kW}$$

In most situations, the electric motor and compressor are paired by the man-ufacturer to supply a specified amount of flow at some pressure. The electric motor is sized by the manufacturer and has enough power to drive the com-pressor at its rated flow and pressure. When selecting a compressor, we are pri-marily concerned with whether the pressure and flow ratings will meet the sys-tem requirements.

The size of the receiver tank is of great importance when selecting a com-pressor unit. It is important because it will determine how much air the sys-tem can store, and therefore what proportion of the time the compressor will have to run. The following equations can be of great help when sizing a receiver tank:

$$V_R = \frac{14.7 \cdot t \cdot (Q_{DEM} - Q_{DEL})}{p_{MAX} - p_{MIN}} \qquad \textbf{(11-2)}$$

$$V_R = \frac{101 \cdot t \cdot (Q_{DEM} - Q_{DEL})}{p_{MAX} - p_{MIN}} \qquad \textbf{(11-2M)}$$

where: V_R = volume of the receiver tank (ft^3, m^3)
$\quad$ t = time (min)
$\quad$ Q_{DEM} = system demand flow (scfm, standard m^3/min)
$\quad$ Q_{DEL} = compressor delivery flow (scfm, standard m^3/min)
$\quad$ p_{MAX} = maximum receiver tank pressure (psig, kPa gauge)
$\quad$ p_{MIN} = minimum receiver tank pressure (psig, kPa gauge)

The demand flow (Q_{DEM}) is the flow out of the receiver tank to the system; the delivery flow (Q_{DEL}) is the flow from the compressor into the receiver tank. The maximum and minimum pressures can be in psig or psia because, when taking a difference between two pressures, the atmospheric pressure cancels out and the resulting difference will be the same either way. It is usually more con-venient to use gauge pressure, however. Equations 11-2 and 11-2M can be used in many situations, as the following examples illustrate.

EXAMPLE 11-2.

A compressor unit is required to provide a flow of 50 scfm to a system for 10 min while the compressor sits idle. What size receiver tank is required? The system must operate between 125 and 90 psig.

SOLUTION:

$$V_R = \frac{14.7 \cdot t \cdot (Q_{DEM} - Q_{DEL})}{p_{MAX} - p_{MIN}} = \frac{14.7 \cdot (10) \cdot (50 - 0)}{125 - 90} = 210 \text{ ft}^3$$

Choosing a receiver tank of this size or larger will mean that the compressor will sit idle for at least 10 minutes.

EXAMPLE 11-2M.

A compressor unit is required to provide a flow of 1.4 standard m³/min to a system for 10 min while the compressor sits idle. What size receiver tank is required? The system must operate between 850 and 650 kPa gauge.

SOLUTION:

$$V_R = \frac{101 \cdot t \cdot (Q_{DEM} - Q_{DEL})}{p_{MAX} - p_{MIN}} = \frac{101 \cdot (10) \cdot (1.4 - 0)}{850 - 650} = 7.07 \text{ m}^3$$

EXAMPLE 11-3.

For the system in Example 11-2, determine the time to recharge the receiver tank from 90 to 125 psig if the compressor delivery flow is 150 scfm. Assume the demand remains constant at 50 scfm.

SOLUTION:

$$t = \frac{V_R \cdot (p_{MAX} - p_{MIN})}{(Q_{DEL} - Q_{DEM})} = \frac{210 \cdot (125 - 90)}{14.7 \cdot (150 - 50)} = 5 \text{ min}$$

EXAMPLE 11-3M.

For the system in Example 11-2M, determine the time to recharge the receiver tank from 650 to 850 kPa gauge if the compressor delivery flow is 4.25 standard m³/min. Assume the demand remains constant at 1.4 standard m³/min.

SOLUTION:

$$t = \frac{V_R \cdot (p_{MAX} - p_{MIN})}{101 \cdot (Q_{DEL} - Q_{DEM})} = \frac{7.07 \cdot (850 - 650)}{101 \cdot (4.25 - 1.4)} = 4.91 \text{ min}$$

Notice that Q_{DEM} and Q_{DEL} were switched in examples 11-3 and 11-3M to avoid getting a negative number. This means that, in this situation, there is a net flow into the receiver tank, while the original equation was written for a net flow out of the tank. The system in the previous two problems will therefore have a cycle in which the compressor runs for 5 minutes and charges the receiver tank from 90 to 125 psig, then sits idle for 10 minutes while the system draws air until the pressure drops again to 90 psig. This analysis assumes that the demand flow will remain constant at 50 scfm. In most systems, however, the flow demand changes considerably throughout the day, depending on which machines are running, at what speed they are running, etc. There may even be brief periods when the demand flow exceeds the delivery flow. The following example shows how Equation 11-3 is used in this situation.

EXAMPLE 11-4.

For the system discussed in Examples 11-2 and 11-3, determine how long the compressor unit can supply a flow of 175 scfm to the system before the pressure drops below 90 psig.

SOLUTION:

$$t = \frac{V_R \cdot (p_{MAX} - p_{MIN})}{14.7 \cdot (Q_{DEL} - Q_{DEM})} = \frac{210 \cdot (125 - 90)}{14.7 \cdot (175 - 0)} = 2.86 \text{ min}$$

EXAMPLE 11-4M.

For the system discussed in Examples 11-2M and 11-3M, determine how long the compressor unit can supply a flow of 5 standard m³ to the system before the pressure drops below 650 kPa gauge.

SOLUTION:

$$t = \frac{V_R \cdot (p_{MAX} - p_{MIN})}{101 \cdot (Q_{DEL} - Q_{DEM})} = \frac{7.07 \cdot (850 - 650)}{101 \cdot (5 - 0)} = 2.80 \text{ min}$$

If the demand in Example 11-4 would continue to be 175 scfm for longer than 2.86 minutes, the pressure will drop below 90 psig because the compressor will not be able to keep up with the demand. If this high demand continues, a larger tank or a higher flow compressor will be needed. Increasing the storage capacity (i.e., a larger tank) is a more efficient way to deal with short-term peak demands. Manufacturers normally give the option of two or more receiver tank sizes with each compressor.

How do we determine what the demand will be in a particular system? This is a difficult task because the demand is dependent upon the number of machines running simultaneously *and* under what conditions they are running, which will certainly vary throughout the day. This is further complicated by the fact that air systems typically lose 20% of their flow to leakage. Learning to properly size a pneumatic system requires both experience and some experimentation to determine the actual flow demands. Flow demand determination is discussed further in Chapter 12.

Air compressors are one of the largest consumers of industrial electricity. By reducing leakage and making proper system sizing a priority, significant savings will result. When putting in a new pneumatic system or upgrading capacity, it is worth bringing in outside expertise if none exists in-house. Many consulting firms provide this service. They also perform audits of existing systems to improve the efficiency by finding leakage points, reducing unnecessary capacity, etc.

11.4 Vacuum Pumps

Vacuum pumps evacuate air from one side of an object, allowing atmospheric pressure to act on the opposite side. Vacuum pumps are basically air compressors that are operated in reverse; the compressor inlet is connected to the system and the outlet is connected to the atmosphere. Industrial vacuum pumps are most commonly of the rotary vane type. Piston and rotary screw designs are also widely available, however. Their construction is very similar to the compressors of the same type.

In addition to vacuum pumps, *vacuum generators* are also commonly used in industry. Vacuum generators use a *venturi* to create a vacuum. A venturi is a pipe with a reduced diameter section in the middle, called the throat. The Bernoulli principle tells us that in the reduced diameter section, the velocity will increase and the pressure will decrease. A simplified vacuum generator is shown in Figure 11-8. Compressed air is fed through the device, which causes a very low pressure at the throat. The throat is designed so that the pressure drops below atmospheric pressure, creating a vacuum. A vacuum port is then connected to the throat section, which can be connected to a suction cup or other device. There are several advantages of vacuum generators:

1. They can be powered by existing pneumatic systems, rather than requiring a separate system utilizing a vacuum pump,
2. They are very reliable,

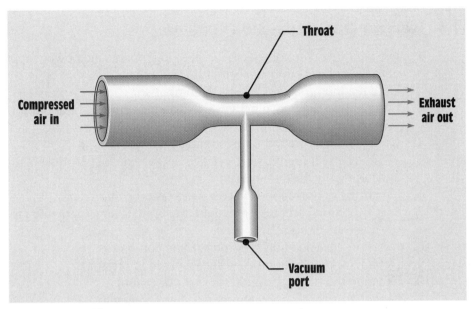

FIGURE 11-8   Vacuum generator.

3. They require no lubrication, and
4. They are virtually maintenance free.

Points 2, 3, and 4 are due to the fact that there are no moving parts within the device.

11.5 Equations

EQUATION NUMBER	EQUATION	REQUIRED UNITS
11-1	$HP_1 = \dfrac{P_{IN} \cdot Q}{65.5 \cdot \eta_O} \cdot \left[\left(\dfrac{P_{OUT}}{P_{IN}} \right)^{0.286} - 1 \right]$	Q in scfm, p in psia, HP_1 in hp
11-1M	$kW_1 = \dfrac{P_{IN} \cdot Q}{17.14 \cdot \eta_O} \cdot \left[\left(\dfrac{P_{OUT}}{P_{IN}} \right)^{0.286} - 1 \right]$	Q in m³/min, p in kPa abs, kW_1 in kW
11-2	$V_R = \dfrac{14.7 \cdot t \cdot (Q_{DEM} - Q_{DEL})}{(p_{MAX} - p_{MIN})}$	t in min, Q in scfm, p in psi
11-2M	$V_R = \dfrac{101 \cdot t \cdot (Q_{DEM} - Q_{DEL})}{(p_{MAX} - p_{MIN})}$	t in min, Q in m³/min, p in kPa

11.6 Review Questions and Problems

1. List the three components that make up a pneumatic power supply unit. Describe the function of each component.
2. Describe the process of cycling a compressor.
3. Describe the difference between a dynamic compressor and a positive displacement compressor.
4. Describe the operation of a centrifugal compressor.
5. What are the three types of positive displacement compressors that are commonly used in industry?
6. Describe the operation of a two-stage piston compressor.
7. Why do piston compressors require multiple stages to generate pressures above 150 psi?
8. Describe the operation of a rotary vane compressor.
9. Describe the operation of a rotary screw compressor.
10. Describe the basic operation of a vacuum pump.
11. Describe the basic operation of a vacuum generator.
12. Determine the actual drive power required to drive a compressor that delivers 100 scfm at 150 psig. The overall efficiency of the compressor is 75%.
13. Determine the drive power required to drive a compressor that delivers 6.25 standard m^3/min at 1000 kPa. The overall efficiency of the compressor is 75%.
14. A compressor unit is required to provide a flow of 100 scfm to a system for 15 minutes while the compressor sits idle. What size receiver tank is required? The system must operate between 90 and 120 psig.
15. A compressor unit is required to provide a flow of 3.0 standard m^3/min to a system for 15 minutes while the compressor sits idle. What size receiver tank is required? The system must operate between 600 and 750 kPa gauge.
16. Determine the time to recharge a 300 ft^3 receiver tank from 100 to 130 psig if the compressor delivery flow is 150 scfm. Assume the demand remains constant at 50 scfm during this time.
17. Determine the time to recharge a 8 m^3 receiver tank from 600 to 700 kPa gauge if the compressor delivery flow is 2.5 standard m^3/min. Assume the demand remains constant at 1.0 standard m^3/min during this time.

CHAPTER 12

Pneumatic Components

OUTLINE

12.1 Introduction

12.2 Pneumatic Cylinders

12.3 Pneumatic Motors

12.4 Other Pneumatic Actuators

12.5 Pneumatic Directional Control Valves

12.6 Pneumatic Flow Control Valves

12.7 Air Preparation

12.8 Air Distribution

12.9 Equations

12.10 Review Questions and Problems

12.1   Introduction

Many of the devices used in hydraulics and pneumatics are very similar. Components such as cylinders, motors, directional control valves, and flow control valves have the same basic construction and perform the same function in both types of systems. Other components, such as pressure regulators, lubricators, and dryers, are unique to pneumatics.

One category of components, air tools, is unique to pneumatics. Air tools include devices such as grinders, impact wrenches, sanders, paint sprayers, drills, screwdrivers, riveting hammers, chipping hammers, hoists, blow guns, saws, nail guns, staple guns, etc. Figure 12-1 shows photographs of some industrial air tools. The large variety of air tools available is one of the reasons for the widespread use of pneumatics in industrial, construction, and automotive applications.

Air consumption is one of the most important specifications for air tools, cylinders, motors, and other pneumatically driven devices. Because of the compressibility of air, the amount of air consumed during the operation of any pneumatic device depends primarily on the air pressure and the magnitude of the load.

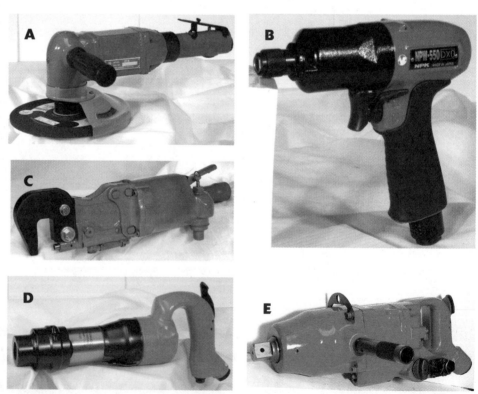

FIGURE 12-1   Air tools. (A) Grinder. (B) Screwdriver. (C) Compression riveter. (D) Chipping hammer. (E) Impact wrench.

(Photos courtesy of Michigan Pneumatic Tool Inc., Detroit, MI.)

Other factors such as mechanical friction and leakage are also significant. Manufacturers often give an air consumption specification under particular conditions. For example, a manufacturer may say that a device will consume 5 scfm at 90 psig under a particular load. 90 psig is the *supply pressure,* or the pressure at the tool. This issue is discussed further in the cylinder and motor sections of this chapter.

12.2 Pneumatic Cylinders

Pneumatic cylinders have the same basic construction as hydraulic cylinders. However, because they are subjected to much lower pressures, they are of a much lighter construction. Pneumatic cylinders are available in many of the same configurations as hydraulic cylinders, such as single- and double-acting models, double-rod models, etc. Figure 12-2 shows a photo of a double-acting pneumatic cylinder.

FIGURE 12-2   Double-acting pneumatic cylinder.
(Photo courtesy of Norgren, Littleton, CO.)

Pneumatic systems are generally much quicker acting, but less precise than hydraulic systems due to the compressibility of air versus the relative incompressibility of hydraulic fluids. Because of the lower pressures used in pneumatic systems, a pneumatic cylinder has a much lower force capability than a hydraulic cylinder of the same size. Pneumatic cylinders are therefore appropriate for applications that require high speed and lower forces, and do not require a high degree of precision. Pneumatic cylinders are often used in automation equipment to assemble, sort, package, and otherwise manipulate parts. They are also commonly used as clamps and low-capacity presses.

Pascal's law applies to pneumatics as well as hydraulics. We can therefore use $F = p \cdot A$ to calculate the force generated when a pressure is applied to a given area. The following equations can be used to calculate the forces generated by a pneumatic cylinder on the extend and retract strokes:

$$F_E = p \cdot A_P \tag{12-1}$$

$$F_R = p \cdot (A_P - A_R) \tag{12-2}$$

where: F_E = extend force (lbs)
F_R = retract force (lbs)
p = pressure (psig)
A_P = area of the piston (in²)
A_R = area of the rod (in²)

These equations are identical to those used in Chapter 4 to calculate the forces generated by hydraulic cylinders. Notice that the pressure should be the gauge pressure, which may be written psig or psi. Recall that the gauge pressure is the pressure above atmospheric pressure, which is the net pressure available to do work.

EXAMPLE 12-1.

A double-acting pneumatic cylinder with a 2 in bore and a 1 in rod is to be used in a system with a supply pressure of 90 psig. What is the force output of the cylinder on the extend and retract strokes?

SOLUTION:

1. Calculate the piston area:

$$A_p = \frac{\pi \cdot D_p{}^2}{4} = \frac{3.142 \cdot (2 \text{ in})^2}{4} = 3.142 \text{ in}^2$$

2. Calculate the rod area:

$$A_p = \frac{\pi \cdot D_R^{\,2}}{4} = \frac{3.142 \cdot (1 \text{ in})^2}{4} = 0.7854 \text{ in}^2$$

3. Calculate the extension force:

$$F_E = p \cdot A_p = 90 \frac{\text{lbs}}{\text{in}^2} \cdot (3.142 \text{ in}^2) = 283 \text{ lbs}$$

4. Calculate the retraction force:

$$F_R = p \cdot (A_p - A_R) = 90 \frac{\text{lbs}}{\text{in}^2} \cdot (3.142 - 0.7854) \text{ in}^2 = 212 \text{ lbs}$$

EXAMPLE 12-1M.

A double-acting pneumatic cylinder with a 50 mm bore and a 20 mm rod is to be used in a system with a supply pressure of 600 kPa gauge. What is the force output of the cylinder on the extend and retract strokes?

SOLUTION

1. Calculate the piston area:

$$A_p = \frac{\pi \cdot D_p^{\,2}}{4} = \frac{3.142 \cdot (0.050 \text{ m})^2}{4} = 0.001964 \text{ m}^2$$

2. Calculate the rod area:

$$A_R = \frac{\pi \cdot D_R^{\,2}}{4} = \frac{3.142 \cdot (0.020 \text{ m})^2}{4} = 0.0003142 \text{ m}^2$$

3. Calculate the extension force:

$$F_E = p \cdot A_p = 600{,}000 \frac{N}{m^2} \cdot (0.001964 \text{ m}^2) = 1178 \text{ N}$$

4. Calculate the retraction force:

$$F_R = p \cdot (A_p - A_R) = 600{,}000 \frac{N}{m^2} \cdot (0.001964 - 0.0003142) \text{ m}^2 = 990 \text{ N}$$

Equations 12-1 and 12-2 are less accurate when used for pneumatic cylinders because they typically have higher internal friction than do hydraulic cylinders, which reduces the force output. The higher internal friction occurs because tighter seals must be used to prevent the air from leaking, and the lubrication is better with hydraulic cylinders. For these reasons manufacturers sometimes supply data that give the actual force output per psi of supply pressure, corrected for frictional losses.

In hydraulics, the speed of a cylinder depends directly upon the flow rate from the pump and can be calculated accurately (see Equations 4-5 and 4-6). Accurate prediction of pneumatic cylinder speed, on the other hand, is extremely difficult due to the compressibility of air. The fact that pneumatics supplies flow to the system from a tank of compressed air causes the speed to depend on the supply pressure and the load on the cylinder. The following situations illustrate this concept: If we punch a hole in the side of a tank of compressed air, how fast will the air flow out? It is obvious that the air will rush out very quickly. What if we force the air to instead go through a small hose to a cylinder? It will, of course, flow out more slowly. What if we put a large weight on the cylinder? The air flow will be even slower, due to the increased resistance. These situations demonstrate that the air flow rate and, consequently, the cylinder speed, are dependant on the load.

A higher tank pressure tends to push the air out of the receiver tank more quickly. This means that the air flow rate is also dependant on the supply pressure. In addition to the load and supply pressure, the cylinder speed will also depend on leakage and friction. All of these constantly changing factors make the determination of pneumatic cylinder speed difficult at best. Obtaining a particular speed with a pneumatic cylinder is most commonly achieved with flow control valves through trial and error.

To determine the air consumption of a pneumatic cylinder in a particular application, we must calculate the volume of air required to extend and retract the cylinder. The volume required to extend a cylinder is the area of the piston (A_p) multiplied by the length of the stroke (S), as given by the following equation:

$$V_E = A_p \cdot S \tag{12-3}$$

The volume required to retract a cylinder is the area of the piston minus the area of the rod ($A_p - A_R$) multiplied by the length of the stroke (S), as given by the following equation:

$$V_R = (A_p - A_R) \cdot S \tag{12-4}$$

EXAMPLE 12-2.

A double-acting pneumatic cylinder with a 3 in bore, 1 in rod, and a 4 in stroke length must cycle (extend and retract) 40 times per minute. The supply pressure is 100 psig. What is the air consumption of the cylinder in scfm?

SOLUTION:

1. Calculate the piston area:

$$A_p = \frac{\pi \cdot D_p^{\,2}}{4} = \frac{3.142 \cdot (3 \text{ in})^2}{4} = 7.069 \text{ in}^2$$

2. Calculate the rod area:

$$A_R = \frac{\pi \cdot D_R^{\,2}}{4} = \frac{3.142 \cdot (1 \text{ in})^2}{4} = 0.7854 \text{ in}^2$$

3. Calculate the total volume per cycle:

$$V_E = A_p \cdot S = 7.069 \text{ in}^2 \cdot 4 \text{ in} = 28.28 \text{ in}^3$$

$$V_R = (A_p \cdot A_R)S = (7.069 \text{ in}^2 - 0.7854 \text{ in}^2) \cdot 4 \text{ in} = 25.13 \text{ in}^3$$

$$V_{TOT} = V_p + V_A = 28.28 \text{ in}^3 + 25.13 \text{ in}^3 = 53.41 \text{ in}^3$$

4. Convert to ft³:

$$53.41 \text{ in}^3 \cdot \left(\frac{1 \text{ ft}^3}{1728 \text{ in}^3} \right) = 0.0309 \text{ ft}^3$$

This is the volume required at 100 psig. Flow rates for pneumatic systems are usually expressed in standard cubic feet per minute (scfm), which is the equivalent flow rate of atmospheric (uncompressed) air. We must therefore convert this volume to an equivalent uncompressed volume using Boyle's law ($p_1 \cdot V_1 = p_2 \cdot V_2$). Recall that absolute pressure must be used in this equation.

5. Calculate the equivalent uncompressed volume:

$$V_1 = \frac{V_2 \cdot p_2}{p_1} = \frac{0.0309 \text{ ft}^3 \cdot (100 + 14.7) \text{ psia}}{(0 + 14.7) \text{ psia}} = 0.2411 \text{ ft}^3$$

6. Calculate the flow rate:

$$0.2411\frac{ft^3}{cycle} \cdot \left(40\frac{cycles}{min}\right) = 9.644 \text{ scfm}$$

EXAMPLE 12-2M.

A double-acting pneumatic cylinder with a 80 mm bore, 25 mm rod, and a 100 mm stroke length must cycle (extend and retract) 40 times per minute. The supply pressure is 700 kPa gauge. What is the air consumption of the cylinder in standard m³/min?

SOLUTION:

1. Calculate the piston area:

$$A_p = \frac{\pi \cdot D_p^2}{4} = \frac{3.142 \cdot (0.080 \text{ m})^2}{4} = 0.005027 \text{ m}^2$$

2. Calculate the rod area:

$$A_R = \frac{\pi \cdot D_R^2}{4} = \frac{3.142 \cdot (0.025 \text{ m})^2}{4} = 0.0004909 \text{ m}^2$$

3. Calculate the total volume per cycle:

$$V_E = A_p \cdot S = 0.005027 \text{ m}^2 \cdot 0.100 \text{ m} = 0.0005027 \text{ m}^3$$

$$V_R = (A_p - A_R) \cdot S = (0.005027 \text{ m}^2 - 0.0004909 \text{ m}^2) \cdot 0.100 \text{ m} = 0.0004536 \text{ m}^3$$

$$V_{TOT} = V_p + V_A = 0.0005027 \text{ m}^3 + 0.0004536 \text{ m}^3 = 0.0009563 \text{ m}^3$$

4. Calculate the equivalent uncompressed volume:

$$V_1 = \frac{V_2 \cdot p_2}{p_1} = \frac{0.0009563 \text{ m}^3 \cdot (700 + 101) \text{ kPa abs}}{(0 + 101) \text{ kPa abs}} = 0.007584 \text{ m}^3$$

5. Calculate the flow rate:

$$0.007584\frac{m^3}{cycle} \cdot \left(40\frac{cycles}{min}\right) = 0.3034 \text{ standard m}^3/\text{min}$$

The graphic symbols used to represent pneumatic cylinders on schematics are the same as those used for the hydraulic cylinder of the same type.

A *slide* is a variant of the standard cylinder design just discussed (Figure 12-3). A slide is essentially a cylinder with guide rods that run parallel to the piston rod. A flat plate is attached to the end of the piston rod and guide rods. Various tooling attachments can be attached to the plate, depending on the application.

FIGURE 12-3   Slides.

(Photograph provided courtesy of PHD, Inc. For additional information, contact us at www.phdinc.com.)

FIGURE 12-4 Rotary vane air motor.

(Photo courtesy of Gast Manufacturing, Inc., Benton Harbor, MI.)

12.3 Pneumatic Motors

Pneumatic motors are similar in construction and function to hydraulic motors. Pneumatic motors are used in applications where low to moderate torque is required. Hydraulic motors, by contrast, are capable of generating much larger torque outputs. Pneumatic motors can operate at speeds in excess of 10,000 rpm, while hydraulic motors are usually confined to speeds below 5000 rpm. Pneumatic motors are most commonly of the rotary vane, radial piston, or axial piston type. A photo of a rotary vane pneumatic motor is shown in Figure 12-4.

The torque and speed capabilities of pneumatic motors put them in direct competition with electric motors, which have similar capabilities. Pneumatic motors are much less efficient than electric motors, but have several advantages that are important in some applications. One advantage is that pneumatic motors can stall under full load for indefinite periods of time, while electric motors will be damaged by heat generation when stalled. For this reason, electric motors are protected by circuit breakers, which cut power to the electric motor if it is overloaded. Another advantage is that pneumatic motors can be used in applications where electric motors would constitute a fire hazard. Pneumatic motors are also lighter and less expensive than electric motors of an equivalent capacity.

Just as with hydraulic motors, the torque of a pneumatic motor depends on displacement and pressure. Larger motors (those with a greater displacement) produce more torque at a given pressure. The compressibility of air, however, makes the direct calculation of torque output inaccurate. Performance graphs are the preferred method of specifying torque output for pneumatic motors. Figure 12-5 shows a typical pneumatic motor performance graph that plots torque versus rotational speed. Notice that the torque increases with increasing speed up to about 300 rpm, at which point the torque begins to decrease with increasing speed. At 100 psi, the maximum torque is about 57 in · lbs.

The torque generated when the motor is rotating at a particular speed is known as the *running torque*. The torque generated when a motor is started from a dead stop under load is known as the *starting torque*. This can be determined from the torque-versus-speed curve by looking at the torque generated at 0 rpm. Figure 12-5 shows that the starting torque for a supply pressure of 100 psi is about 33 in · lbs. The *stall torque*, the torque required to stop a rotating motor at a particular supply pressure, is also important. The stall torque is always higher than the starting torque because when the motor is rotating, the load must overcome the momentum built up by the motor and the pressure in order to stop the motor.

The speed of a pneumatic motor is difficult to predict accurately due to the compressibility of air. Just as with cylinders, the speed depends primarily on the

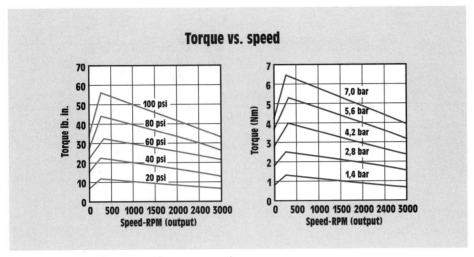

FIGURE 12-5   Air motor performance graph.

(Courtesy of Gast Manufacturing, Inc., Benton Harbor, MI.)

FIGURE 12-6   Air-powered gear motor.

(Photo courtesy of Gast Manufacturing, Inc., Benton Harbor, MI.)

load and supply pressure. If the load is increased on the motor, the torque will increase to match the load and the speed will automatically decrease. If the torque is greater than the stall torque, the motor will stop. Reducing the load causes the speed to increase. Supply pressure has the opposite effect. Increasing the supply pressure causes the speed to increase, while decreasing the supply pressure causes the speed to decrease.

Pneumatic motors are also available with internal gear-reducer mechanisms that increase torque and decrease rotational speed. For example, an air-powered gear motor with a 20:1 gearing ratio will increase torque by a factor of 20 and decrease speed by a factor of 20, as compared to the equivalent pneumatic motor alone. A cut-away of an air powered gear motor is shown in Figure 12-6.

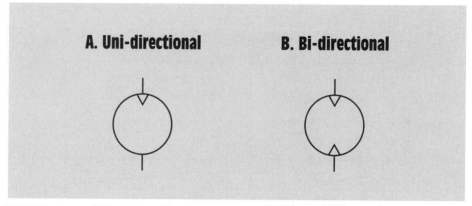

FIGURE 12-7   Air motor graphic symbols. (A) Uni-directional. (B) Bi-directional.

Manufacturers specify the air consumption of each pneumatic motor model under particular conditions; for example, a motor may be said to consume 50 scfm at 100 psi and 10,000 rpm. At lower pressures or lower speeds, the air consumption will be lower. Manufacturers also provide the *free speed* and *design speed* of a motor. The free speed is the speed under no load at a particular supply pressure. The design speed is the rotational speed at the maximum power output.

The graphic symbols that represent pneumatic motors are the same as those for the corresponding hydraulic motor, except the arrows are not shaded (Figure 12-7).

12.4 Other Pneumatic Actuators

Other actuators are commonly used in pneumatic circuits, including *grippers* and *escapements*. A gripper is used to pick up and manipulate parts in automated manufacturing (Figure 12-8). In this actuator, a piston is connected to the jaws via a pivot and connecting rod. The motion of the piston causes the jaws to open and close to grip and manipulate parts. Various jaw configurations can be used depending on the application.

An escapement is an actuator that allows parts to be released one at a time along an assembly line (Figure 12-9). It functions by extending and retracting two rods. Both rods are initially extended. The back rod is retracted first to allow one part to be fed in. Next, the back rod is extended to trap the part between the rod fixtures. The front rod is then retracted, allowing one part to proceed. Finally, the front rod extends and another cycle is initiated.

FIGURE 12-8 Gripper.
(Photograph provided courtesy of PHD, Inc. For additional information, contact us at www.phdinc.com.)

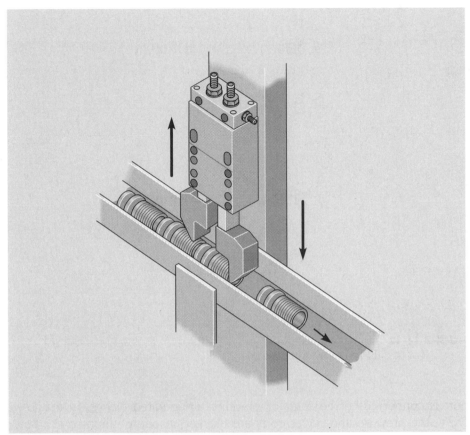

FIGURE 12-9 Escapement.

(Photograph provided courtesy of PHD, Inc. For additional information, contact us at www.phdinc.com.)

12.5 Pneumatic Directional Control Valves

Pneumatic directional control valves (DCVs) are very similar to their counterparts in hydraulics. Pneumatic DCVs are primarily of the spool type, as is the case with hydraulics. Because they operate at lower pressures, however, pneumatic DCVs are of a lighter construction.

Like hydraulic DCVs, pneumatic DCVs can be divided into three main categories: *two-way, three-way,* and *four-way* valves. The number of "ways" is the number of ports. The basic construction of pneumatic two-way and three-way DCVs is the same as that found in hydraulics. The graphic symbols are also the same (Figure 12-10). Both valves are two-position models that have pushbutton actuation and spring return.

A two-way has a pressure port (*P*) and an outlet port (*A*). The pressure port is connected to a compressed air source (i.e., the receiver tank) and the outlet

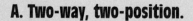

A. Two-way, two-position.

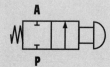

B. Three-way, two-position.

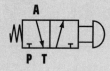

FIGURE 12-10 Two-way and three-way pneumatic DCVs. (A) Two-way, two-position. (B) Three-way, two-position.

port is connected to the actuator or other air-operated device. In a two-way DCV, the pressure and outlet ports are blocked in one position so the flow is stopped. They are connected in the other position so that flow can go from *P* to *A*. The symbol for this particular valve shows that it is normally closed because the spring is on the closed side of the valve symbol. Two-way valves can be used to simply start or stop flow in a particular line.

A three-way valve has a pressure port (*P*), an outlet (*A*) and an exhaust port (*E*). The pressure port is connected to a compressed air source and the outlet port is connected to the actuator or other air-operated device. The exhaust port is open directly to the atmosphere. It normally has a small filter to remove lubricating oil and other particles that may be in the air. In a three-way DCV, the outlet and exhaust ports are connected in one position and the pressure port is blocked, so flow will go from *A* to *E*. In the other position, the pressure and outlet ports are connected and the exhaust port is blocked, so flow will go from *P* to *A*. In hydraulics, the exhaust port is replaced by the tank port (*T*) because the fluid is returned to a tank rather than to the atmosphere. Perhaps the most common application for a three-way valve is controlling a single-acting cylinder (Figure 12-11). This simple circuit can be used as a clamp, as a press, or for numerous other applications.

Another common application for a three-way DCV is to shut-off and vent a branch of a pneumatic system (Figure 12-12). Part A shows the valve when the

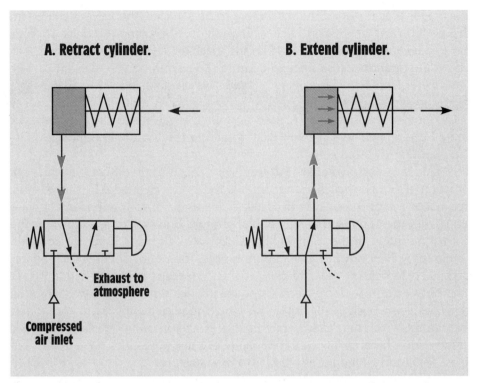

FIGURE 12-11 Three-way, two-position DCV controlling a single-acting cylinder. (A) Retract cylinder. (B) Extend cylinder.

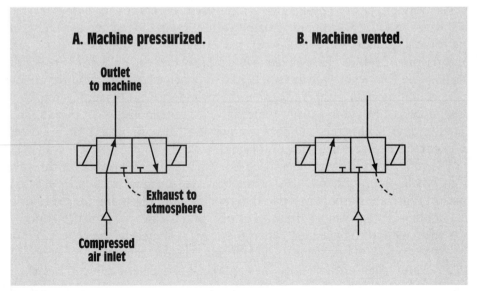

FIGURE 12-12 Shut-off and vent application. (A) Machine pressurized. (B) Machine vented.

machine is pressurized. If the machine is to be serviced, the valve is shifted into the position shown in part B. In this position, the compressed air supply is blocked *and* the compressed air from the machine is vented to the atmosphere. This is important because trapped compressed air can be very dangerous when servicing equipment. This valve is a *double-solenoid* actuated valve. A solenoid is a device that uses electrical current to shift a valve. A momentary electrical signal to one of the solenoids is used to shift the valve into the pressurized or vented positions. The valve will then remain in the last position indicated until the opposing solenoid is shifted.

Pneumatic and hydraulic four-way valves are slightly different in their construction, although their basic function is the same. Figure 12-13 shows simplified cut-aways of a four-way hydraulic valve and a four-way pneumatic valve, along with their corresponding graphic symbols. Both valves are shown in their normal positions. In the four-way hydraulic valve, the tank ports are combined within the valve body into one common port. The return flow for each outlet is exhausted through a separate port in the pneumatic valve. In hydraulics, the fluid must be returned to a reservoir, so having two separate ports would require two lines going back to the tank. This would require buying more components and fittings, which may leak. In pneumatics, the return flow is exhausted directly to the atmosphere, so having two separate exhaust ports does not add any extra lines. This configuration is also easier to manufacture.

Notice that the four-way pneumatic valve actually has five ports. It is still called a four-way valve, however, because the two exhaust ports are considered to have the same function.

Pneumatic four-way DCVs are most commonly used to control double-acting cylinders. Figure 12-14 shows a four-way DVC controlling a double-acting cylinder. This DCV is solenoid actuated with spring return. When the DCV is in the normal position, the cylinder will retract and remain retracted under pressure (Figure 12-14A). The cylinder will extend when the solenoid is energized (Figure 12-14B). Whenever the solenoid is de-energized, the spring returns the valve to the normal position, causing the cylinder to automatically retract. In other applications we may want the cylinder to remain in the last position indicated, and not returned to some normal position. This can be accomplished with the double solenoid valve shown in Figure 12-15.

There are also five-way pneumatic DCVs. The graphic symbol for this valve is shown in Figure 12-16A. This valve has two pressure inlets and only one exhaust port and is used when two different pressure levels are desired to operate a double-acting cylinder (Figure 12-16B). In this case, the pressure at port P_1 is used to extend the cylinder. When the valve is shifted to the other position, the pressure at P_2 retracts the cylinder. These valves can be used for control of an air clamp. High pressure may be required in the extend direction to clamp the workpiece, while only low pressure is required to return the cylinder to the

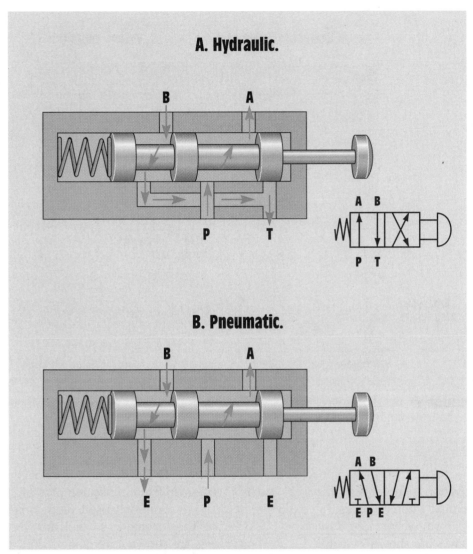

A. Hydraulic.

B   A

P   T

A B

P T

B. Pneumatic.

B   A

E   P   E

A B

E P E

FIGURE 12-13 Four-way hydraulic and pneumatic DCVs. (A) Hydraulic. (B) Pneumatic.

retracted position under no load. Using a lower pressure is always desirable because less air is consumed, reducing the cost of operation. The five-way valve is essentially the same as the four-way valve; it is just hooked up differently. The two exhausts on the four-way become P_1 and P_2, while the pressure port on the four-way becomes the exhaust on the five-way. Some five-port DCVs are designed to be used as four-way or five-way valves, while others are exclusively four-way. As always, follow the manufacturer's recommendations.

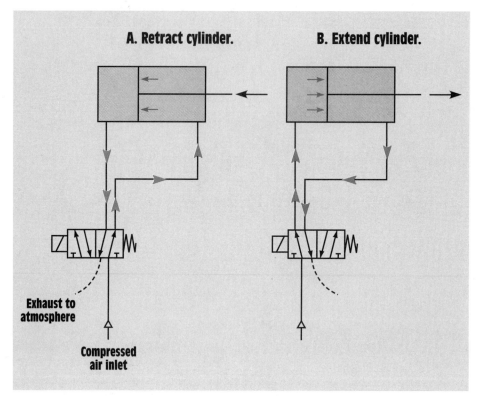

FIGURE 12-14 Single-solenoid, four-way DCV controlling a double-acting cylinder. (A) Retract cylinder. (B) Extend cylinder.

In addition to two-position pneumatic DCVs, DCVs can also have a third position (neutral) that allows the actuator to be held fixed in position or allowed to float when not in use. Two types of neutrals are commonly used in pneumatic four-way valves: *closed center* and *float center*. The schematic symbols for these valves are shown in Figure 12-17. These valves are shown controlling a double-acting cylinder. The closed-center valve will cause the cylinder to hold position when in neutral because air is trapped between the DCV and the cylinder. It will not hold position with high precision, however, due to the compressibility of air. The float-center valve will allow the cylinder to float because both ports on the cylinder are connected to the exhaust ports. This will allow any external force to move the cylinder back and forth. For both types of neutrals, the pressure port must be blocked to prevent compressed air from escaping from the supply line. Closed-center and float-center valves have a similar effect when used with pneumatic motors.

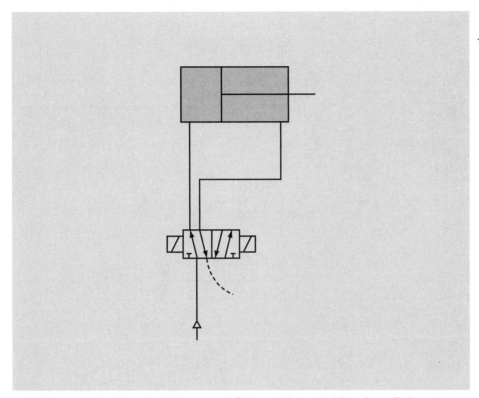

FIGURE 12-15 Double-solenoid, four-way DCV controlling a double-acting cylinder.

Both valves shown controlling the cylinders in Figure 12-17 are *double-solenoid, spring-centered* valves. These valves cause the cylinder to extend or retract as long as the corresponding solenoid is energized. Whenever neither solenoid is energized, the valve automatically returns to the neutral position.

Like hydraulic directional control valves, pneumatic DCVs can be actuated in a variety of ways (Figure 12-18). All of the valves are shown as two-position, spring-returned valves for the sake of comparison. Figure 12-18 A, B, and C show valves that are manually actuated by an operator. Part D shows a cam-actuated valve, which is shifted by contact with another device, such as a cylinder. Part E shows a pilot-operated valve, which is commonly used in pneumatics. Pilot-operated valves are shifted by system pressure. Figure 12-19 shows a cylinder being controlled by a double-pilot-actuated DCV. In this circuit, the two pushbutton-actuated, three-way valves are pilot valves. Their only function is to shift the main valve. In this circuit, actuating-pilot DCV #1 will cause the main

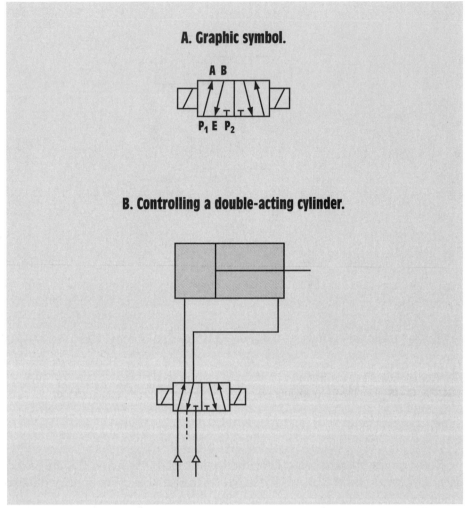

FIGURE 12-16 Five-way DCV. (A) Graphic symbol. (B) Controlling a double-acting cylinder.

valve to shift so that the cylinder will extend, while actuating pilot DCV #2 will cause the main valve to shift so that the cylinder will retract. This valve will maintain position whenever neither pilot valve is actuated. Pilot control is commonly used because it allows the pushbuttons to be remotely located from the main valve.

Figure 12-18F shows a solenoid-actuated valve, which is shifted by electrical current. Figure 12-18G shows a pilot-operated solenoid valve. These valves are actually two valves in one package. In these DCVs, a small pilot valve that is controlled by a solenoid uses air pressure to shift a larger DCV.

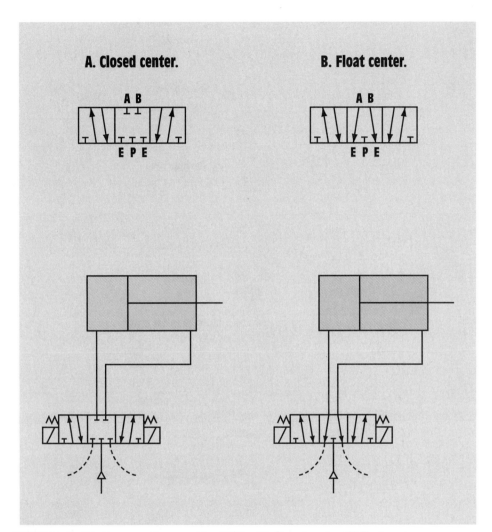

FIGURE 12-17 Four-way, three-position DCVs. (A) Closed center. (B) Float center.

Figure 12-20 shows the most commonly used positioning methods for pneumatic DCVs. The spring-return and spring-centered valves return to a normal position when not actuated. The valve in Figure 12-20C maintains the last position indicated. It is held in place by the friction between the seals and the valve spool. In hydraulics, valves are held in position by a detent, a mechanism that holds a valve securely in the current position. The friction of the valve is insufficient to hold it in position against the higher pressures used in hydraulics. In pneumatics, however, the lower pressures and tighter seals allow the valve to maintain position with friction alone.

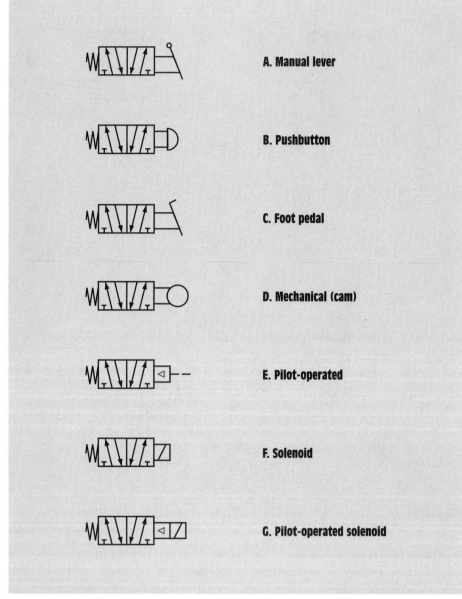

FIGURE 12-18 Pneumatic DCV actuation types. (A) Manual lever. (B) Pushbutton. (C) Foot pedal. (D) Mechanical (cam). (E) Pilot-operated. (F) Solenoid. (G) Pilot-operated solenoid.

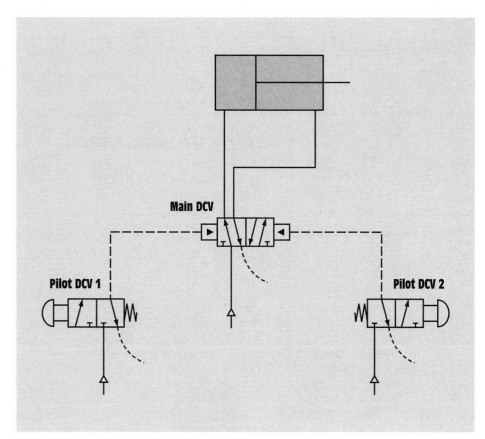

FIGURE 12-19 Pilot control of a four-way valve.

The exhaust air that is returned from the actuator is sent directly to the atmosphere through the DCV. This high-velocity exhaust air can be very loud, especially when several machines are operating in the same area. Exhaust silencers, also called mufflers, are frequently used to lessen the noise. They are screwed directly into the exhaust ports of the DCV. Some actual mufflers are shown in Figure 12-21, along with their graphic symbol.

In addition to the two-way, three-way, and four-way spool valves, other directional control valves are used in pneumatics. These include check valves, shuttle valves, AND valves, and quick-exhaust valves, among others (Figure 12-22). Hydraulic check valves and shuttle valves and their counterparts in pneumatics have the same function (see Chapter 6). A check valve allows flow in one direction (*A* to *B*), but not the other (*B* to *A*). A shuttle valve allows flow from either of two inlets (*P*1 or *P*2) to be sent to a common outlet (*A*). It is often called an

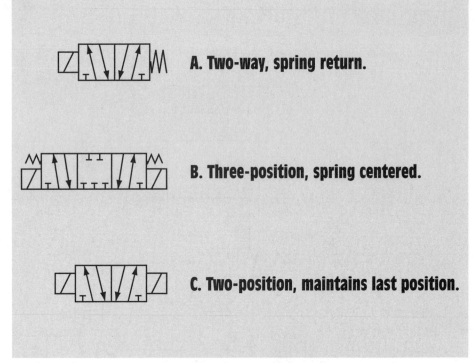

A. Two-way, spring return.

B. Three-position, spring centered.

C. Two-position, maintains last position.

FIGURE 12-20 Pneumatic DCV positioning. (A) Two-position, spring return. (B) Three-position, spring-centered. (C) Two-position, maintains last position.

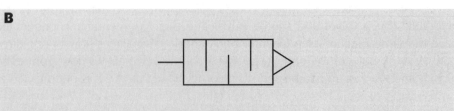

FIGURE 12-21 (A) Exhaust port silencers. (B) Exhaust port silencer graphic symbol.
(Photo courtesy of Norgren, Littleton, CO)

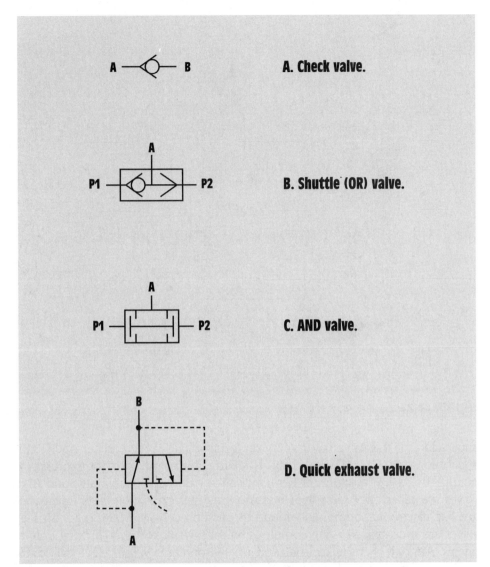

FIGURE 12-22 Other pneumatic directional control valves. (A) Check valve. (B) Shuttle (OR) valve. (C) AND valve. (D) Quick exhaust valve.

OR valve because receiving pressure at *P*1 or *P*2 causes flow to be sent to outlet *A*. Figure 12-23 shows the most basic application for a shuttle valve. In this circuit, either of the three-way DCVs will actuate the single-acting cylinder. This could be used on a press to provide two places where the operator could activate the machine.

Figure 12-22C shows the symbol for an AND valve. An AND valve requires pressure at both *P*1 and *P*2 simultaneously for flow to be sent to outlet *A*. One

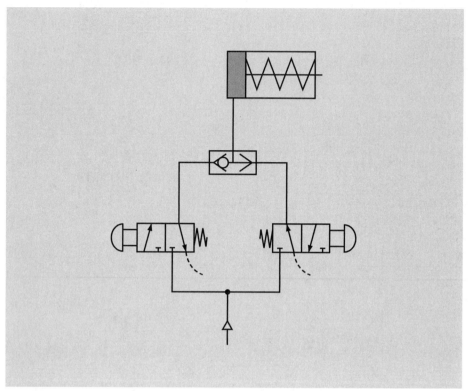

FIGURE 12-23 Shuttle valve application.

application for this valve is for safety on a pneumatic press. A schematic for this application is shown in Figure 12-24. This circuit forces the operator to press both buttons, thereby keeping his or her hands away from the press. Shuttle and AND valves are called air logic valves because they perform simple logic operations, just like electronic components found in microprocessors. Many other air logic valves are available, allowing complex control systems to be built using only air signals. Air logic systems can be used in environments where electrical control presents a fire hazard. They are also used in applications where there is extreme heat or other factors that can cause quick failure of an electrical control system.

Figure 12-22D shows the symbol for a quick exhaust valve. These valves are used to increase cylinder speed by speeding up the exhaust flow out of a cylinder. In the basic cylinder circuit shown in Figure 12-14, the exhaust air must go back to the DCV before being exhausted to the atmosphere. This slows the cylinder down because the resistance of the lines and the DCV put a backpressure on the piston. If a quick exhaust valve is installed, such as in the circuit shown in Figure 12-25, the exiting air will be exhausted at the cylinder. A quick exhaust valve is basically a three-way valve with internal piloting that causes the air coming from the DCV to be sent to the cylinder, while allowing flow out of

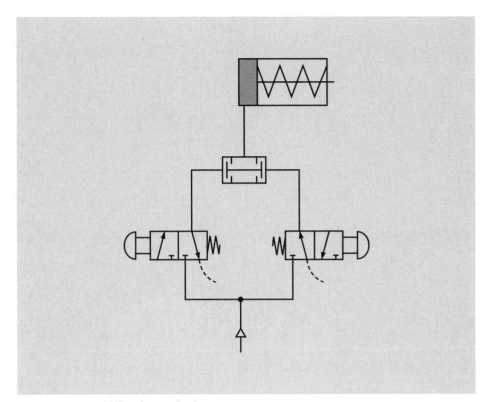

FIGURE 12-24 AND valve application.

the cylinder to be passed directly to the atmosphere. These valves read pressure on either side and shift to the appropriate position depending on the direction of flow. They are screwed directly into the cylinder ports. Quick exhaust valves are used in applications where speed is critical, such as in high-speed assembly machinery.

Just as with hydraulic DCVs, the pressure drop characteristics of pneumatic DCVs are of primary importance when selecting a valve. This information can be provided with pressure drop-versus-flow graphs, such as those that were shown for hydraulic valves in Chapter 6. It is more common, however, to use the flow coefficient (C_v) for pneumatic valves. The flow coefficient is a parameter that relates the pressure drop across a valve and the flow rate through it (see Chapter 8). It is determined experimentally by the valve manufacturer. Due to the compressibility of air, the relationship between flow and pressure is more complicated than that used for hydraulics. The following equations can be used to determine the flow coefficient for pneumatic valves:

$$C_V = \frac{Q}{22.48}\sqrt{\frac{T}{\Delta p \cdot p_{OUT}}}$$

(12-5)

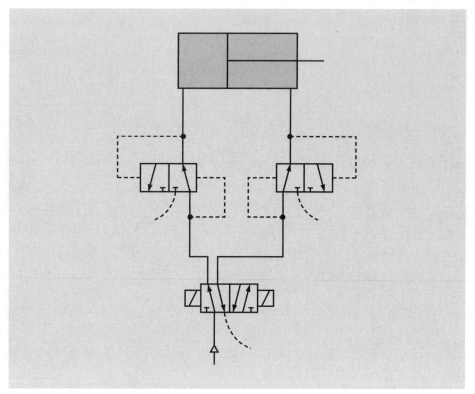

FIGURE 12-25 Quick exhaust valve application.

$$C_V = \frac{Q}{68.7} \sqrt{\frac{T}{\Delta p \cdot p_{OUT}}}$$ **(12-5M)**

where: Q = flow rate (scfm, l/min)
  T = air temperature (°R, K)
  Δp = pressure drop (psi, kPa)
  p_{OUT} = outlet pressure (psia, kPa abs)

The C_V value is a valve-sizing tool. A valve with a larger C_V will have a smaller pressure drop across the valve for a particular flow rate. The following examples illustrate how the C_V value is used.

EXAMPLE 12-3.

An air-powered machine consumes 25 scfm at 90 psig. If the directional control valve that will control this machine can have a pressure drop of no more than 5 psi, what is the minimum C_V? Assume the temperature of the air is 80°F.

SOLUTION:

Assume that the outlet pressure of the DCV is 90 psig, as this is the pressure required at the machine.

1. Convert to absolute values:

$$p_{OUT} = 90 + 14.7 = 104.7 \text{ psia}$$

$$T = 80 + 460 = 540°R$$

2. Calculate the C_v:

$$C_v = \frac{Q}{22.48}\sqrt{\frac{T}{\Delta p \cdot p_{OUT}}} = \frac{25}{22.48} \cdot \sqrt{\frac{540}{5 \cdot (104.7)}} = 1.13$$

Choosing a valve with a C_v of 1.13 or higher will ensure that the pressure drop will be no more than 5 psi.

EXAMPLE 12-3M.

An air-powered machine consumes 700 l/min at 620 kPa gauge. If the directional control valve that will control this machine can have a pressure drop of no more than 35 kPa, what is the minimum C_v? Assume the temperature of the air is 27°C.

SOLUTION:

1. Convert to absolute values:

$$p_{OUT} = 620 + 101 = 721 \text{ kPa abs}$$

$$T = 27 + 273 = 300 \text{ K}$$

2. Calculate the C_v:

$$C_v = \frac{Q}{68.7}\sqrt{\frac{T}{\Delta p \cdot p_{OUT}}} = \frac{700}{68.7}\sqrt{\frac{300}{35 \cdot (721)}} = 1.11$$

EXAMPLE 12-4.

A valve with a C_v of 1.25 is used in a system with a line pressure of 100 psig. The pressure drop across the valve is 10 psi. What is the flow through this valve? Assume the temperature of the air is 80°F.

SOLUTION:

The line pressure is the inlet pressure to the valve. The outlet pressure will then be $100 - 10 = 90$ psig.

1. Convert to absolute values:

$$p_{OUT} = 90 + 14.7 = 104.7 \text{ psia}$$

$$T = 80 + 460 = 540°R$$

2. Calculate the flow:

$$Q = 22.48 \cdot C_V \cdot \sqrt{\frac{\Delta p \cdot p_{OUT}}{T}} = 22.48 \cdot (1.25) \cdot \sqrt{\frac{10 \cdot (104.7)}{540}} = 39.1 \text{ scfm}$$

When solved for the flow rate, as in the previous example, Equation 12-4 shows that a larger pressure drop across a valve will result in a higher flow rate through the valve. This relationship has a limiting condition, however. The air flow rate through a fixed-size orifice cannot exceed the speed of sound. This condition occurs when $p_{OUT} / p_{IN} = 0.533$, or when the outlet pressure is about half the inlet pressure. The value of 0.533 is known as the *critical pressure ratio*. Decreasing this ratio to less than 0.533 will result in no further increase in flow, and the valve is said to be *choked*. The following example illustrates what to do when this situation occurs.

EXAMPLE 12-5.

A valve with a C_V of 1.5 has an inlet pressure of 80 psig and an outlet pressure of 20 psig. What is the flow through this valve? Assume the temperature of the air is 80°F.

SOLUTION:

1. Convert to absolute values:

$$p_{IN} = 80 + 14.7 = 94.7 \text{ psia}$$

$$p_{OUT} = 20 + 14.7 = 34.7 \text{ psia}$$

$$T = 80 + 460 = 540°R$$

2. Calculate the ratio:

$$\frac{p_{OUT}}{p_{IN}} = \frac{34.7}{94.7} = 0.366$$

The ratio is less than 0.533, so the valve is choked. To determine the flow rate through this valve, use an outlet pressure that would correspond to a ratio of 0.533: $p_{OUT} = 0.533 \cdot (94.7) = 50.5$ psia. The pressure drop in this situation is $94.7 - 50.5 = 44.2$ psi. Using these values will give us the maximum achievable flow rate.

3. Calculate the flow:

$$Q = 22.48 \cdot C_V \cdot \sqrt{\frac{\Delta p \cdot p_{OUT}}{T}} = 22.48 \cdot (1.5) \cdot \sqrt{\frac{44.2 \cdot (50.5)}{540}} = 68.6 \text{ scfm}$$

Valve choking is seldom an issue in industrial pneumatic systems. A valve that is choked is undersized and should be replaced with a larger valve.

12.6 Pneumatic Flow Control Valves

Pneumatic flow control valves (FCVs), like those in hydraulics, are most commonly needle valves with a return check for free flow in the reverse direction. Figure 12-26 shows their basic construction, function, and graphic symbol. The flow is controlled from A to B, while from B to A the flow is relatively unrestricted because it can pass through the low-pressure check. Flow control valves often have an arrow on the valve body that indicates the controlled flow direction.

Recall from Chapter 8 that flow controls may be connected in a meter-in or meter-out configuration. Meter-in means that the flow rate into the actuator is being controlled, while meter-out means that the flow rate out of the actuator is being controlled. Both types of control are illustrated with a double-acting cylinder in Figure 12-27. Trace the flow path when each cylinder is extending and retracting to be sure that you understand the operation of the flow control valves. If necessary, go back and review their operation in Chapter 8.

How do we decide whether to use meter-in or meter-out flow control? Meter-in is preferred in hydraulics for resistive loads (those that oppose the movement), while meter-out is preferred for tractive loads (those that are in the direction of movement). In pneumatics, however, *meter-out is preferred in*

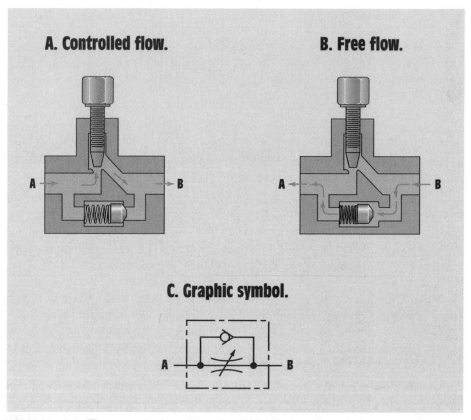

FIGURE 12-26 Flow control valve. (A) Restricted flow. (B) Free flow. (C) Graphic symbol.

all situations. This is because meter-out flow control places a stabilizing back-pressure on the piston, which smoothes out the jumpy motion of pneumatic actuators.

In addition to the standard flow control valve that has threaded female connections on either end, it is also common to use a valve with a threaded male connector on one end so that it can be screwed directly into the cylinder port. This eliminates the need for a line and fitting between the cylinder and the flow control. Some cylinder manufacturers also offer flow controls that are integrated into the cylinder itself.

In pneumatics, it is also common to use flow controls that are at the exhaust port of the directional control valve, rather than between the directional control valve and the actuator (Figure 12-28). No return flow check valve is required with these valves because the air flows out only at the exhaust ports. Exhaust port flow controls are most commonly small cylindrical valves that are screwed directly into the exhaust port of the directional control valve. They are also sometimes built into the directional control valve itself.

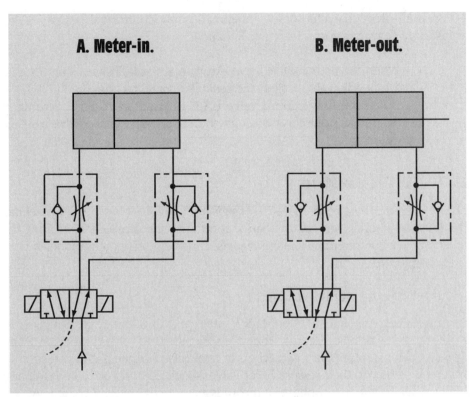

FIGURE 12-27 Meter-in versus meter-out flow control. (A) Meter-in. (B) Meter-out.

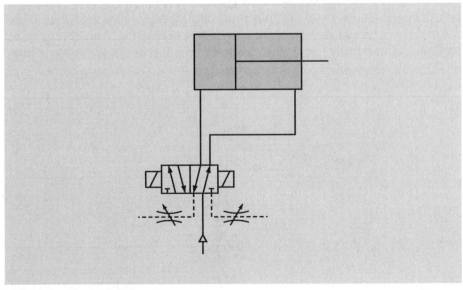

FIGURE 12-28 Exhaust port flow controls.

Pressure drop characteristics of pneumatic flow control valves are usually supplied by C_v values for at least two situations:

1. When the flow control valve is fully open and flow is going through the needle valve (called the *controlled-flow* direction), and
2. When the flow control valve is fully closed and fluid is flowing in the reverse direction through the check valve (called the *free-flow* direction).

12.7 Air Preparation

Compressed air must be prepared, or *conditioned,* before it is sent to a machine. This includes filtering out contaminants, controlling the pressure, lubricating the air, and removing moisture. The following sections discuss the components used to accomplish these tasks.

12.7.1 PRESSURE RELIEF VALVES

Pressure relief valves in a pneumatic system perform the same function as those used in hydraulic systems. They are spring-loaded valves that open when a preset maximum pressure is reached to prevent a further rise in pressure. In pneumatic systems, the pressure relief valve is located at the receiver tank and exhausts air to the atmosphere if the pressure rises above its setting. As we know from our discussion on compressor operation in Chapter 11, the compressor is automatically cycled to keep the pressure in the receiver tank between a minimum and maximum pressure. The pressure relief valve is set to a higher pressure than the maximum and, therefore, should theoretically never open. It is still necessary, however, to prevent the tank pressure from becoming dangerously high if the electronics that cycle the compressor should fail. The graphic symbol for a pneumatic pressure relief valve is shown in Figure 12-29.

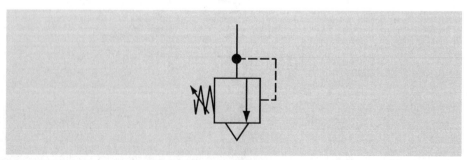

FIGURE 12-29 Pneumatic pressure relief valve symbol.

12.7.2 PRESSURE REGULATORS

Pressure regulators are used to adjust the supply pressure to an appropriate level for a particular machine. Their operation is very similar to the operation of a pressure-reducing valve, which is a hydraulic pressure control valve (see Chapter 7). Regulators are used because the line pressure may be higher than is required by a given machine. Using a lower pressure means using less air, which is always desirable because it reduces the cost of operation. In addition to reducing air consumption, regulators provide a machine with a more consistent supply pressure. Supplying a machine with air directly from the distribution lines will not give a consistent supply pressure because the line pressure in a facility will fluctuate. Fluctuating air pressure will cause the machine performance to be inconsistent and more difficult to predict and control. For these reasons, a pressure regulator should be located at each pneumatic machine.

Figure 12-30 shows a cut-away of a pressure regulator and its graphic symbol. The air flows from the inlet to the outlet past the main poppet, which is shown closed in the figure. The main poppet stem has a light spring on the bottom and it contacts the diaphragm at the top. The pressure from the outlet is applied to the diaphragm through the pilot opening. As the pressure increases, the diaphragm deflects upward against the main spring. The main poppet follows the motion of the diaphragm due to the light spring on the bottom of the

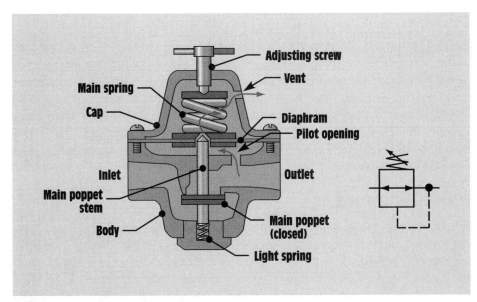

FIGURE 12-30 Pressure regulator.
(From *Industrial Fluid Power*, Vol. 1, Womack Educational Publications.)

stem. The main poppet will close completely when the pressure at the outlet becomes high enough. The pressure setting of the regulator can be modified by changing the compression of the main spring with the adjusting screw. If a further increase in pressure were to occur at the outlet of the regulator after the main poppet is closed, the diaphragm would continue to deflect upward and pull away from the main poppet stem. This causes a small hole to be uncovered in the diaphragm, allowing excess pressure to be relieved through the vent. Because of this last feature, this type of regulator is known as a *self-relieving* regulator. Excess pressure is also relieved if the pressure setting is reduced after the valve is already closed. Self-relieving regulators are the most commonly used type, although nonrelieving types are available.

12.7.3 FILTERS

Contamination in pneumatic systems causes damage to components and greatly decreases their performance and efficiency, just as it does in hydraulics systems. Filtration is therefore an important issue in pneumatics as well as in hydraulics. Filtration begins at the compressor inlet with an intake filter. This filter removes contaminants from the atmospheric air that could damage the compressor. Intake filters are most commonly of the paper element type, similar to those found in automobile engines. An intake filter is not enough, however, to adequately protect pneumatic components because the air will pick up other contaminants as it moves through the distribution network. Each machine should be preceded by an *air-line filter* to protect it from damage from contamination.

A typical air-line filter is shown in Figure 12-31. Air enters through the inlet and flows through the *deflector plate,* causing the air to swirl. Centrifugal force causes water and larger solid particles to be hurled against the interior wall. Gravity then carries them down to the bottom, which is called the *quiet zone* because there is no air flow in this area. The *baffle* separates the quiet zone and prevents these particles from re-entering the air stream. The air then passes through the filter element where smaller particles are removed. The *shroud* blocks the air from getting to the filter element before the larger particles are removed by the swirling action.

12.7.4 LUBRICATORS

The viscosity of a fluid largely determines its lubricating capability. Air has a very low viscosity and, consequently, has no lubricating capability. Many pneumatic components are permanently lubricated by the manufacturer and can therefore use unlubricated air. A *lubricator* must be used for other components. The function of a lubricator is to spray a fine oil mist into the airflow that will be carried downstream and precipitate out onto the components.

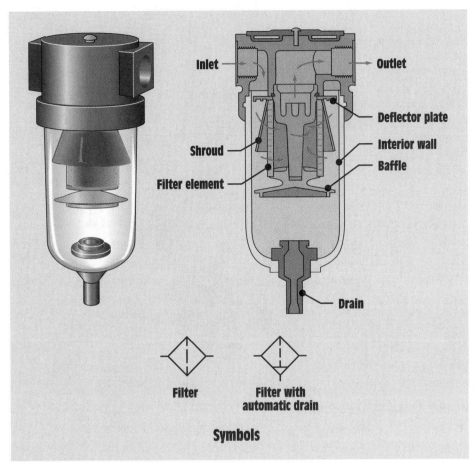

Inlet — Outlet

Deflector plate

Shroud

Interior wall

Filter element — Baffle

Drain

Filter

Filter with
automatic drain

Symbols

FIGURE 12-31 Air-line filter.
(Courtesy of Parker Hannifin Corporation. From Norvelle, *Fluid Power Technology:* West, 1993, p. 528.)

A typical air-line lubricator is shown in Figure 12-32. Pressurized air entering at the inlet applies a pressure to the oil through the pilot line. This causes oil to be forced up into the feed tube and to the needle valve, which controls the flow rate of the oil. The throat causes a small pressure drop so that the pressure at the inlet and at the surface of the oil is higher than the pressure at the outlet, allowing the oil to flow up the feed tube. Drops of oil fall into the air stream and are broken up and carried downstream to the components to be lubricated. The sight glass allows the drop rate of the oil to be observed.

A *micromist lubricator* is a variant on the preceding design. These lubricators have a mechanism that allows only the smaller droplets to enter the air stream. Larger

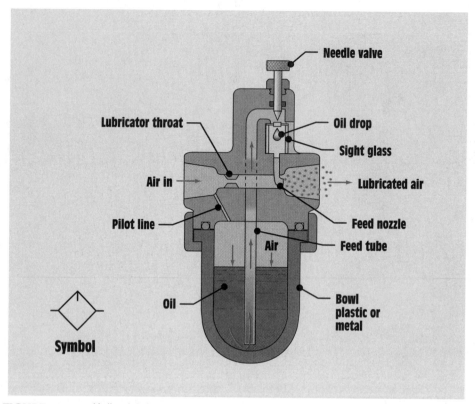

FIGURE 12-32 Air-line lubricator.

(From *Industrial Fluid Power,* Vol. 1, Womack Educational Publications.)

droplets are recirculated back to the reservoir. Micromist lubricators allow the oil to remain suspended in the air longer before precipitating out onto the components.

Each machine that requires lubrication should have its own lubricator. Lubricators should be located as close as possible to the machine they supply so that the oil actually gets to the components, rather than just precipitating out onto the plumbing. Micromist lubricators can be located farther away from a machine due to the finer oil mist they produce.

The amount of lubricating oil that is required by a given machine depends on many factors, such as the number of components to be lubricated, the operating speed, etc. Only a thin coating of oil on the components should be provided. Excess oil tends to gum up components and degrade performance.

When the air is exhausted back into the atmosphere, even a small amount of lubricating oil can cause a breathing hazard. Filters should be located on the exhaust ports to remove most of the oil from the exiting air. Oil removal filters that remove nearly all of the oil from the air are available.

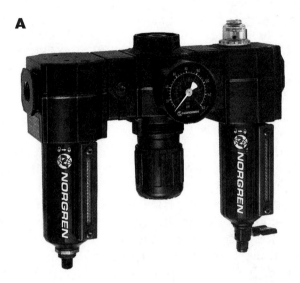

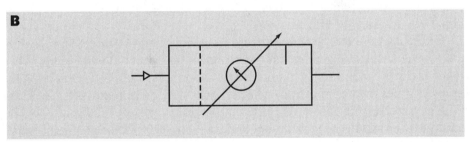

FIGURE 12-33 Filter regulator lubricator (FRL) unit. (A) Actual FRL unit. (B) Graphic symbol. (Photo courtesy of Norgren, Littleton, CO.)

Many pneumatic components are permanently lubricated by the manufacturer and therefore require no external lubrication. Industry trends are moving toward these components because they eliminate the expense, required maintenance, and possible breathing hazards associated with lubricators.

12.7.5 COMBINED CONDITIONING UNITS

Combined conditioning units that include a filter, regulator, and lubricator combined into one unit are offered by many manufacturers. The combined unit, called an *FRL* or *trio,* and its graphic symbol are shown Figure 12-33. *FR* units, or *duos,* that combine the filter and regulator are available for systems that do not require external lubrication. Each machine should have its own conditioning unit for the reasons that were discussed previously.

12.7.6. WATER REMOVAL

Water exists in a pneumatic system in two forms: as water vapor and as liquid water. Vapor is water in its gaseous form and causes no harm to the system because it behaves like air. Water does become a problem, however, when it condenses from vapor to liquid water. Liquid water in a pneumatic system is undesirable because it causes components to rust and reduces the effectiveness of lubricants.

To understand the water problem in a pneumatic system we must first define two terms: *relative humidity* and *dew point*. Relative humidity is the percentage of water in the air as compared to the maximum amount it could hold. For example, air at 100% relative humidity is *saturated* and cannot hold any more water. The amount of water that air can hold changes with temperature and pressure. As the temperature increases, air can hold more water. This means that 30% relative humidity at 90°F represents more water in the air than 30% humidity at 70°F. As the pressure increases, air can hold less water. This means that 30% relative humidity at 0 psig represents more water in the air than 30% humidity at 100 psig.

Dew point is another way to measure the amount of moisture in air. The dew point is the temperature at which the air becomes saturated with water. When the temperature is at the dew point, condensation will result. A higher dew point represents more water in the air, while a lower dew point represents less water. If the dew point of the air is 50°F, the temperature can drop all the way to 50°F before the air will become saturated and condensation will occur. This is significant for a pneumatic system because as long as we stay above the dew point, there will be no condensation.

Because air cannot hold as much water at higher pressures, it is important to know whether the dew point being referred to is the *atmospheric dew point* or the *pressure dew point*. The atmospheric dew point is the dew point at atmospheric pressure, while the pressure dew point is the dew point at system pressure. The pressure dew point is higher because pressurized air cannot hold as much water. When dealing with pneumatic systems, the pressure dew point is of greater concern because the air is under pressure.

Water becomes a problem in a pneumatic system because when the compressor increases the pressure of the air, it increases the relative humidity because air at a higher pressure cannot hold as much water. In fact, after the air is compressed, it will normally not be able to hold the amount of water contained in atmospheric air. This means that after compression, water will have condensed out and the air will be at 100% relative humidity.

Compressors may have an *aftercooler* between the compressor and receiver tank that removes some of the heat of compression. Aftercoolers are heat exchangers that pass cool air or water over pipes through which the compressed air flows. Cooling the air causes even more water to condense out because cooler air cannot hold as much water. The air is now cooler, but still at 100% relative humidity. The receiver tank must be drained regularly to remove the condensed water. This is important to prevent the receiver tank from rusting, which can be dangerous.

Due to the reasons just discussed, air in the receiver tank is generally saturated. This is undesirable because if the temperature drops at any time, which it almost certainly will, water will condense out. The only way to prevent condensation from forming is to reduce the relative humidity of the air to well below 100%. *Air dryers* are used to accomplish this. Two types of dryers are prevalent in industry: *refrigeration dryers* and *absorption dryers*. Refrigeration dryers, also called *chillers,* work on the common refrigeration cycle like that used in a commercial refrigerator or air conditioner. In this cycle, a refrigerant is alternately compressed and heated, then expanded and cooled. The compressed air is passed by the cold side of the cycle, where it is cooled to temperatures as low as 35°F. This causes most of the moisture to condense out. The air is then either allowed to come up to room temperature on its own, or heated to bring it back quickly to room temperature. The air then has a pressure dew point of 35°F. Again, this means that the air can drop all the way down to 35°F before any water will condense out. Figure 12-34 shows a photo of an actual refrigeration dryer.

FIGURE 12-34
Refrigeration-type air dryer.
(Photo courtesy of Norgren, Littleton, CO.)

Absorption dryers use a gel material, called a *desiccant,* which absorbs water as the air is passed through it. It basically acts as a water filter. Because the desiccant eventually becomes saturated with water, it must be dried periodically. Two sections are used so that the dryer can operate continuously. While one section is drying the compressed air, the other is being dried. A small portion of the compressed air is expanded and used to dry the section that is not in use. These are called *regenerative* absorption dryers. Absorption dryers can typically bring the pressure dew point down to as low as −40°F. Because temperatures in just about any manufacturing facility will never reach this low level, the water problem is eliminated. Figure 12-35 shows a photo of a regenerative dryer.

FIGURE 12-35
Regenerative-type air dryer.
(Photo courtesy of Norgren, Littleton, CO.)

The graphic symbol for a dryer applies to both the refrigerated and absorptive type dryers (Figure 12-36).

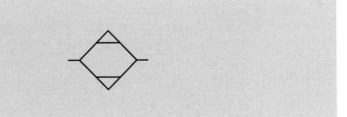

FIGURE 12-36 Air-dryer graphic symbol.

12.8 Air Distribution

Pneumatic systems normally use a centrally located compressor and receiver tank that supplies multiple machines (see Chapter 11). The compressed air is distributed throughout the facility through a *distribution header* (conduits located overhead). Lines are then dropped down to supply each machine, which has, at minimum, a filter and a regulator, and possibly a lubricator if the machine is not internally lubricated.

The previous section addressed the problems caused by water in a pneumatic system. In addition to using air dryers, the distribution system can also be designed to keep water away from the machines. Three design features accomplish this:

1. Drop legs to the machines are taken upward, then down to the machine. This allows gravity to prevent liquid water from reaching the machine.
2. Water drop legs with automatic drains are placed at various locations along the distribution header. These are taken straight down off the header.
3. The distribution lines are sloped slightly downward (½° to 1°) to keep any liquid water that may be present moving along until it reaches a water drop leg.

Just as with hydraulic systems, pneumatic systems can use pipe, tubing, or hose as conduits. Schedule 40 steel pipe is the most durable and is commonly used for permanent, stationary connections. Tubing is also used for stationary connections but is not as durable as pipe. Tubing may be plastic (i.e., polyethylene, nylon), copper, or aluminum. Synthetic rubber hose is used for connections between moving components or between components that would be to difficult to plumb with a rigid conduit.

12.9 Equations

EQUATION NUMBER	EQUATION	REQUIRED UNITS
12-1	$F_E = p \cdot A_p$	Any consistent units
12-2	$F_R = p \cdot (A_p - A_R)$	Any consistent units
12-3	$V_E = A_p \cdot S$	Any consistent units
12-4	$V_R = (A_p - A_R) \cdot S$	Any consistent units
12-5	$C_v = \dfrac{Q}{22.48} \sqrt{\dfrac{T}{\Delta p \cdot p_{OUT}}}$	Q in scfm, T in °R, Δp in psi, p_{OUT} in psia
12-5M	$C_v = \dfrac{Q}{68.7} \sqrt{\dfrac{T}{\Delta p \cdot p_{OUT}}}$	Q in l/min, T in K, Δp in kPa, p_{OUT} in kPa abs

12.10 Review Questions and Problems

1. List some advantages of pneumatic motors over electric motors.
2. Define the terms *running torque, starting torque,* and *stall torque.*
3. Draw the graphic symbol for a two-way, two-position DCV. What are the two ports on this valve? What happens to the flow in each position?
4. Draw the graphic symbol and label the ports for a pneumatic three-way, two-position DCV. What happens to the flow in each position?
5. Draw the graphic symbol and label the ports for a pneumatic four-way, two-position DCV. Describe the flow paths in each position.
6. Draw the graphic symbol and label the ports for a pneumatic five-way, two-position DCV. Describe the flow paths in each position.
7. Draw the graphic symbol and label the ports for the following pneumatic directional control valves:
   a. Two-way, two-position, normally open with pushbutton actuation
   b. Three-way, two-position, normally closed with solenoid actuation
   c. Four-way, two-position, with double solenoid actuation
   d. Four-way, three-position (float neutral), spring-centered with double-pilot actuation
8. What is the purpose of a shuttle valve? Draw the graphic symbol, label the ports, and describe its operation.
9. What is the purpose of an AND valve? Draw the graphic symbol, label the ports, and describe its operation.
10. What is the purpose of a quick exhaust valve? Where are these valves installed?

11. Describe the difference between meter-in and meter-out flow control. Which is preferred for pneumatics? Why?
12. Draw a cylinder with meter-out flow control of both strokes.
13. What is the purpose of a pressure regulator?
14. What devices make up an FRL unit?
15. Where should an FRL unit be located in a pneumatic circuit?
16. Define the terms *relative humidity* and *dew point*.
17. What is the purpose of a dryer in a pneumatic system? Why is it necessary?
18. Describe the operation of a refrigeration-type air dryer.
19. Describe the operation of a absorption-type air dryer.
20. A double-acting pneumatic cylinder with a 2.5 in bore and a 0.625 in rod is to be used in a system with a supply pressure of 80 psig. What is the force output of the cylinder on the extend and retract strokes?
21. A double-acting pneumatic cylinder with a 63 mm bore and a 20 mm rod is to be used in a system with a supply pressure of 650 kPa gauge. What is the force output of the cylinder on the extend and retract strokes?
22. Why is it difficult to accurately predict a pneumatic cylinder's speed?
23. What factors determine a pneumatic cylinder's speed?
24. A double-acting pneumatic cylinder with a 2.5 in bore, a 1 in rod, and a 5 in stroke length must cycle (extend and retract) 60 times per minute. The supply pressure is 100 psig. What is the air consumption of the cylinder in scfm?
25. A double-acting pneumatic cylinder with a 63 mm bore, a 20 mm rod, and a 120 mm stroke length must cycle (extend and retract) 60 times per minute. The supply pressure is 600 kPa gauge. What is the air consumption of the cylinder in standard m^3/min?
26. An air-powered machine consumes 40 scfm at 80 psig. If the directional control valve that will control this machine can have a pressure drop of no more than 5 psi, what is the minimum C_v? Assume the temperature of the air is 80 °F.
27. An air-powered machine consumes 1000 lpm at 550 kPa gauge. If the directional control valve that will control this machine can have a pressure drop of no more than 40 kPa, what is the minimum C_v? Assume the temperature of the air is 30 °C.
28. A valve with a C_v of 1.2 is used in a system with a supply pressure of 80 psig. The pressure drop across the valve is 10 psi. What is the flow through this valve? Assume the temperature of the air is 90°F.
29. A valve with a C_v of 1.35 is used in a system with a supply pressure of 600 kPa gauge. The pressure drop across the valve is 70 kPa. What is the flow through this valve? Assume the temperature of the air is 25°C.

CHAPTER 13

Electronic Control of Fluid Power

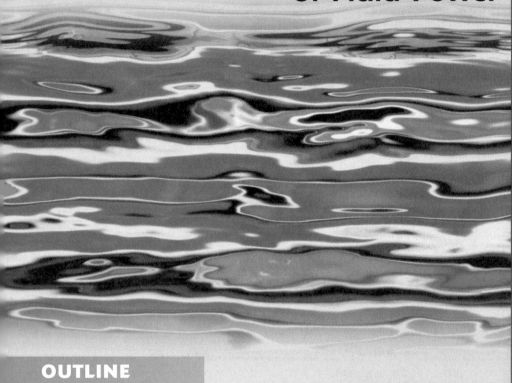

OUTLINE

13.1 Introduction

13.2 Solenoid Valves

13.3 Proportional and Servo Valves

13.4 Pump Controls

13.5 Review Questions

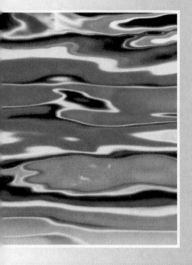

13.1   Introduction

Electronic control is prevalent in all types of industrial manufacturing systems, and fluid power is no exception. Electronic control provides many benefits, but perhaps the most significant are a high degree of precision and decision-making capabilities. Electronic control of fluid power is achieved through the use of several devices. At the highest level is the central processing unit, usually a *programmable logic controller* (*PLC*). A PLC is an industrial computer that has sophisticated input/output capabilities. PLCs give the control system decision-making capabilities by monitoring input signals and generating appropriate output signals based on a predefined program. At the other end of the system is the control valve, which may simply control actuator direction or also other variables such as position, velocity, acceleration, and force. In between the PLC and the control valve may be various other electronic devices, depending on the sophistication of the system.

Sensors are the interface between the physical components and the electronics. Sensors are devices that take some physical parameter (i.e., position, speed, acceleration, pressure, flow, temperature, etc.) and convert it into an electrical signal so that it can be processed by the electronics. The electronics then monitor and act upon the data received from the sensors. Most sensors fall into a category of devices known as *transducers*. A transducer is any device that converts energy from one form into another. A *pressure transducer,* for example, takes fluid pressure energy and converts it into electrical energy (voltage). Because the voltage signal produced by a pressure transducer is proportional to the pressure of the fluid, the signal can be used to sense the pressure.

There are two basic types of control systems: *open loop* and *closed loop*. A basic schematic of an open loop control system is shown in Figure 13-1. In open loop control, a *command signal* (*C*) is sent to the control valve. The command signal is the input that instructs the valve to produce a desired output (*O*). The output may be a certain pressure or flow rate, for example. Open loop control

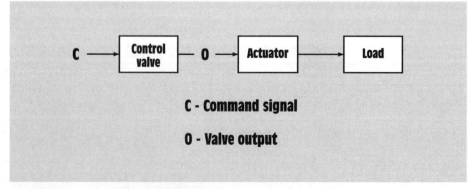

FIGURE 13-1   Open-loop control system.

systems *do not* have sensors that determine what is really happening at the load. In open loop control, we must trust the fact that the control valve causes the desired result to be produced at the actuator.

In closed loop control, the actual result being produced by the actuator is read by a sensor. This allows the control system to correct any error between the desired result and the actual result. Closed loop control is shown schematically in Figure 13-2. The sensor produces a *feedback signal* (F) that is compared with the command signal at the *comparator.* The comparator determines if there is any difference between the desired result (the command signal) and the actual result (the feedback signal). The comparator produces an *error signal* (E), which is sent to the control valve via the *amplifier.* The amplifier converts the weak voltage error signal to an amplified current signal that is sufficient to drive the control valve. This signal instructs the control valve to correct any difference between the desired result and the actual result being produced by the actuator at the load.

Open and closed loop control are concepts that apply to any control system, not just those found in fluid power. The heating system in our homes is an example of closed-loop control. The command signal (the desired temperature) is dialed in at the thermostat. A sensor inside the thermostat reads the temperature in the room. When the temperature in the room drops below the command signal by a certain amount, an error signal instructs the furnace to turn on.

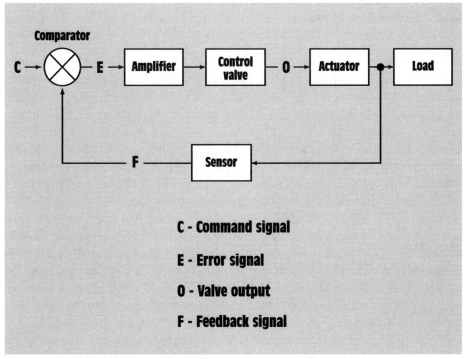

C - Command signal

E - Error signal

O - Valve output

F - Feedback signal

FIGURE 13-2   Closed-loop control system.

When the temperature again reaches the desired level, the error signal becomes zero and the furnace is turned off.

Response time, accuracy, and *precision* are of great importance in with electronic control systems. Response time refers to the difference in time between when a desired result is commanded and when it is actually achieved. For example, if a hydraulic motor is commanded to turn at 500 rpm, that speed will certainly not be attained instantaneously. Did it require 1 second or 0.1 seconds to achieve this speed? Accuracy is a measure of how close the actual result is to the desired result. Did the hydraulic motor just discussed achieve a final speed of 501 rpm or 500.1 rpm? Because the desired result can never be achieved exactly, we must always answer the following question: How close to the desired result is close enough? Precision, often confused with accuracy, is a measure of repeatability. It measures the amount of variance in the actual result when the same command signal is sent on successive trials. We might find that in ten trials, the motor achieved a final speed of between 499 rpm and 501 rpm. The precision or repeatability is therefore ± 1 rpm (± 0.2%).

The reader is encouraged to pursue further readings on PLCs, sensors, and control system theory. A good, basic understanding of control system components and theory is critical to understanding today's industrial workplace. This text focuses primarily on electrohydraulic and electropneumatic valves.

13.2 Solenoid Valves

Solenoids are switches that are shifted by electrical current. They are commonly used as actuation devices for hydraulic and pneumatic valves and in many other applications, such as in automobile ignition systems. Solenoids operate on the principle that when electrical current is sent through a coil of wires, a magnetic field is generated within the coil. The magnetic field is then used to attract a movable element called a *plunger,* which is made of a magnetized material such as iron. The attraction between the coil and the plunger produces a shifting action that can be used for actuation. Solenoid-actuated valves can be found in virtually every industrial manufacturing facility due to the widespread use of electronically controlled hydraulic and pneumatic circuits.

Figure 13-3 shows a simplified cut-away of a four-way, two-position, single solenoid, spring-return directional control valve. The primary components of the solenoid are the coil and the plunger. Figure 13-3A shows the solenoid coil de-energized (no current), which allows the spring to hold the valve in its normal position. Part B shows the solenoid coil energized (current is applied), which causes a magnetic field to be generated within the coil. The magnetic field causes the iron plunger to be pulled into the coil, shifting the valve spool into the position shown. A cut-away of an actual valve of this type is shown in Figure 13-4.

The graphic symbols for some common solenoid actuated hydraulic four-way valves are shown in Figure 13-5. Part A shows the two-position, single solenoid,

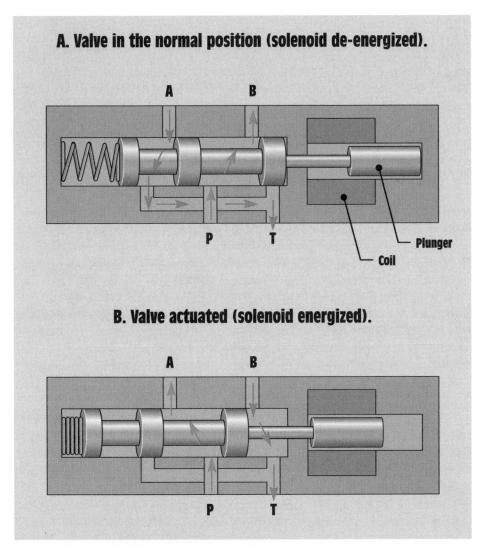

A. Valve in the normal position (solenoid de-energized).

A   B

P   T

— Plunger

— Coil

B. Valve actuated (solenoid energized).

A   B

P   T

FIGURE 13-3   Single solenoid valve with spring return. (A) Valve in the normal position. (B) Valve actuated (solenoid energized).

spring-return directional control valve. Part B shows a double solenoid, two-position detented valve. A detent is a mechanism that holds a valve firmly in position, thus preventing the valve from being inadvertently shifted by hydraulic pressure. This valve is shifted by momentarily energizing one of the solenoids and maintains position until the opposing solenoid is energized. Part C shows a double solenoid, three-position, spring-centered valve that returns to neutral automatically whenever neither solenoid is energized. Figure 13-6 shows a photo of an actual valve of this type. Solenoids are also commonly used to shift two-way and three-way directional control valves, as well as some types of pressure control valves.

FIGURE 13–4   Cutaway of a four-way, two-position, single solenoid, spring-return directional control valve.

(Photo courtesy of Continental Hydraulics.)

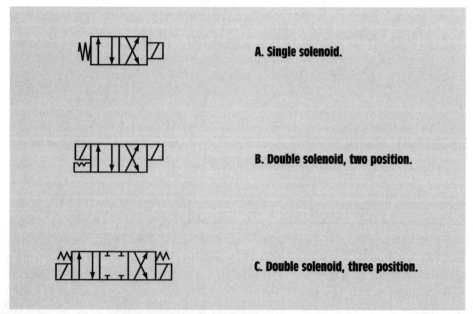

FIGURE 13–5   Hydraulic solenoid DCV symbols. (A) Single solenoid, two-position, spring return. (B) Double solenoid, two-position, detented. (C) Double solenoid, three-position, spring centered.

FIGURE 13-6   Four-way, two-position, double solenoid, spring-centered directional control valve.
(Courtesy of Continental Hydraulics, Savage, Minnesota.)

Hydraulic solenoid valves are usually of the *wet armature* type, which means that the hydraulic fluid is allowed to circulate through the solenoid case. This carries away the heat generated by the current flowing through the coil.

The solenoid valves discussed previously are called direct-acting valves because the solenoid shifts the valve spool directly. Solenoids can also be *pilot-operated*. Pilot-operated solenoid valves are essentially two valves in one package. The solenoid is used to actuate a small pilot valve, which in turn uses the pressure of the system to shift the main valve. This method of actuation is necessary on large valves that operate in systems at high pressures because the solenoid alone cannot generate enough force to shift a large valve against high pressure. The solenoid can, however, generate enough force to shift the small pilot valve, which can then use the pressure of the system to shift the main valve. Figure 13-7A shows a photo of an actual hydraulic pilot-operated solenoid valve; Part B shows the graphic symbol.

Solenoid valves are available in both AC and DC models. Each type is also available in a variety of voltages. It is important to match the type of solenoid with the power supply in the application when selecting a solenoid valve. For AC solenoids, the power supply specification will include voltage and line frequency (e.g., 120V/60 Hz). For DC solenoids, only a voltage is required (e.g., 24 V). AC solenoids generally have a faster response time than DC solenoids.

Many solenoid valves have a small indicator light that turns on when there is current in the solenoid coil. This aids in troubleshooting a circuit because no light at the solenoid indicates a problem with the electrical circuit, rather than a mechanical problem such as the valve spool sticking. A manual override pin that allows the solenoid to be shifted manually also aids in troubleshooting.

Solenoids are often used to actuate pneumatic valves as well as hydraulic valves. Most of the concepts discussed in this section apply equally to pneumatic solenoid valves.

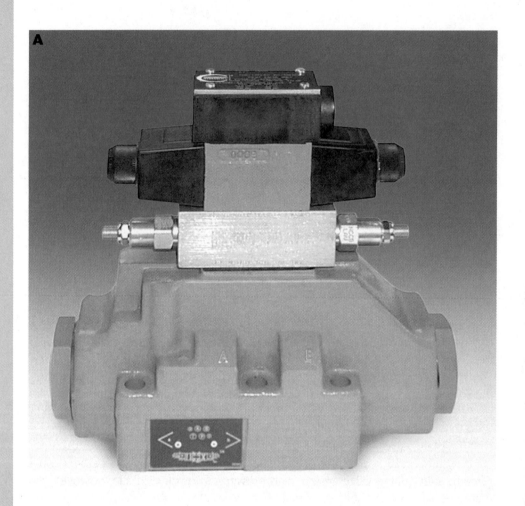

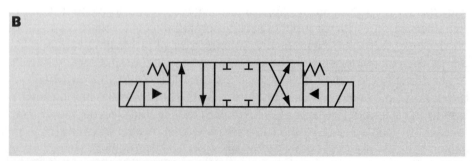

FIGURE 13-7   Pilot-operated solenoid directional control valve. (A) Photo of a pilot-operated solenoid directional control valve. (B) Graphic symbol.
(Courtesy of Continental Hydraulics, Savage, Minnesota)

13.3 Proportional and Servo Valves

The solenoid valves discussed in the previous section have only two states: shifted or not shifted. This means that a solenoid valve can only be shifted open to allow flow or closed to block flow, but nowhere in between. *Proportional valves* are similar to solenoid valves in that they are electronically controlled. They are different in that they are able to precisely position the valve spool between the open and closed positions. This allows the flow to be throttled (metered) through the valve and provides precise flow control as well as directional control. The outlet pressure can also be controlled by using the valve spool position to control the pressure drop across the valve.

The simplest type of proportional valve uses a variable force solenoid and a spring to oppose its motion. It operates on the principle that the magnetic force generated within the solenoid coil depends on the current through the coil. At higher current input levels, the solenoid acts with more force. The greater the solenoid force, the farther the spring will deflect. Therefore, for each particular current input there is a particular spool position. A particular spool position corresponds to a particular flow rate, so the flow rate is *proportional* to the input current.

The simple proportional valve design just discussed can be improved by adding a position transducer that senses the spool position and provides feedback for error correction. The control diagram for this method of spool position control is shown in Figure 13-8. The presence of feedback makes this a closed-loop control system. Proportional valves are readily available with a position transducer and other associated electronics integrated into the valve package. The most commonly used position transducer for proportional valves is a *linear variable differential transformer* (LDVT). These transducers consist of a stationary element and a movable element. When the movable element is displaced from the reference position, a voltage is induced that is proportional to the distance moved. The movable element is attached to the valve spool, allowing its position to be sensed through the induced voltage. Figure 13-9 shows a proportional valve with an LDVT position transducer. The LDVT is the device on the left side of the valve.

The control loop shown in Figure 13-8 is called the *inner control loop* because it is contained within the control valve itself. The control system that was diagrammed in Figure 13-2 is the *outer control loop*. The inner control loop is contained within the block marked *control valve* in Figure 13-2. The inner control loop is only aware of the spool position, not what is happening at the load. The outer loop is providing feedback and error correction on what is actually happening at the load (position, velocity, force, etc.). The outer control loop may or may not be necessary, depending on the application.

The amount of *spool overlap* is an important aspect of a proportional valve. Spool overlap is the amount that the spool land overlaps the pressure port when in the closed neutral position (Figure 13-10). Spool overlap is important because when the command voltage is increased from zero and the spool begins to shift, there will be no flow until the spool land begins to uncover the pressure port.

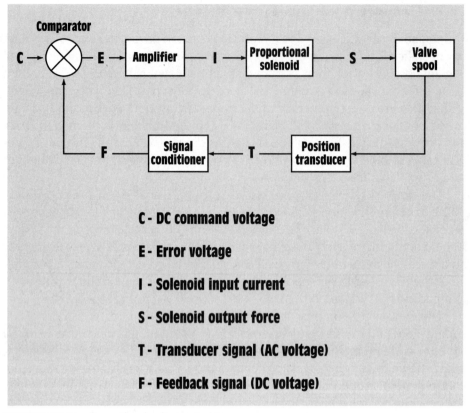

C - DC command voltage

E - Error voltage

I - Solenoid input current

S - Solenoid output force

T - Transducer signal (AC voltage)

F - Feedback signal (DC voltage)

FIGURE 13-8   Proportional valve control loop.
(Adapted from *Introduction to Design of Hydraulic Circuits Using Servo and Proportional Technology,* IDAS Engineering, Inc.)

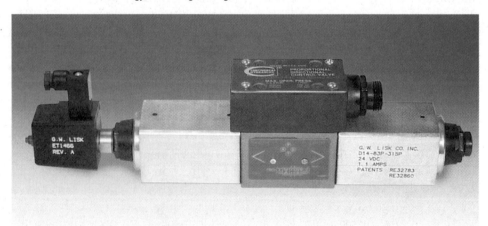

FIGURE 13-9   Proportional valve with LDVT.
(Photo courtesy of Continental Hydraulics)

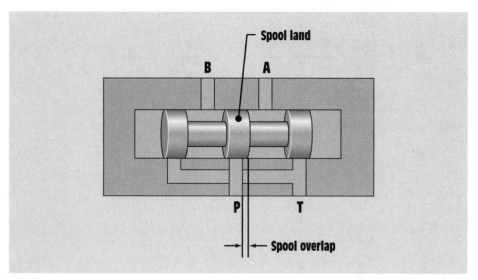

FIGURE 13-10 Spool overlap.

This is best illustrated with a flow-versus-command voltage graph, which is normally supplied by the manufacturer for each model at a specified pressure drop. A typical graph for a proportional valve with overlap is shown in Figure 13-11. As the command voltage is increased from 0 to 1 volt, there is no resulting flow through the valve. This lack of response at low input levels is due to the overlap and is called the *deadband* of the valve. As the command voltage is increased beyond 1 volt, there is a proportional increase in flow through the valve.

Deadband in a proportional valve with overlap can be eliminated by incorporating electronics that compensate for the overlap. The electronic compensation, often called a *deadband eliminator,* causes the valve to respond immediately when a command voltage is applied. *Zero lap* proportional valves, which eliminate deadband by eliminating the overlap of the spool land and pressure port, are also available.

A *servo valve* is another type of electronically controlled valve that produces output proportional to an electrical input signal. In general, the term *servo* refers to any mechanism that uses feedback for self-regulation. In the past, three features differentiated servo valves from proportional valves:

1. The use of feedback for self-regulation of the spool position,
2. Zero lap, and
3. The position of the spool is controlled by a *torque motor,* rather than a proportional solenoid.

Torque motors are electromechanical devices that provide precise positioning. Like solenoids, they produce a force that is proportional to input current. It has become more difficult to differentiate between proportional and servo valves, however,

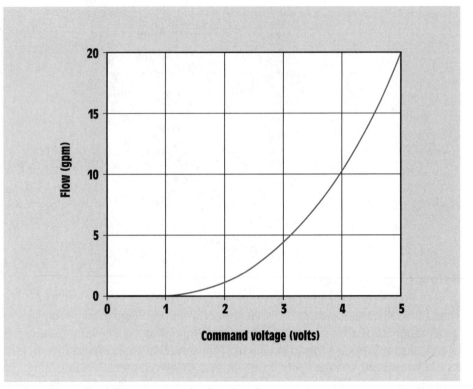

FIGURE 13-11 Flow versus command voltage graph.

because zero lap and feedback are now commonly found in proportional valves. Some manufacturers even refer to valves that use proportional solenoids as *servo solenoid* valves because they have feedback and zero lap. The confusion can be eliminated if one focuses on the valve's performance specifications, rather than on what it is called. Both proportional and servo valves have the same objective: to precisely control the actuator's position, force, velocity, and acceleration, or any combination of these by controlling the pressure and flow rate to the actuator.

In systems that use proportional or servo valves, a *motion controller* may be used in addition to a PLC, or it may replace the PLC altogether. Motion controllers provide more sophisticated control when continuous, closed-loop control of position, velocity, or acceleration is required. This is due to the difference in operation between a PLC and motion controller. A PLC scans the status of its input signals (from sensors and other devices) every few milliseconds and then generates output signals based on a predefined program. Because a few milliseconds may elapse between an input change and a resulting change in output, response time and accuracy may suffer. Motion controllers continuously monitor sensor input signals and can react faster to changes at the load, increasing response time and accuracy.

Proportional and servo valves, like other directional control valves, have specifications that state the viscosity range of the fluid, maximum pressure, maximum flow, filtration level, etc. In addition to these, several other important specifications are related to the performance of the control functions. *Frequency response* measures how fast a valve responds to a changing command signal. It is tested by applying a sinusoidal command signal that quickly shifts the valve spool back and forth. For example, a frequency response of 50 Hz means that the valve spool was able to keep pace with a signal that moved it back and forth at a rate of 50 times per second. Frequency response is normally provided at different amplitudes of spool travel. For example, the frequency response of 50 Hz may be at 10% spool travel relative to the maximum travel.

Hysteresis error of a control valve is the difference in flow output when the valve is responding to an increasing versus a decreasing signal. It is expressed as a maximum variation. For example, hysteresis error may be specified as a maximum of 1% over the operating range of the valve. *Repeatability* is the difference in flow output between successive trials using the same input signal. Repeatability is also specified as a percentage variation.

The deadband (overlap) and flow-versus-input signal data are also important. Manufacturer's catalogs are an excellent resource for learning how the data are presented. Many manufacturers supply catalog information at their Internet sites.

The basic graphic symbols for hydraulic proportional and servo valves are shown in Figure 13-12. The arrows through the solenoids of the proportional valve indicate that they have a variable force output depending on the input current. The lines above and below both valve symbols indicate that the spool position is infinitely variable between the closed neutral and open positions.

Other types of valves, such as pressure relief valves and pressure-reducing valves, may also be equipped with proportional technology. Proportional pressure relief valves allow their setting to be controlled by an electronic signal, rather than by a manual control knob. This allows the electronics to set an appropriate maximum pressure according to a predefined program, depending on what is happening in the circuit.

In the past, proportional technology was limited to hydraulics because the compressibility of air made it too difficult to provide precision at a reasonable cost in pneumatic applications. Recent advances in electronics and the development of high-precision pneumatic valves, however, have made proportional and servo technology available to pneumatic circuit designers. Most of the concepts discussed previously apply equally to pneumatic servo and proportional valves, although the precision of the output is less due to the compressibility of air.

13.4 Pump Controls

The previous section discussed electronically controlled directional control valves that are capable of controlling flow and pressure, in addition to direction. Controlling flow and pressure with a directional control valve is wasteful because it is obtained by controlling the pressure drop across the valve. Pressure drops represent wasted energy that is transformed into heat. A more efficient way to control flow and pressure is to control it at the source: the pump. Electrohydraulic proportional

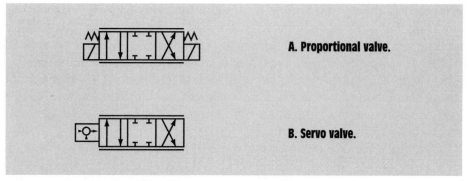

A. Proportional valve.

B. Servo valve.

FIGURE 13-12 Proportional and servo valve graphic symbols. (A) Proportional valve. (B) Servo valve.

controls are also available to control pump output. In electrohydraulic pump control, the pressure and flow output of the pump are adjusted to match the load requirements. This ensures that only the power required by the load is produced at the pump, thereby reducing waste. Pump control is a much more efficient way to control pressure and flow.

The disadvantage to electrohydraulic pump control is that it is much less responsive than valve control. While electrohydraulic valves can commonly achieve frequency responses of 100 to 200 Hz, electrohydraulic pumps are currently capable of frequency responses of only 10 to 12 Hz, making them much slower to respond to command signal changes. For this reason valve control is the predominantly used method for proportional control of hydraulic circuits, in spite of the inherent inefficiency.

13.5 Review Questions

1. What is a PLC?
2. What is the purpose of sensors in a control system?
3. Describe the difference between open loop and closed loop control.
4. Describe the operation of a closed loop control system. Describe in your own words the function of each component.
5. Define the terms response time, accuracy, and precision.
6. Describe the operation of a solenoid.
7. What is the purpose of a proportional valve? How is it different from a solenoid valve?
8. What is valve overlap? How does it affect the performance of a proportional valve?
9. What is the purpose of a servo valve?
10. Define the terms frequency response, hysteresis, and repeatability.
11. Describe the advantages and disadvantages of using a proportional valve to control flow and pressure versus using pump controls.

Appendix A

NOMENCLATURE AND COMMON UNITS

SYMBOL	DEFINITION	U.S. CUSTOMARY UNIT	METRIC UNIT
F	Force	lb	N
A	Area	in^2	m^2
p	Pressure	lb/in^2 (psi)	N/m^2 (Pa), kPa, bar
p_{GAUGE}	Gauge pressure	psig	kPa gauge
p_{ABS}	Absolute pressure	psia	kPa abs
p_{ATM}	Atmospheric pressure	psia	kPa abs
π	Pi (constant)	3.1416	3.1416
g	Gravity (constant)	32.2 ft/s^2	9.81 m/s^2
D	Diameter	in	m
D_P	Piston diameter	in	m
D_R	Rod diameter	in	m
d	Distance	in, ft	m
h	Height	in, ft	m
L	Length	in, ft	m
t	Time	s	s
v	Velocity	ft/s, in/s, in/min	m/s, m/min
v_E	Extend velocity (cylinder)	ft/s, in/s, in/min	m/s, m/min
v_R	Retract velocity (cylinder)	ft/s, in/s, in/min	m/s, m/min
T	Temperature	°F, °R	°C, K
T	Torque	in · lbs, ft · lbs	N · m
T_T	Theoretical torque	in · lbs, ft · lbs	N · m
T_A	Actual torque	in · lbs, ft · lbs	N · m
Q	Flow rate	gpm, scfm	lpm, m^3/min
Q_A	Theoretical flow rate	gpm	lpm
Q_T	Actual flow rate	gpm	lpm
W	Work	in · lbs, ft · lbs	N · m
P	Power	hp	kW
HP_I	Input horsepower	hp	–
HP_H	Hydraulic horsepower	hp	–

SYMBOL	DEFINITION	U.S. CUSTOMARY UNIT	METRIC UNIT
HP_O	Output horsepower	hp	—
kW_I	Input power (in kilowatts)	—	kW
kW_H	Hydraulic power (in kilowatts)	—	kW
kW_O	Output power (in kilowatts)	—	kW
w	Weight	lbs	N
m	Mass	slugs	kg
ρ	Density	slug/ft^3	kg/m^3
V	Volume	in^3, ft^3	m^3
V_P	Pump displacement	in^3/rev	cm^3/rev
V_M	Motor displacement	in^3/rev	cm^3/rev
γ	Specific weight	lb/ft^3	N/m^3
S_g	Specific gravity (liquid)	Unitless	Unitless
G	Specific gravity (gas)	Unitless	Unitless
μ	Dynamic viscosity	lb · s/ft^2	N · s/m^2
y	Film thickness	in, ft	m
v	Kinematic viscosity	ft^2/s, SSU	m^2/s,cSt
B	Bulk modulus	psi	Pa
Re	Reynolds number	Unitless	Unitless
N	Rotational speed	rev/min (rpm)	rev/min (rpm)
η_V	Volumetric efficiency	Unitless	Unitless
η_M	Mechanical efficiency	Unitless	Unitless
η_O	Overall efficiency	Unitless	Unitless
C_V	Flow coefficient	$gpm/\sqrt{psi}$	$lpm/\sqrt{kPa}$
β_X	Beta ratio for particle size X	Unitless	Unitless

Appendix B

METRIC (SI) PREFIXES AND CONVERSION FACTORS

B.1 Metric (SI) Prefixes

PREFIX	MULTIPLICATION FACTOR	SYMBOL
pico	$0.000\ 000\ 000\ 001\ (10^{-12})$	p
nano	$0.000\ 000\ 001\ (10^{-9})$	n
micro	$0.000\ 001\ (10^{-6})$	μ
milli	$0.001\ (10^{-3})$	m
centi	$0.01\ (10^{-2})$	c
deci	$0.1\ (10^{-1})$	d
kilo	$1000\ (10^{3})$	k
mega	$1\ 000\ 000\ (10^{6})$	M
giga	$1\ 000\ 000\ 000\ (10^{9})$	G
tera	$1\ 000\ 000\ 000\ 000\ (10^{12})$	T

B.2 Conversion Factors

QUANTITY	CONVERSIONS
Force or weight	1 lb = 4.448 N = 0.004448 kN
	1 N = 0.2248 lb
Mass	1 slug = 14.594 kg
	1 kg = 0.06852 slug
Distance	1 in = 0.0254 m = 25.40 mm
	1 ft = 12 in = 0.3048 m
	1 m = 3.281 ft = 39.37 in
	1 μm = 0.000001 m = 0.00003937 in
Area	1 in^2 = 0.0006452 m^2 = 645.16 mm^2
	1 m^2 = 1550 in^2

(continued)

QUANTITY	CONVERSIONS
Pressure	1 psi = 6895 Pa = 6.895 kPa = 0.06895 bar
	$1\text{ Pa} = 1\dfrac{N}{m^2} = 0.0001450\text{ psi}$
	1 kPa = 1000 Pa = 0.1450 psi
	1 bar = 14.50 psi
Temperature	$°F = \dfrac{9}{5} \cdot °C + 32$
	$°C = \dfrac{5}{9} \cdot (°F - 32)$
	°R = °F + 460
	K = °C + 273
Volume	1 gal = 231 in³ = 3.7854 l
	1 in³ = 0.01639 l = 0.00001639 m³ = 16.39 cm³
	1 l = 0.001 m³ = 1000 cm³ = 61.02 in³ = 0.2642 gal
	1 cm³ = 0.000001 m³ = 0.06102 in³
Power	$1\text{ hp} = 550\dfrac{ft \cdot lbs}{s} = 745.7\text{ W} = 0.7457\text{ kW}$
	$1\text{ W} = 1\dfrac{N \cdot m}{s} = 0.001341\text{ hp}$
	1 kW = 1000 W = 1.341 hp
Vacuum	1 in Hg = −0.491 psi = 25.4 mm Hg
	1 mm Hg = − 0.1333 kPa = 0.03937 in Hg
Viscosity	$1\,St = 100\text{ cSt} = 1\dfrac{cm^2}{s} = 0.0001\dfrac{m^2}{s}$
	$\text{cSt} = 0.2253 \cdot \text{SUS} - \dfrac{194.4}{\text{SUS}}\qquad (32 \le \text{SUS} \le 100)$
	$\text{cSt} = 0.2193 \cdot \text{SUS} - \dfrac{134.6}{\text{SUS}}\qquad (100 \le \text{SUS} \le 240)$
	$\text{cSt} = \dfrac{\text{SUS}}{4.635}\qquad (\text{SUS} > 240)$

B.3 How to Use Conversion Factors

The correct use of conversion factors is of upmost importance when working in any technical field. Mistakes in unit conversions are a common source of errors in all areas of industry and can be very costly in terms of lost time and money. Mistakes can be avoided if the method outlined in this section is used.

What are we doing when we convert a quantity from one unit to another? We use the conversion from feet to inches to answer this question:

$$1 \text{ ft} = 12 \text{ in}$$

This conversion factor is simply an equation that states that 1 ft is equivalent to 12 in. Because this is an equation, we may multiply or divide both sides of the equation by the same factor without changing its validity. We divide both sides of the equation by 12 in to obtain:

$$\frac{1 \text{ ft}}{12 \text{ in}} = \frac{12 \text{ in}}{12 \text{ in}} = 1$$

This tells us that when we multiply a quantity by the factor $\frac{1 \text{ ft}}{12 \text{ in}}$, we are in fact multiplying by 1 because the numerator and denominator are equal. We now use this conversion factor to convert 100 inches to feet:

$$100 \text{ in} \cdot \left(\frac{1 \text{ ft}}{12 \text{ in}}\right) = 8.333 \text{ ft}$$

We multiplied by the factor $\frac{1 \text{ ft}}{12 \text{ in}}$, which was shown to be equal to 1. Inches cancel out and we are left with feet. Whenever we do a conversion, we are multiplying by 1 and changing the units. Including the units is important because it verifies that the conversion has been done correctly. If the units cancel out and we are left with the desired unit, we can be fairly certain that we have done the conversion correctly. This all but eliminates the errors resulting from dividing when we should have multiplied or vice versa. We now go the opposite direction and convert 10 feet to inches:

$$10 \text{ ft} \cdot \left(\frac{12 \text{ in}}{1 \text{ ft}}\right) = 120 \text{ in}$$

The conversion is simply inverted. We know this was necessary because we wanted feet to cancel out and leave us with inches.

Two conversions are done simultaneously in the following example. What if we want to convert 5 feet per second to inches per minute? This is done as follows:

$$5 \frac{\text{ft}}{\text{s}} \cdot \left(\frac{12 \text{ in}}{1 \text{ ft}}\right) \cdot \left(\frac{60 \text{ s}}{1 \text{ min}}\right) = 3600 \frac{\text{in}}{\text{min}}$$

The units cancel to leave us with the desired units.

Appendix C

EQUATIONS

EQUATION NUMBER	EQUATION	REQUIRED UNITS
	Chapter 2	
2-1	$p = \dfrac{F}{A}$	Any consistent units
2-2	$A = \dfrac{\pi \cdot D^2}{4}$	Any consistent units
2-3	$F_{OUT} = \dfrac{A_{OUT}}{A_{IN}} \cdot F_{IN}$	Any consistent units
2-4	$W = F \cdot d$	Any consistent units
2-5	$d_{IN} = \dfrac{A_{OUT}}{A_{IN}} \cdot d_{OUT}$	Any consistent units
2-6	$v_{IN} = \dfrac{A_{OUT}}{A_{IN}} \cdot v_{OUT}$	Any consistent units
2-7	$w = m \cdot g$	Any consistent units
2-8	$\rho = \dfrac{m}{V}$	Any consistent units
2-9	$\gamma = \dfrac{w}{V}$	Any consistent units
2-10	$sg_X = \dfrac{\gamma_X}{\gamma_{WATER}}$	Any consistent units
2-11	$\mu = \dfrac{F \cdot y}{v \cdot A}$	Any consistent units
2-12	$v = \dfrac{\mu}{\rho}$	Any consistent units
2-13	$B = \dfrac{-\Delta p}{\Delta V / V}$	Any consistent units
2-14	$Q = v \cdot A$	Any consistent units
2-15	$v_1 \cdot A_1 = v_2 \cdot A_2$	Any consistent units
2-16	$h_1 + \dfrac{p_1}{\gamma} + \dfrac{v_1^2}{2 \cdot g} = h_2 + \dfrac{p_2}{\gamma} + \dfrac{v_2^2}{2 \cdot g}$	Any consistent units

EQUATION NUMBER	EQUATION	REQUIRED UNITS
	Chapter 2 (continued)	
2-17	$v_2 = \sqrt{\dfrac{2 \cdot g \cdot (p_1 - p_2)}{\gamma \cdot \left[1 - \left(\dfrac{A_2}{A_1}\right)\right]}}$	Any consistent units
2-18	$v_2 = \sqrt{2 \cdot g \cdot h}$	Any consistent units
2-19	$Re = \dfrac{v \cdot D}{v}$	Any consistent units
2-20	$p_H = \gamma \cdot h$	Any consistent units
2-21	$p = \dfrac{W}{t} = \dfrac{F \cdot d}{t}$	Any consistent units
2-22	$P = F \cdot v$	Any consistent units
2-23	$HP = \dfrac{F \cdot v}{550}$	F in lb, v in ft/s, HP in hp
2-23M	$kW = \dfrac{F \cdot v}{1000}$	F in N, v in m/s, kW in kW
2-24	$P_H = p \cdot Q$	Any consistent units
2-25	$HP_H = \dfrac{p \cdot Q}{1714}$	p in psi, Q in gpm, HP_H in hp
2-25M	$kW_H = \dfrac{p \cdot Q}{60,000}$	p in kPa, Q in lpm, kW_H in kW
	Chapter 3	
3-1	$Q_T = V_p \cdot N$	Any consistent units
3-2	$Q_T = \dfrac{V_p \cdot N}{231}$	V_p in in³/rev, N in rpm, Q_T in gpm
3-2M	$Q_T = \dfrac{V_p \cdot N}{1000}$	V_p in cm³/rev, N in rpm, Q_T in lpm
3-3	$T = F \cdot d$	Any consistent units
3-4	$T_T = \dfrac{p \cdot V_p}{2 \cdot \pi}$	Any consistent units

EQUATION NUMBER	EQUATION	REQUIRED UNITS
	Chapter 3 (continued)	
3-5	$HP_1 = \dfrac{T \cdot N}{63{,}025}$	T in in · lbs, N in rpms, HP_1 in hp
3-5M	$kW_1 = \dfrac{T \cdot N}{9550}$	T in N · m, N in rpms, kW_1 in kW
3-6	$HP_H = \dfrac{p \cdot Q}{1714}$	Q in gpm, p in psi, HP_H in hp
3-6M	$kW_H = \dfrac{p \cdot Q}{60{,}000}$	Q in lpm, p in kPa, kW_H in kW
3-7	$\eta_V = \dfrac{Q_A}{Q_T}$	Any consistent units
3-8	$Q_A = \eta_V \cdot \dfrac{V_P \cdot N}{231}$	V_P in in³/rev, N in rpm, Q_A in gpm
3-8M	$Q_A = \eta_V \cdot \dfrac{V_P \cdot N}{1000}$	V_P in cm³/rev, N in rpm, Q_A in lpm
3-9	$\eta_M = \dfrac{T_T}{T_A}$	Any consistent units
3-10	$\eta_O = \dfrac{HP_H}{HP_1}$	HP_H in hp, HP_1 in hp, η_O is unitless
3-10M	$\eta_O = \dfrac{kW_H}{kW_1}$	kW_H in kW, kW_1 in kW, η_O is unitless
3-11	$\eta_O = \eta_M \cdot \eta_v$	All quantities are unitless
	Chapter 4	
4-1	$F_E = p \cdot A_P$	Any consistent units
4-2	$F_R = p \cdot (A_P - A_R)$	Any consistent units
4-3	$v = \dfrac{Q}{A}$	Any consistent units

EQUATION NUMBER	EQUATION	REQUIRED UNITS
	Chapter 4 (continued)	
4-4	$v = \dfrac{231 \cdot Q}{A}$	Q in gpm, A in in², v in in/min
4-5	$v_E = \dfrac{231 \cdot Q}{A_p}$	Q in gpm, A_p in in², v_E in in/min
4-5M	$v_E = \dfrac{Q}{1000 \cdot A_p}$	Q in lpm, A_p in m², v_E in m/min
4-6	$v_R = \dfrac{231 \cdot Q}{A_p - A_R}$	Q in gpm, A_X in in², v_E in in/min
4-6M	$v_R = \dfrac{Q}{1000 \cdot (A_p - A_R)}$	Q in lpm, A_X in m², v_E in m/min
4-7	$HP_O = \dfrac{F \cdot v}{550}$	F in lbs, v in ft/s, HP_O in hp
4-7M	$kW_O = \dfrac{F \cdot v}{1000}$	F in N, v in m/s, kW_O in kW
4-8	$Q_{RET,E} = \dfrac{Q_{PUMP} \cdot (A_p - A_R)}{A_p}$	Any consistent units
4-9	$Q_{RET,R} = \dfrac{Q_{PUMP} \cdot A_p}{A_p - A_R}$	Any consistent units
4-10	$F_{CYL} = F_{LOAD} \cdot \cos a$	Any consistent units
4-11	$F_{BEAR} = F_{LOAD} \cdot \sin a$	Any consistent units
4-12	$F_{LEV} = F_{CYL} \cdot \cos a$	Any consistent units
4-13	$T = F_{LEV} \cdot L$	Any consistent units
	Chapter 5	
5-1	$T_T = \dfrac{V_M \cdot \Delta p}{2 \cdot \pi}$	Any consistent units
5-2	$Q_T = V_M \cdot N$	Any consistent units
5-3	$Q_T = \dfrac{V_M \cdot N}{231}$	Q_T in gpm, N in rpm, V_M in in³/rev

EQUATION NUMBER	EQUATION	REQUIRED UNITS
	Chapter 5 (continued)	
5-3M	$Q_T = \dfrac{V_M \cdot N}{1000}$	Q_T in lpm, N in rpm, V_M in cm³/rev
5-4	$HP_O = \dfrac{T \cdot N}{63,025}$	T in in · lbs, N in rpms, HP_O in hp
5-4M	$kW_O = \dfrac{T \cdot N}{9550}$	T in N · m, N in rpms, kW_O in kW
5-5	$\eta_M = \dfrac{T_A}{T_T}$	Any consistent units
5-6	$T_A = \dfrac{V_M \cdot \Delta p \cdot \eta_M}{2 \cdot \pi}$	Any consistent units
5-7	$\eta_V = \dfrac{Q_T}{Q_A}$	Any consistent units
5-8	$Q_A = \dfrac{V_M \cdot N}{231 \cdot \eta_V}$	Q_A in gpm, N in rpm, V_M in in³/rev
5-8M	$Q_A = \dfrac{V_M \cdot N}{1000 \cdot \eta_V}$	Q_A in lpm, N in rpm, V_M in cm³/rev
5-9	$\eta_O = \dfrac{HP_O}{HP_H}$	HP_H in hp, HP_O in hp, η_O is unitless
5-9M	$\eta_O = \dfrac{kW_O}{kW_H}$	kW_H in kW, kW_O in kW, η_O is unitless
5-10	$\eta_O = \eta_M \cdot \eta_V$	All quantities are unitless

EQUATION NUMBER	EQUATION	REQUIRED UNITS
	Chapter 6	
6-3	$v_{REGEN} = \dfrac{231 \cdot Q}{A_R}$	Q in gpm, A_R in in², v_{REGEN} in in/min
6-3M	$v_{REGEN} = \dfrac{Q}{1000 \cdot A_R}$	Q in lpm, A_R in m², v_{REGEN} in m/min
6-5	$F_{REGEN} = p \cdot A_R$	Any consistent units
	Chapter 8	
8-1	$Q = C_V \cdot \sqrt{\dfrac{\Delta p}{S_g}}$	Any consistent units
	Chapter 9	
9-1	$p_{INT} = p_{PUMP} \cdot \dfrac{A_{PISTON}}{A_{ROD}}$	Any consistent units
9-2	$Q_{INT} = Q_{PUMP} \cdot \dfrac{A_{ROD}}{A_{PISTON}}$	Any consistent units
9-3	$p_{INT} = p_{PUMP} \cdot IR$	Any consistent units
9-4	$Q_{INT} = \dfrac{Q_{PUMP}}{IR}$	Any consistent units
9-5	$\beta_X = \dfrac{N_U}{N_D}$	Any consistent units
9-6	$\eta_X = 1 - \dfrac{1}{\beta_X}$	Any consistent units
9-7	$ID = OD - 2 \cdot t$	Any consistent units
	Chapter 10	
10-1	$p_{ABS} = p_{GAUGE} + p_{ATM}$	Any consistent units
10-3	$p_1 \cdot V_1 = p_2 \cdot V_2$	Any consistent units (p in absolute)
10-4	$\dfrac{p_1}{T_1} = \dfrac{p_2}{T_2}$	Any consistent units (p, T in absolute)

EQUATION NUMBER	EQUATION	REQUIRED UNITS
10-5	$\dfrac{V_1}{T_1} = \dfrac{V_2}{T_2}$	Any consistent units (*T* in absolute)
10-6	$\dfrac{p_1 \cdot V_1}{T_1} = \dfrac{p_2 \cdot V_2}{T_2}$	Any consistent units (*p, T* in absolute)
10-7	$p_1 \cdot V_1^{14} = p_2 \cdot V_2^{14}$	Any consistent units (*p* in absolute)
10-8	$\dfrac{p_1}{T_1^{3.5}} = \dfrac{p_2}{T_2^{3.5}}$	Any consistent units (*p, T* in absolute)
Chapter 11		
11-1	$HP_1 = \dfrac{p_{IN} \cdot Q}{65.5 \cdot \eta_O} \cdot \left[\left(\dfrac{p_{OUT}}{p_{IN}} \right)^{0.286} - 1 \right]$	*Q* in scfm, *p* in psia, HP_I in hp
11-1M	$kW_1 = \dfrac{p_{IN} \cdot Q}{17.14 \cdot \eta_O} \cdot \left[\left(\dfrac{p_{OUT}}{p_{IN}} \right)^{0.286} - 1 \right]$	*Q* in m³/min, *p* in kPa abs, kW_I in kW
11-2	$V_R = \dfrac{14.7 \cdot t \cdot (Q_{DEM} - Q_{DEL})}{P_{MAX} - P_{MIN}}$	*t* in min, *Q* in scfm, *p* in psi, V_R in ft³
11-2M	$V_R = \dfrac{101 \cdot t \cdot (Q_{DEM} - Q_{DEL})}{P_{MAX} - P_{MIN}}$	*t* in min, *Q* in m³/min, *p* in kPa, V_R in m³
Chapter 12		
12-1	$F_E = p \cdot A_P$	Any consistent units
12-2	$F_R = p \cdot (A_P - A_R)$	Any consistent units
12-3	$V_E = A_P \cdot S$	Any consistent units
12-4	$V_R = (A_P - A_R) \cdot S$	Any consistent units
12-5	$C_V = \dfrac{Q}{22.48} \sqrt{\dfrac{T}{\Delta p \cdot p_{OUT}}}$	*Q* in scfm, *T* in °R, Δp in psi, p_{OUT} in psia
12-5M	$C_V = \dfrac{Q}{68.7} \sqrt{\dfrac{T}{\Delta p \cdot p_{OUT}}}$	*Q* in l/min, *T* in K, Δp in kPa, p_{OUT} in kPa abs

Appendix D

Graphic Symbols

Pump symbols

 Fixed displacement pump

 Variable displacement pump

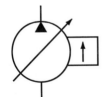 Pressure compensated pump

 Bi-directional pump

Air compressor

Cylinder symbols

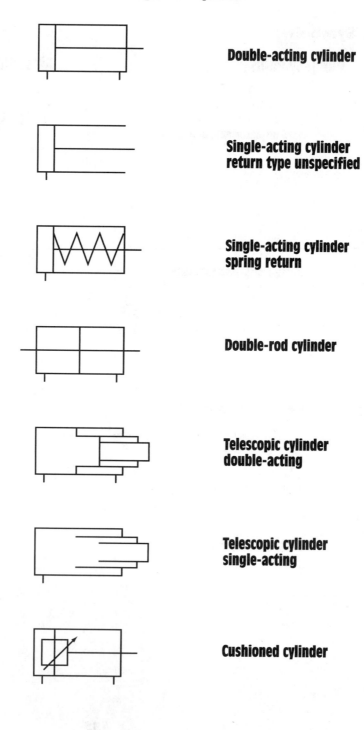

Double-acting cylinder

Single-acting cylinder
return type unspecified

Single-acting cylinder
spring return

Double-rod cylinder

Telescopic cylinder
double-acting

Telescopic cylinder
single-acting

Cushioned cylinder

Hydraulic motor symbols

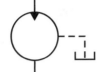 **Fixed displacement
uni-directional**

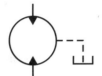 **Fixed displacement
bi-directional**

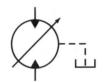 **Variable displacement
bi-directional**

 **Limited rotation
bi-directional**

Pneumatic motor symbols

 **Fixed displacement
uni-directional**

 **Fixed displacement
bi-directional**

Directional control valves

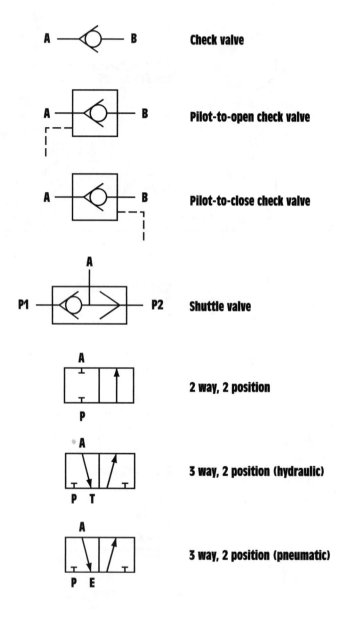

Check valve

Pilot-to-open check valve

Pilot-to-close check valve

Shuttle valve

2 way, 2 position

3 way, 2 position (hydraulic)

3 way, 2 position (pneumatic)

Directional control valves
Hydraulic

 4 way, 2 position

 4 way, 3 position closed neutral

 4 way, 3 position tandem neutral

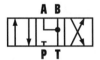 4 way, 3 position float neutral

 4 way, 3 position open neutral

 4 way, 3 position regenerative neutral

Directional control valves
Pneumatic

 4 way, 2 position

 4 way, 3 position closed neutral

 4 way, 3 position float neutral

Electrohydraulic

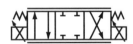 a. Proportional valve

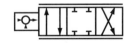 b. Servo valve

Directional control valve actuation types

Manual lever

Pushbutton

Foot pedal

Mechanical (cam)

Pilot-operated

Solenoid

Pilot-operated solenoid

DCV positioning methods

Spring return

Detent

Friction

Hydraulic pressure control valves

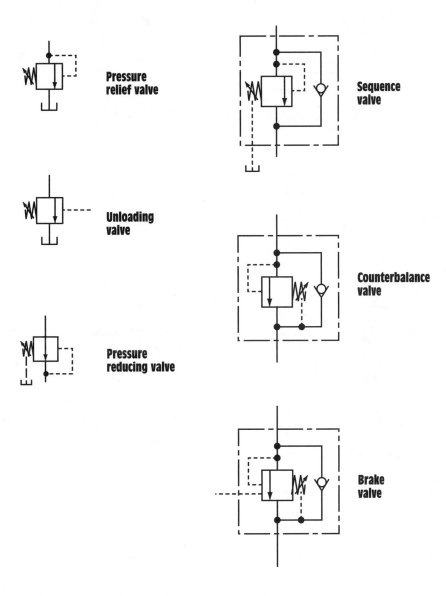

Pressure
relief valve

Unloading
valve

Pressure
reducing valve

Sequence
valve

Counterbalance
valve

Brake
valve

Flow control valves

 A ⊘ B   **Needle valve**

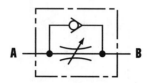 A — B   **Flow control**

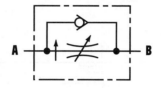 A — B   **Flow control pressure compensated**

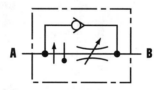 A — B   **Flow control pressure and temperature compensated**

    Flow divider

Miscellaneous hydraulic components

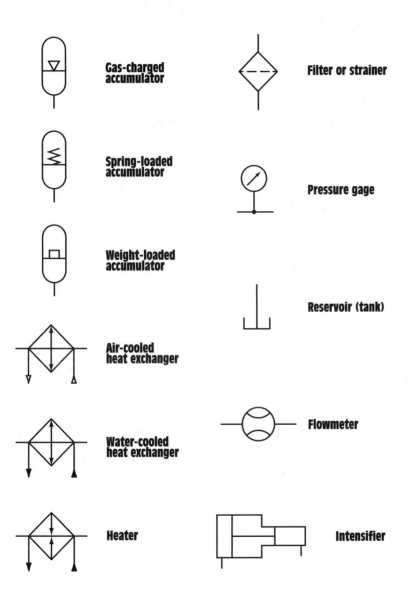

Gas-charged accumulator

Spring-loaded accumulator

Weight-loaded accumulator

Air-cooled heat exchanger

Water-cooled heat exchanger

Heater

Filter or strainer

Pressure gage

Reservoir (tank)

Flowmeter

Intensifier

Miscellaneous pneumatic components

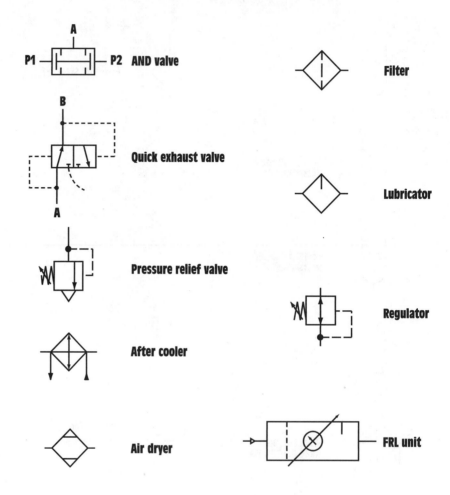

AND valve

Quick exhaust valve

Pressure relief valve

After cooler

Air dryer

Filter

Lubricator

Regulator

FRL unit

Glossary

A

Accumulator a device that stores energy in a hydraulic system by storing hydraulic fluid under pressure.

Adiabatic a process in which no heat is transferred to the surrounding atmosphere.

Aeration atmospheric air leaking into a hydraulic system.

Atmospheric pressure the pressure exerted by the weight of the atmosphere above the point of measurement (14.7 *psia* at sea level).

B

Beta ratio the ratio of the particles per milliliter of a given size or larger before a filter to the particles of that size that are present after passing through the filter.

Boyle's law states that the absolute pressure of a confined gas is inversely proportional to its volume, provided its temperature remains constant.

Brake valve a type of hydraulic pressure control valve that prevents hydraulic motors from accelerating uncontrollably due to an overrunning load.

Bulk modulus a measure of a fluid's incompressibility or stiffness.

C

Cavitation the formation of oil vapor and air bubbles on the inlet side of the pump due to very low pressure (high vacuum).

Charles' law states that the volume of a confined gas is proportional to its temperature, provided its pressure remains constant.

Check valve a valve that allows flow in one direction, but prevents flow in the opposite direction.

Closed loop control a method of electronic control in which sensors are used to provide feedback of actual system behavior for error correction.

Compressor an air pump that compresses air into a receiver tank.

Counterbalance valve a type of hydraulic pressure control valve that prevents hydraulic cylinders from accelerating uncontrollably when lowering a suspended weight.

Cracking pressure the pressure at which a pressure relief valve begins to open.

Cylinder a device used to convert fluid power into mechanical power in the form of linear motion.

D

Density the mass of a substance per unit volume.

Desiccant a substance, such as a silica gel, that absorbs water.

Detent a mechanism that holds a valve in its current position.

Dew point the temperature at which the air becomes saturated with water and condensation results.

Directional control valve a valve used to control the direction of flow in a hydraulic circuit.

Displacement for pumps, the volume output per revolution of the pump shaft—for motors, the volume required to turn a motor one revolution.

Double-acting cylinder a cylinder that can act under pressure in both directions (extend and retract) to move a load.

Double-rod cylinder a cylinder with a rod on both sides of the piston.

Drain line a line used to remove the leakage from a valve or other hydraulic device.

F

Filter a device used to remove contamination from a fluid.

Flow coefficient (C_v) an experimentally determined value that relates the pressure drop across a valve to the flow rate through it.

Flow control valve a valve that controls the flow rate in a hydraulic circuit. It is used to control actuator speed.

Flow divider a device used to divide the flow from a pump into fixed proportions, usually two equal streams, regardless of the pressure on each stream.

Flow meter a device used to measure flow rate.

Fluid power the use of a fluid (liquid or gas) to transmit power from one location to another.

Full flow pressure the pressure at which a pressure relief valve is fully open and returning all of the pump flow to the tank.

G

Gay-Lussac's law states that the absolute pressure of a confined gas is proportional to its temperature, provided its volume remains constant.

H

Heat exchanger a device used to add or remove heat from a system.

Hydraulics the use of a liquid flowing under pressure to transmit power from one location to another.

I

Intensifier a device used to generate pressures greater than those achievable with a standard hydraulic pump alone.

ISO cleanliness code a standard that determines how a contaminant sensitivity for a hydraulic component should be specified.

Isothermal a process in which the temperature remains constant (any added heat is transferred to the surrounding atmosphere).

L

Laminar flow smooth and layered fluid flow.

Lubricator a device used to spray an oil mist into the air stream in a pneumatic system.

M

Mechanical efficiency for pumps, the ratio of the theoretical drive torque to the actual drive torque—for motors, the ratio of the actual torque output to the theoretical torque output.

Micron (μm) one millionth of a meter.

Motor a device that converts fluid power into mechanical power in the form of rotational motion.

N

NFPA (National Fluid Power Organization) an organization that sets design and performance standards and promotes the fluid power industry.

O

Open loop control a method of electronic control that does not use feedback.

Overall efficiency the power output of a device divided by the power input required to drive that device.

P

Pascal's law states that the pressure exerted on a confined fluid is transmitted undiminished in all directions and acts perpendicular to the containing surfaces.

Pilot line a line used for control purposes in a hydraulic or pneumatic circuit.

PLC (programmable logic controller) an industrial computer that has sophisticated input/output capabilities.

Pneumatics the use of a gas flowing under pressure to transmit power from one location to another.

Power a measure of the amount of energy expended per unit time.

Pressure compensated pump a pump equipped with a device that limits the maximum pressure in a hydraulic circuit by reducing the displacement of the pump when a preset firing pressure is reached.

Pressure override the difference between the full flow pressure and cracking pressure of a pressure relief valve.

Pressure reducing valve a type of hydraulic pressure control valve that controls the maximum pressure in a branch of a circuit.

Pressure regulator a type of pneumatic pressure control valve that controls the maximum pressure in a branch of a circuit.

Pressure relief valve a type of pressure control valve that limits the maximum pressure in a hydraulic or pneumatic circuit.

Pressure transducer a device that converts fluid pressure into an electrical signal.

Prime mover a device, such as an electric motor or gas engine, used to drive a pump or compressor.

Proportional valve an electronically controlled valve that is able to precisely position the valve spool between the open and closed positions. It allows the flow rate through the valve and the pressure drop across it to be precisely controlled with an electronic signal.

Pump a device used to create flow in a hydraulic system.

R

Ram a large capacity single-acting cylinder.

Receiver tank a device that holds the compressed air in a pneumatic system.

Relative humidity the percentage of water in the air as compared to the maximum amount it could hold at a specific temperature.

Reservoir the tank that holds the fluid in a hydraulic system.

S

scfm standard cubic feet per minute.

Sequence valve a type of hydraulic pressure control valve that is used to force two actuators to be operated in sequence.

Servo valve an electronically controlled valve that is able to precisely position the valve spool between the open and closed positions. It allows the flow rate through the valve and the pressure drop across it to be precisely controlled with an electronic signal.

Shuttle valve a valve that has two inlets and one outlet. The outlet receives flow from whichever inlet is at a higher pressure.

Single-acting cylinder a cylinder that acts under pressure in one direction only and returns automatically when the pressure is released.

Solenoid a switching device that uses the magnetic field generated by an electrical current for actuation.

Specific gravity for liquids, the ratio of the specific weight of a liquid to the specific weight of water—for gases, the ratio of the specific weight of a gas to the specific weight of air.

Specific weight the weight per unit volume of a substance.

Static head pressure pressure due to the weight of a fluid above the point of measurement.

Strainer a coarse filter usually made of a wire mesh.

T

Tandem cylinder a cylinder that has two pistons attached to a single piston rod. It provides increased output force by increasing the surface area over which the pressure is applied.

Telescopic cylinder a cylinder that consists of multiple stages that actuate in sequence. It is used when a long stroke length and a short retracted length are required.

Turbulent flow rough and uneven fluid flow.

U

Unloading valve a type of hydraulic pressure control valve that opens when a preset pressure is reached in an external pilot line.

V

Vacuum containing a gas at a pressure less than atmospheric pressure.

Vacuum pump a pump used to remove air from a container.

Viscosity a measure of a fluid's thickness or resistance to flow.

Viscosity index a measure of how much a fluid's viscosity changes with temperature.

Volumetric efficiency for pumps, the ratio of the actual flow rate produced to the theoretical flow rate when driven at a particular speed—for motors, the ratio of the theoretical flow to the actual flow required to achieve a particular speed.

W

Work a measure of expended energy. For a physical system it is the force multiplied by the distance moved due to that force.

Index

Absolute pressure, 378
Absolute temperature, 378
Absolute zero, 378
Access plate, 327
Accumulators, 310–320
   bladder, 312
   diaphragm, 312
   discharge valve, 320
   gas-charged, 310–312
   spring-loaded, 313–314
   supplementing pump flow, 317–318
   used with a press, 318, 320
   weight-loaded, 315, 317
Accuracy, 462
Actuator, 51, 104
Adiabatic process, 387
Aeration, 88
Aftercooler, 454
Air, 376
   See also Pneumatics principles
Air preparation, 448–455
   air distribution, 456
   combined conditioning units, 453
   cooling, 454–455
   filters, 450
   lubricators, 450–453
   pressure regulators, 449–450
   pressure relief valves, 448
   refrigeration-type air dryer, 455
   regenerative-type air dryer, 455
   water removal, 454–455
Air-cooled heat exchanger, 329
Air-line lubricator, 451
Air-powered gear motor, 425
American National Standards Institute
   (ANSI), website, 3
American Society for Testing and Materials
   (ASTM), website, 3
ANSI. See American National Standards
   Institute
Assembly guidelines, for hose, 356
ASTM. See American Society for Testing and
   Materials
Atmospheric dew point, 454
Atmospheric pressure, 378, 390–393
   See also Pneumatics principles
Axial piston pump, 73, 76

Baffle, 450
Baffle plate, 327
Balanced vane motor, 154–155

Balanced vane pumps, 81, 83, 85
Ball valve, 197
Bar, 10
Barrel, 104
Bearings. See Seals and bearings
Bernoulli's equation, 34–40
   See also Torricelli's theorem
Beta ratio, 337–339
Bi-directional limited rotation motor, 174
Bladder accumulators, 312
Blind end, 104
Bourdon tube gauge, 342
Boyle's law, 379–381, 386–388
Brake valves, 268–269
Bulk modulus, 25–26

Cartridge-type valves, 236, 238
Cavitation, 88
Centistokes (cSt), 24
Charles's law, 383–385
Check valves, 188–190
CIR, 177
Circuits
   directional control valves (DCVs),
      224–234
   flow control, 287–298
Clamp and bend circuit, 262–263, 264
Cleanliness code (ISO), 96–97
Closed system, 16
Closed-loop control system, 460–462
Combined units, 453
Component manufacturers, 4
Compound seal, 365
Compound or two-stage pressure relief
   valve, 248
Compressor, 395
   See also Pneumatic power supply
Conduits and fittings, 347–361
   hose, 355–356
   pipe, 252, 349–350
   tubing, 352–353
Consistent units, 49
Constant horsepower, 174
Constant torque, 174
Contamination, 333
   See also Filters
Continuity equation, 31–33
Control, 51
Control valves, pneumatic, 394–395
Cooling, 454–455
Corrosion, 370

Counterbalance circuit, 224–225
Counterbalance valves, 265–267
Counterflow, 329
Crimping, 355
Critical pressure ratio, 444
Cross-over pressure relief, 252
cSt. See Centistokes
Cushioned cylinders, 298–299
Cut-out pressure, 400
Cylinder lifting a large weight, 294
Cylinder rotating a lever arm, 355
Cylinder synchronization, 301, 304
Cylinders
   with bleed-off flow control, 295–296
   cushioned, 298–299
   See also Hydraulic cylinders; Pneumatic
      cylinders

DCVs. See Hydraulic; Pneumatic directional
   control valves
Deflector plate, 450
Delivery, 60
Density, 21
Detent, 202–203
Dew point, 454
Diaphragm accumulators, 312
Differential flow, of cylinders, 121–124
Differential pressure flowmeter, 343
Diffuser, 401
Directional control valves (DCVs). See
   Hydraulic directional control
   valves; Pneumatic directional
   control valves
Discharge valve, 320
Displacement, 60, 90–91, 154
Distributors, 4
Diverter valve, 203–204
Double-acting hydraulic cylinders, 104
Double-acting pneumatic cylinder, 417–422
Double-rod cylinder, 127
Draining a vessel, 197, 203–204
Drive speed, 90–91
Dynamic compressors (nonpositive
   displacement pumps), 401

Eccentric piston block, 76
Efficiency
   hydraulic pump, 67–73, 92–95
   motor, 167–172
Elastomers, 366
Elastoplastics, 366

Electrical power transmission, 2
Electromechanical devices, 343
Electronic control of fluid power, 459–472
    closed-loop control system, 460–462
    open-loop control system, 460–461
    proportional valves, 467–471
    pump controls, 471–472
    servo valves, 469–471
    solenoid valves, 462–463, 465
Emulsions, 369–370
    oil in water, 370
    water-in-oil (invert), 370
End caps, 104
Equations
    ancillary hydraulic components, 371
    directional control valves (DCVs), 240
    flow control, 306
    hydraulic cylinders, 146–147
    hydraulic motors, 182–183
    hydraulic pumps, 98–99
    hydraulics principles, 52–53
    pneumatic, 396, 413, 457
Escapement, 426
External gear pump, 79
Extrusion, 366

Fastness, 48
Filler/breather, 327
Filling a vessel, 193, 197, 203–204
Filter pressure drop indicators, 340
Filters, 333–342, 450
Filtration, and pumps, 95–97
Fire point, 371
Fire-resistant fluid, 369
Firing pressure, 87
Fittings. See Conduits and fittings
Fixed displacement, 63, 172
Fixed displacement, bi-directional
        motor, 172
Fixed displacement, uni-directional
        motor, 172
Flared connections, 353
Flareless connections, 353
Flash point, 371
Filtration, motors, 182
Flow, 58
    continuity of, 31–33
    laminar versus turbulent, 42–45
Flow control. See Hydraulic flow control;
        Pneumatic flow control valves
Flow dividers, 300–304
Flow rate, 90–91
    versus flow velocity, 26–31
Flow velocity, versus flow rate, 26–31
Flowing fluid, 47

Flowmeters, 343–347
    differential pressure, 343
    headland, 344–346
    orifice, 343
    rotameter (variable area flowmeter), 344
    temperature gauges, 347
    turbine, 346
Fluid power, 2–5
    applications, 3–4
    defined, 2
    hydraulics versus pneumatics, 2–3
    industry, 4
    standards, 3
    units, 5
    websites, 3
Fluid transfer, 3–4
Fluids
    hydraulic, 369–371
    motor, 182
    pumps, 95
Fluorocarbon (viton), 366–367
Foot pedal actuation, 219
Force
    of cylinders, 107–112

Gas flow, 389–390
Gas laws. See Pneumatics principles
Gas-charged accumulators, 310–312
Gay-Lussac's law, 381–383
Gear motor, 154
Gear pumps, 79–81
General gas law, 385–386, 389
    See also Pneumatics principles
Graphic symbols
    cylinders, 130, 132
    motors, 172, 174
    pilot-operated check valves, 189–190
    pneumatic, 404
    pumps, 88–90
Gravity, specific, 21
Gravity return-type single-acting cylinder, 125
Gripper, 426
Gauges
    pressure, 342–343
    temperature, 347
Guide bushings, 362

Headland flowmeter, 344–346
Heat exchangers, 329, 331–332
High pressure filter, 335
High-pressure standby mode, 273
Horsepower (HP), 48
    constant, 174
Hose, 355–356
    assembly guidelines, 356

HP. See Horsepower
Hydraulic cylinders, 103–151
    applications, 132–142
    differential flow, 121–124
    double-acting, 104
    double-rod, 127
    equations, 146–147
    force, 107–112
    graphic symbols, 130, 132
    gravity return-type single-acting, 125
    hydraulic rams, 129–130
    mounting styles, 143
    parts of, 104
    piston area, 106
    power, 118–121
    pressure spikes, 146
    safety factor, 146
    single-acting, 124–125
    size, 142
    speed, 112–118
    spring return-type single-acting, 126
    stroke lengths, 142
    tandem, 127, 129
    telescopic, 126–127
    tie rod construction, 104
Hydraulic directional control valves (DCVs),
        187–242
    check valves, 188–190
    circuits, 224–234
        counterbalance, 224–225
        jump, 230
        junction, 230
        multispool directional control
            valve, 230
        parallel, 228
        pressure line, 230
        regenerative, 228
        shuttle valve, 225, 228
        tank line, 230
    equations, 240
    four-way, 204–216
        closed neutral, 209–210
        controlling a bi-directional motor, 208
        controlling a double-acting
            cylinder, 207
        float neutral, 210
        open neutral, 211–212
        regenerative neutral, 212–215
        tandem neutral, 210
    mounting, 234–238
    shuttle valves, 190–191
    three-way, 198–204
        normally closed, 198, 200
        normally open, 200
        pressure relief valve, 200, 202

two-way, 193–198
  ball valve, 197
  filling and draining a vessel, 193, 197
  normally closed, 193
  normally open, 193
valve actuation, 218–224
valve specifications, 238
pressure drop-versus-flow graph, 238
Hydraulic flow control, 279–308
  circuits, 287–298
    extending, 287
    meter-in control of both strokes, 289
    meter-in flow, extend stroke,
      287–288
    meter-out control of both strokes,
      290–292
    meter-out, extend stroke, 288–289
  cushioned circuits, 298–299
  cylinders
    bleed off, 295–296
    lifting a large weight, 294
    meter-out control of the retract
      stroke, 294, 296
  equations, 306
  flow coefficient, 284–287
  flow dividers, 300–304
    balanced spool and rotary, 300
    cylinder synchronization, 301, 304
  metering valve, 281
  needle valve, 280–281
  pressure-compensated valve, 283–284
specifications, 304–305
Hydraulic fluids, 369–371, 369–391
  types of, 369
Hydraulic motors
  applications, 174–176
  balanced vane, 154–155
  equations, 182–183
  filtration, 182
  fluids, 182
  gear, 154
  performance, 177–181
  piston, 155
  rack and pinion rotary actuator, 157
  rotary actuator, 155–156
  size and physical dimensions, 177
  speed and pressure ratings, 181–182
  torque, 161–163
  two-vane rotary actuator, 156–157
  vane-type rotary actuator, 156–157
Hydraulic press circuit, 270–272
Hydraulic pumps, 57–101
  cavitation and aeration, 88
  displacement, flow rate, and drive speed,
    90–91
  efficiency, 67–73, 92–95

equations, 98–99
filtration, 95–97
fluids, 95
gear pumps, 79–81
graphic symbols, 88–90
inlet pressure, 98
noise level, 97
physical dimensions, 97
piston pumps, 73–79
pressure, 91
pressure-compensated pumps, 87
pump drive torque and power, 63–67
pump flow and pressure, 58–63
shaft rotation, 98
vane pumps, 81–86
Hydraulic rams, 129–130
Hydraulic systems, 51
Hydraulics, versus pneumatics, 2–3, 395
Hydraulics principles, 7–56
  equations table, 52–53
  hydraulic fluids properties, 19–26
    bulk modulus, 25–26
    density, specific weight, and specific
      gravity, 21
    mass versus weight, 19–20
    viscosity, 22–25
  hydraulic systems, 51
  liquid flow, 26–45
    Bernoulli's equation, 34–40
    continuity equation, 31–33
    flow rate versus flow velocity, 26–31
    laminar versus turbulent flow, 42–45
    Torricelli's theorem, 40–42
  multiplication of force, 15–19
  Pascal's Law, 8–13
  power, 48–51
  pressure losses, 47
  static head pressure, 45–47
  transmission of force, 14
Hydrodynamic system, 8
Hydrolysis, 367
Hydrostatic transmission system, 174
Hydrostatics, 8
Hydraulic motors
  rack and pinion rotary actuator, 157
  vane-type rotary actuator, 156–157

Immersion heaters, 331
Industrial hydraulic, 4
Industrial pneumatic, 4
Inline check valve, 188
Intensifiers, 321–325
Internal gear pump, 79–80
International Standards Organization (ISO)
  cleanliness code, 337
  website, 3

ISO. See International Standards Organization
Isothermal process, 387

Jump, 230
Junction, 230

Laminar flow, versus turbulent flow, 42–45
Linear actuator (cylinder), 51
Liquid flow. See Hydraulic principles
Load, 51
Load sensing, 273–274
Loaded u-ring, 364
Low pressure filter, 334
Low-speed, high-torque (LSHT), 155, 174,
  181
LSHT. See Low-speed, high-torque
Lubricators, 450–453

Manifold, 234–236
Manual lever, 202, 219
Manual shut-off valve, 197
Mass, versus weight, 19–20
Mechanical power transmissions, 2
Mechanically (cam) actuated, 219
Mercury barometer, 390
Mesh, 79
Metering notches, 284
Metering valves, 281
Metric (SI) units, 5, 9
Micromist lubricator, 451–452
Millimeters of mercury (mm Hg0), 392
Mobile hydraulic, 4
Motion, 8
Motors. See Hydraulic motors; Pneumatic
  motors
Mounting
  directional control valves (DCVs),
    234–238
  pressure control valve, 275–276
Multiplication of force, 15–19
Multispool directional control valve, 230

National Fluid Power Association (NFPA),
  website, 3
Needle valve, 280–281
NFPA. See National Fluid Power Association
Nitrile (NBR or buna-N), 366
Nonpositive displacement pumps (dynamic
  compressors),
    58, 401
NPTF threads, 350
Nylon, 368

O-rings, 350, 353, 362
Occupational Safety and Health
  Administration (OSHA), 97

OEMs. *See* Original Equipment Manufacturers
Offline filtration, 335–336
Open-loop control system, 460–461
Orifice flowmeter, 343
Original Equipment Manufacturers (OEMs), 4
OSHA. *See* Occupational Safety and Health Administration
Output, 51, 60
  pneumatic, 394
Oxidation, 370–371

Pa. *See* Pascal
Parallel circuit, 228, 273
Particulate contamination, 333
Pascal (Pa), 10
Pascal's Law, 8–13, 14, 51, 376, 418
Perfect gas laws, 379
Petroleum-based hydraulic fluid, 369
Phosphate ester, 366, 370
Pilot lines, 189–190
Pilot-operated check valves, 189–190
Pilot-operated solenoid valve, 219, 222, 224, 465
Pilot-operated valve, 219
Pipe, 349–350, 352
  NPTF threads, 350
  schedule 40 specifications
  straight threads, 350
Piston, 104
Piston motor, 155
Piston pumps, 73–79
Piston rod, 104, 106
Plastics, 366
Plunger, 462
Pneumatic cylinders, 417–423
  double-acting, 417–422
  slides, 423
Pneumatic directional control valves (DCVs), 427–445
  AND valve, 439–44
  double-solenoid, four-way controlling a double-acting cylinder, 430
  exhaust port silencers, 437
  five-way controlling a double-acting cylinder, 430–431
  four-way hydraulic and pneumatic, 430
  four-way, three-position, 432–433
  pilot control of a four-way valve, 433–434
  pneumatic actuation types, 433–434
  positioning methods, 435
  quick exhaust valve, 440–441
  shut-off and vent application, 428, 430
  shuttle (OR) valve, 439

single-solenoid, four-way controlling a double-acting cylinder, 430
three-way, two position controlling a single-acting cylinder, 428
Pneumatic flow control valves, 445, 445–448, 446, 448
  double-acting cylinder, 445
  exhaust port, 446
  meter-in versus meter-out, 445–446
Pneumatic motors, 424–426
  air powered gear motor, 425
  rotary vane, 424
Pneumatic power supply, 399–414
  background, 400–401
  compressor sizing, 407–412
  equations, 413
  graphic symbol, 404
  large rotary screw compressor, 404
  rotary screw compressor, 403–404
  rotary vane compressor, 403
  small rotary screw compressor, 404
  two-stage piston compressor, 401
  vacuum pumps, 412–413
Pneumatics, versus hydraulics, 2–3, 395
Pneumatics principles, 375–398
  absolute pressure and temperature, 378
  equations, 396, 457
  gas flow, 389–390
  gas laws, 379–389
    Boyle's law, 379–381, 386–388
    Charles's law, 383–385
    Gay-Lussac's law, 381–383
    General gas law, 385–386, 389
  operation, 376
  pneumatic systems, 394–395
  vacuum, 390–394
    and atmospheric pressure, 390–393
    mercury barometer, 390
    suction cup, 390
Polyol ester, 370
Polytetrafluoroethylene. *See* PTFE
Polyurethane, 367–368
Portable filter carts, 336–337
Positive displacement pumps, 58, 401
Pounds per square inch. *See* psi
Pour point, 371
Power, 48–51
  of cylinders, 118–121
  motor, 166–167
Power supply
  hydraulic, 51
  pneumatic, 394–395
Precision, 462
Pressure, 8, 91
Pressure compensated pumps, 87

Pressure control, 244–278
  brake valves, 268–269
  counterbalance valves, 265–267
    remote sensing, 266–267
  pressure reducing valves, 255, 258
  pressure-compensated pumps, 269–275
    high-pressure standby mode, 273
    hydraulic press circuit, 270–273
    load sensing, 273–274
    parallel circuit, 273
  relief valves, 244–253
    compound or two-stage, 248
    cross-over pressure relief, 252
    pilot-operated, 244, 246, 248, 250
  sequence valves, 258–262, 264–265
    clamp and bend circuit, 262, 264
    remote sensing, 264
  uploading valves, 253, 255
  valve mounting, 275–276
  valve specifications, 276
Pressure dew point, 454
Pressure drop-versus-flow graph, Directional Control Valves (DCVs), 238
Pressure gauges, 342–343
  bourdon tube gauge, 342
Pressure line, 230
Pressure losses, 47
Pressure relief valve, 200, 448
  See also Pressure control
Pressure switch, 400
Pressure transducers, 343
Pressure-compensated flow control valve, 283–284
Pressure-compensated pumps, 269–275
Prime mover, 51, 58
Proportional valves, 467–471
psi (pounds per square inch), 10
psia (kPa abs), defined, 378
psig (kPa gauge), defined, 378
PTFE (polytetrafluoroethylene, Teflon), 368
Pull type single-acting cylinder, 124
Pump controls, 471–472
Pump inlet, 327
Pumps, 25, 51
  vacuum, 412–413
  See also Hydraulic pumps
Push type single-acting cylinder, 124
Push-button actuation, 219

Quick disconnect socket and plug, 355
Quiet zone, 450

Rack and pinion rotary actuator, 157
Radial-piston pump, 78
  See also Rotating cam radial-piston pump
Rankine, 378

Receiver tank, 395
Refrigeration-type air dryer, 455
Regenerative circuit, 228
Regenerative-type air dryer, 455
Relative humidity, 454
Remote sensing, 264, 266–267
Reservoirs, 325, 327–329
Response time, 462
Return line, 327
Reynolds number, 43
Rod end, 104
Rod seal, 364
Rod wiper seal, 366
Rotameters (variable area flowmeters), 344
Rotary actuator, 51, 155–156
Rotary flow divider, 301
Rotary screw compressors, 403–404
    large, 404
    small, 404
Rotary vane compressors, 403
Rotary vane pneumatic motor, 424
Rotating cam radial-piston pump, 78
Rotational seal, 366
Running torque, 179, 424

SAE. *See* Society of Automotive Engineers
St. *See* Stoke
Saturation, 454
Saybolt Second Universal (SSU), 24, 371
Seals and bearings, 361–369
    cut-away of a hydraulic cylinder, static and
        dynamic seals, 361–362
    rod seal, 364
Self-priming pumps, 58
Sequence valves, 258–262, 264–265
Servo valves, 469–471
Shroud, 450
Shuttle valve circuit, 225, 228
Shuttle valves, 190–191
Single-acting cylinders, 124–125
Slides, 423
Society of Automotive Engineers (SAE),
    website, 3
Socket and plug, 355
Solenoid valves, 462–463, 465
Solenoid-actuated DCV, 219
Specific gravity, 21
Specific weight, 21
Specifications
    flow control valves, 304–305
    pressure control valve, 276

Speed
    cylinders, 112–118
    motor, 163–166
Spin-on canister filter, 335
Spool land, 193
Spool overlap, 467–469
Spring return-type single-acting cylinders, 126
Spring-loaded accumulators, 313–314
SSU. *See* Saybolt Seconds Universal
Stall, 166
Stall torque, 424
Standards organizations and their websites, 3
Starting torque, 179, 424
Static head pressure, 45–47
Stoke (St), 24
Stop, 156
Straight threads, 350
Strainers, 334
Stroke lengths, 142
Suction cup, 390
Suction strainer, 335
SUS. *See* Saybolt seconds universal
Swash plate, 76
Synthetic hydraulic fluid, 369–370

Tandem cylinder, 127, 129
Tank line, 230
Teflon (PTFE), 368
Telescopic cylinders, 126–127
Temperature gauges, 347
Temperature-compensated flow control
    valve, 284
Terminal pressure drop, 340
Throat, 38
Throttle, 284
Tie rod construction, 104
Torque, 63–67, 161–163, 424
    constant, 174
    running, 179
    starting, 179
Torricelli's theorem, 40–42
Transmission of force, 14
Tubing, 352–353
Turbine flowmeters, 346
Turbulent flow, versus laminar flow, 42–45
Two-stage piston compressor, 401
Two-vane rotary actuator, 156–157

U-rings, 362–364
Unbalanced vane pumps, 81
Uni-directional motor, 172

Units, 5
Unloading valves, 253
U.S. customary unit, 5, 9

V-packing seal, 366
Vacuum, and pneumatics, 390–394
Vacuum pumps, pneumatic power supply,
    412–413
Valve subplate, 234
Valves. *See* Flow control; Hydraulic;
        Pneumatic directional control
        valves (DCVs); Pressure control;
        Electronic control of fluid power
Vane pumps, 81–86
Vane-type rotary actuator, 156–157
Variable area flowmeters (rotameters), 344
Variable displacement, 63
Variable displacement, bi-directional
        motor, 172
Vegetable-based hydraulic fluid, 369–370
Venturi, 38
Vessels, draining and filling, 193, 197
V.I. *See* Viscosity index
Viscosity, 22–25, 182, 371
Viscosity Index (V.I.), 25
Viton (fluorocarbon), 366–367
Viton seal, 370
Volumetric efficiency, 67

Water removal, 454–455
Water wheel, 8
Water-cooled heat exchanger, 329, 331
Water-glycol hydraulic fluid, 369–370
Wear bands, 362
Websites
    American National Standards Institute
        (ANSI), 3
    American Society for Testing and
        Materials
    International Standards Organization
        (ISO), 3
    National Fluid Power Association
        (NFPA), 3
    Society of Automotive Engineers
        (SAE), 3
Weight
    specific, 21
    versus mass, 19–20
Weight-loaded accumulators, 315, 317
Wet armature solenoid valve, 465
Work, 16